Green Energy and Technology

Climate change, environmental impact and the limited natural resources urge scientific research and novel technical solutions. The monograph series Green Energy and Technology serves as a publishing platform for scientific and technological approaches to "green"—i.e. environmentally friendly and sustainable—technologies. While a focus lies on energy and power supply, it also covers "green" solutions in industrial engineering and engineering design. Green Energy and Technology addresses researchers, advanced students, technical consultants as well as decision makers in industries and politics. Hence, the level of presentation spans from instructional to highly technical.

Indexed in Scopus.
Indexed in Ei Compendex.

Hongbing Ding · Chuang Wen · Yan Yang

Supersonic Separators for Sustainable Gas Purification

Springer

Hongbing Ding
School of Electrical and Information Engineering
Tianjin University
Tianjin, China

Chuang Wen
School of the Built Environment
University of Reading
Reading, UK

Yan Yang
Faculty of Environment, Science and Economy
University of Exeter
Exeter, UK

ISSN 1865-3529 ISSN 1865-3537 (electronic)
Green Energy and Technology
ISBN 978-3-031-85212-1 ISBN 978-3-031-85213-8 (eBook)
https://doi.org/10.1007/978-3-031-85213-8

This Springer imprint is published by the registered company Springer Nature Switzerland AG
The registered company address is: Gewerbestrasse 11, 6330 Cham, Switzerland

Preface

As the world faces the growing urgency of environmental degradation and the increasingly evident effects of climate change, the global energy landscape is undergoing a profound transformation. At the core of this shift is the need for cleaner, more efficient energy that meets growing demands while reducing climate risks. With its relatively low-carbon footprint, natural gas has emerged as a key transitional fuel in the journey toward a sustainable, low-carbon future. However, unlocking the full potential of natural gas requires advanced technologies capable of enhancing its purity and reducing its carbon intensity. Supersonic separators have garnered increasing attention for their ability to dehydrate and decarbonize natural gas efficiently. With their low energy consumption, simplified processes, and compact design, supersonic separators offer a promising solution for the energy sector, both in terms of performance optimization and ecological responsibility.

This book is an exploration of the science, engineering, and practical applications behind supersonic separation technology. By delving into the complex flow dynamics, heat and mass transfer mechanisms, and coupling phenomena within supersonic separators, this work seeks to bridge the gap between theoretical research and industrial application. Through a combination of the computational fluid dynamics (CFD)numerical modeling, experimental methodologies, and optimization techniques, a comprehensive analysis aimed at improving the efficiency of natural gas purification is presented in detail. The book also introduces new strategies to enhance separation efficiency through heterogeneous condensation enhancement and multi-objective optimization techniques, providing readers with a deep understanding of the latest research in gas purification. In addition to the technical aspects, this book offers a broader perspective on the role of supersonic separators within the global energy market, highlighting their potential as a crucial tool in addressing climate change. The findings and discussions reflect the urgent need for innovations that can simultaneously advance energy security and environmental protection. To achieve this, the book is organized into four distinct yet interconnected sections.

Chapters 1–3, introduce a brief overview of the development history of natural gas processing and supersonic separation technology. This part provides an overview of the development of natural gas processing and supersonic separation technologies, highlighting their evolution and advantages in terms of efficiency, sustainability, and resource utilization. The natural gas industry has developed various processing technologies suited to specific applications and operating conditions. These technologies, such as membrane separation, absorption, adsorption, and cryogenic processes, each have advantages and limitations. The selection of the most suitable method is driven by environmental considerations and economic feasibility, with hybrid solutions that combine multiple techniques often providing the best results for complex processing challenges. Supersonic separation technology uses supersonic flow to induce condensation and liquefaction, making it highly efficient and environmentally friendly. With benefits like reduced emissions, no need for chemical additives, and low noise and vibration levels, supersonic separators offer a promising solution for the future of natural gas processing.

Chapters 4 and 5, discuss the homogeneous condensation of water vapor in supersonic flow with monodispersed and polydisperse models. This discussion primarily focuses on the fundamental theory of condensation and separation in supersonic flow, particularly non-equilibrium condensation and droplet size distribution. By developing a corresponding numerical model, the heat and mass transfer characteristics of homogeneous condensation of water vapor in supersonic flow are revealed, the distribution of droplet sizes is predicted, and the energy losses during non-equilibrium condensation are clarified. This provides a crucial foundation for developing the theory of strong swirling condensation coupled with flow in supersonic separators. Thus, Chap. 4 establishes a non-equilibrium condensation model, examining the effects of inlet subcooling and supersaturation on condensation behavior. It reveals the flow, heat transfer, and mass transfer characteristics during the homogeneous condensation of water vapor. More importantly, it provides an analytical solution predicting the location of the condensation Wilson point and the local thermophysical properties within the supersonic flow. Building on this, Chap. 5 further investigates the distribution of condensation droplets and develops a polydisperse model using the Quadrature Method of Moments (QMOM). By integrating the transitional SST model, moment representation, and entropy generation analysis, the droplet radius's probability density function (PDF) is reconstructed. The chapter also elucidates the energy losses and erosion rates during droplet flow, emphasizing the accuracy of the polydisperse model in predicting droplet spectra and energy losses in supersonic flows.

Chapters 6 and 7, aim at homogeneous and heterogeneous phase coupling in multiphase systems. Based on the condensation theory presented in Chaps. 4 and 5, this serves as the core framework for explaining the phase and field behavior evolution in strong swirling supersonic flow. It also acts as a crucial method for analyzing the coupling of condensation and separation in supersonic separators. Two numerical models are developed from distinct perspectives, comprehensively accounting for homogeneous and heterogeneous condensation coupling. These models clarify the coupled flow processes among the gas phase, droplets, and liquid film and the complex heat and mass transfer mechanisms occurring within the supersonic separator. (i) Euler-Euler-Euler three-field two-phase model. It treats the gas phase, liquid droplet phase, and liquid film phase within the supersonic separator as distinct continuous phases. It elucidates the complex heat-mass transfer characteristics inside the supersonic separator and reveals the evolution law of the separation performance under multivariate conditions. (ii) Euler-Lagrangian-Euler three-field two-phase model. This model comprehensively considers both homogeneous and heterogeneous condensations. It uncovers the interaction between these two types of condensation in supersonic condensation flow, revealing the generation of strong swirling flow fields within supersonic separators, as well as the dynamic evolution of droplet condensation and deposition, the formation of liquid films, and other related phenomena. It provides a more detailed explanation of how inlet operating conditions affect separation performance and identifies practical strategies for optimizing the efficiency of supersonic separators. The two methods provide valuable insights for optimizing performance and supporting future applications of supersonic separators.

Chapters 8 and 9, analyze the condensation behaviors and separation performances for supersonic separators for high-pressure natural gas, especially for CO_2-rich natural gas (NG) offshore processing. This part aims to extend the previously developed theory of the supersonic separator unit to applications in natural gas processing under high-pressure environments. It clarifies effective methods for enhancing separation efficiency and minimizing energy losses when employing supersonic separators in these processes. Furthermore, it highlights significant advantages in processing efficiency and economic benefits associated with using supersonic separators for natural gas dehydration and decarbonization. Two major process applications are considered: (i) high-pressure natural gas supersonic dehydration process. The complete high-pressure natural gas supersonic dehydration processes are analyzed, including the Joule–Thomson effect under various operating conditions and the optimal droplet parameters. Additionally, structural parameters' sensitivity to separation performance and energy conversion is examined, particularly separation efficiency and energy conversion under different swirling strengths. (ii) CO_2 separation process of CO_2-rich natural gas. A predictive model for Wilson points, supercooling, pressure, and expansion rate is established to aid in forecasting CO_2 separation in supersonic flows. It is recommended that in practical operation, increasing the inlet heterogeneous droplet mass concentration, lowering the temperature, and raising the pressure are effective strategies to enhance CO_2 separation efficiency and reduce energy consumption.

This book serves as an invaluable resource for researchers, engineers, and industry professionals involved in natural gas processing and gas purification. It illuminates the potential of supersonic separators in enhancing the efficiency of natural gas dehydration and decarbonization. Given the urgent challenges posed by climate change and environmental degradation, the innovative technologies discussed are crucial for advancing cleaner energy solutions. Readers are encouraged to critically engage with the research presented, considering its implications for real-world applications, particularly in offshore platforms where CO_2-rich natural gas is processed. By bridging theoretical insights with practical implementation, the book aims to inspire dialogue among diverse stakeholders, including policymakers, academia, and industry leaders. The knowledge shared within these pages is intended to provoke reflection on the collaborative efforts required to address global energy challenges, ultimately ensuring a healthier planet for future generations.

Tianjin, China Hongbing Ding
Reading, UK Chuang Wen
Exeter, UK Yan Yang

Contents

Abbreviations

1D	One dimensional
2D	Two dimensional
3D	Three dimensional
AGRU	Acid Gas Removal Unit
BTEX	Benzene, toluene, ethylbenzene, and xylene
BTU	British thermal unit
BTX	Benzene–toluene–xylene
CCS	Carbon capture and storage
C–F	Carbon-fluorine
CFD	Computational Fluid Dynamics
CNT	Classical nucleation theory
COS	Carbonyl sulfide
DE	Dehydration efficiency
DEA	Diethanolamine
DEG	Diethylene glycol
DGA	Diglycolamine
DPD	Dew-point depression
EOR	Enhanced oil recovery
EOS	Equation of state
ERDS	Expansion ratio of divergent section
GCI	Grid convergence index
GHG	Greenhouse gases
HAPs	Hazardous air pollutants
HCDPA	Hydrocarbon dew-point specifications
HeRU	Helium recovery unit
IGM	Ideal gas model
IRIB	Inlet radius of inner body
J-T	Joule–Thomson
LCS	Length of convergent section
LNG	Liquefied natural gas
LTS	Low-temperature separation

MD	Molecular dynamics
MDEA	Methyldiethanolamine
MEA	Monoethanolamine
MEG	Monoethylene glycol
MMscf	Million Standard Cubic Feet
MMscfd	Million Standard Cubic Feet per Day
MOFs	Metal-organic frameworks
MP	Membrane permeation
MRE	Mean relative error
NG	Natural gas
NGCC	Natural Gas Combined Cycle
NGLs	Natural gas liquids
NPV	Net present value
NRU	Nitrogen rejection unit
OPEX	Operational expenses
PDF	Probability density function
PLR	Pressure loss ratio
PSA	Pressure swing adsorption
psi	Pounds per square inch
PTFE	Polytetrafluoroethylene
QMOM	Quadrature Method of Moments
R^2	R-Square
RANS	Reynolds-averaged Navier-Stokes
RGM	Real gas model
RMSE	Root mean square error
rRMSE	Relative root mean square error
SP	Separation performance
SRU	Sulfur recovery unit
SS	Supersonic separator
TEA	Triethanolamine
TEG	Triethylene glycol
TRIB	Throat radius of inner body
TSA	Temperature swing adsorption
UDF	User-defined functions
VOCs	Volatile organic compounds
WDPA	Water dew-point specifications

Nomenclature

Terms

A	Surface area (m^2)
a_B	Baumann factor (–)
ΔB	Roughness correction (–)
C	Mass coefficient (–)
C_c	Cunningham correction (–)
C_{con}	Condensation constant (kg m^{-2} s^{-1})
C_D	Drag coefficient (–)
C_{evap}	Evaporation constant (kg m^{-3} s^{-1})
C_{phase}	Phase change constant (kg m^{-2} s^{-1})
C_s	Roughness constant (–)
C_V	Coefficient of variation (–)
c_p	Specific heat of fluid (J kg^{-1} K^{-1})
D	Cell-center-to-wall distance (m)
D_i	Diffusion coefficient (m^2 s^{-1})
$D_{i,m}$	Mass diffusion coefficient for species i (m^2 s^{-1})
$D_{T,i}$	Thermal diffusion coefficient for species i (m^2 s^{-1})
d	Diameter (mm)
d_{32}	Sauter diameter (μm)
d_{in}	Diameter of inlet (m)
d_{th}	Diameter of throat (m)
E	Total energy (J kg^{-1})
EN	Source terms of entropy (J K^{-1} m^{-3} s^{-1})
Er	Erosion rate (–)
e_D	Exergy destruction (J kg^{-1})
e_x	Specific flow exergy (J kg^{-1})
$\vec{F}_D$	The drag force (kg m^{-2} s^{-2})
f_{Tu}	Function of the effective turbulence intensity (–)
$f_A f_\Lambda$	Correction function (–)

G_k	Generation of k due to mean velocity gradient (J)
G_ω	Generation of ω due to mean velocity gradient (kg m^{-3} s^{-2})
$\vec{g}$	Acceleration of gravity (m s^{-2})
g_j	j-wise acceleration of gravity (m·s^{-2})
H_v	Constant for Vickers hardness (–)
h	Enthalpy (J kg^{-1})
h_f	Enthalpy of film (J kg^{-1})
h_i	Enthalpy of species i (J kg^{-1})
h_{lg}	Latent heat (J kg^{-1})
h_v	Latent heat of vapor (J kg^{-1})
I_{te}	Unit tensor (–)
J	Nucleation rate (m^{-3} s^{-1})
$\vec{J}$	Diffusion flux (kg m^{-2} s^{-1})
Kn	Knudsen number (–)
Kr	Kronecker delta number (–)
K_s	Average height of sand-grain roughness element (μm)
k	Turbulence kinetic energy (J kg^{-1})
k_e	Erosion constant (–)
k_B	Boltzmann's constant (1.38065 × 10^{-23} J K^{-1})
k_p	Expansion rate (s^{-1})
L	Length (mm)
L_{con}	Length of convergent section (mm)
M_w	Molecular weight (kg)
M_v	Mass diffusivity of vapor (m^2 s^{-1})
m	Mass (kg)
$\dot{m}$	Mass changing rate (kg s^{-1})
m_m	Mass of single water molecule (2.99 × 10^{-26} kg)
m_v	Liquid mass per unit volume in unit time (kg m^{-3} s^{-1})
$\dot{m}_{col}$	Mass source term due to collision/coalescence (kg m^{-3} s^{-1})
$\dot{m}_d$	Deposition rate (kg m^{-2} s^{-1})
$\dot{m}_f$	Film phase change rate (kg m^{-2} s^{-1})
N	Droplet number density (m^{-3})
$\dot{N}_{col}$	The source term due to collision (m^{-3} s^{-1})
$\vec{n}$	Film area normal (–)
n_{fa}	The cumulative count of facets (–)
O	Mass transfer source term (kg m^{-3} s^{-1})
Pr	Prandtl number (–)
Pr_t	Turbulent Prandtl number for energy (–)
p	Pressure (Pa)
Q	Energy transfer source term (J kg^{-1} s^{-1})
q	Mass flow rate (kg s^{-1})
R_g	Specific gas constant (J kg^{-1} K^{-1})
Re	Reynolds number (–)
RH	Humidity (–)

r Radius (mm)
r_{10} Number-average radius (μm)
r_{32} Sauter radius (μm)
S Supersaturation (–)
$SR_{i,j}$ Mean strain rate (s^{-1})
Sc Turbulent Schmidt number (–)
$S_{ma,g}$ Mass source term of gas phase (kg m^{-3} s^{-1})
$S_{mo,f}$ Momentum source term of film (Pa m^{-1}, kg m^{-2} s^{-2})
S_w Swirl strength (–)
s Entropy (J kg^{-1} K^{-1})
T Temperature (K)
Tu Turbulence intensity (–)
ΔT Subcooling (K)
T_d Dew-point temperature (K)
ΔT_d Dew-point depression (K)
ΔT_H Superheating (K)
T_{mid} Temperature of the film half-depth (K)
T_{wall} Temperature of the film wall (K)
U Momentum transfer source term (Pa m^{-1})
u Velocity (m s^{-1})
u_j j-wise velocity (m s^{-1})
u_t Film tangential velocity vector (m^{-1} s^{-2})
V_{cell} Cell volume (m^3)
υ Volume of single molecule (m^3)
We Weber number (–)
w_i Quadrature weight (–)
x Mole fraction (–)
x_j j-wise Cartesian coordinate (m)
Y_i Mass fraction of species i (–)
Y_k Dissipations of k produced by turbulence (J)
Y_ω Dissipations of ω produced by turbulence (kg m^{-3} s^{-2})
y Distance (m)
y^+ Normalized distance (m)
∇_{sur} Surface gradient (–)

Greek Letters

α Thermal diffusivity (m^2 s^{-1})
α_g The volume fraction of gas phase (–)
α_k Heat transfer coefficient (W m^{-1} K^{-1})
α_t Turbulence thermal diffusivity (m^2 s^{-1})
β Wet fraction (–)

γ	Pressure loss ratio (–)
γ_c	Constant for hardness (–)
γ_g	Specific heat ratio (–)
ε	Turbulence dissipation (J s^{-1})
η	Separation efficiency (–)
η_{dry}	Dry total isentropic efficiencies (–)
η_p	Foreign particles removal rate (–)
η_v	Water vapor removal rate (–)
λ	Thermal conductivity (W m^{-1} K^{-1})
λ_{eff}	Effective thermal conductivity (W m^{-1} K^{-1})
λ_{mg}	Mean free path of a gas molecule (nm)
λ_v	Thermal conductivity of vapor (W m^{-1} K^{-1})
ν	Kinematic viscosity (Pa s)
ξ	Volume fraction (–)
ξ_η	Efficiency ratio (–)
ρ	Density (kg m^{-3})
ρ_{in}	Inlet mass concentration (kg m^{-3})
τ_e	Stress tensor (Pa)
τ_{gf}	Viscous shear stress (N)
τ_i	Droplet relaxation time (s)
τ_w	Wall shear stress (N)
Φ	Relative humidity (–)
ϕ	Diameter of a circle (mm)
φ	Angle (°)
δ	Film thickness (μm)
δ_1	Displacement thickness (μm)
ζ_D	Exergy destruction ratio (–)
θ	Kantrowitz correction (–)
κ	Von Kármán constant (–)
μ	Dynamic viscosity (Pa s)
μ_0	Number of droplets per unit mass of the mixture (kg^{-1})
μ_3	Total volume of droplets per unit mass of mixture (m^3 kg^{-1})
μ_{eff}	The effective molecular viscosity (Pa s)
μ_k	The k-th order moment of number density (m^k kg^{-1})
μ_t	Turbulent viscosity (Pa s)
σ	Surface tension (N m^{-1})
σ_ε	Turbulent Prandtl number for k (–)
σ_g	Standard deviation (m)
σ_k	Turbulent Prandtl number for ε (–)
χ_v	Mole fraction of vapor (–)
Ω	Strain rate (s^{-1})
ω	Specific dissipation rate (s^{-1})
ω_{log}	Friction velocity in logarithmic near-wall region (m s^{-1})
ω_{vis}	Friction velocity in linear near-wall region (m s^{-1})
ω_{wf}	Wave frequency (Hz)

Superscripts

*	Stagnation condition
in	Inlet
out	Outlet

Subscripts

A	Due to aerodynamic losses
C	Due to heat conduction
c	Critical
con	Condensation
D	Due to viscosity
$\overline{D}$	Viscous dissipation
de	Deposition
eff	Effective
evap	Evaporation
f	Film
g	Gas
gen	Generation
het	Heterogeneous
hom	Homogeneous
i	Species index
in	Inlet
is	Isentropic condition
L	Due to phase change
l	Liquid
m	Molecule
mix	Mixture
o	Optimum
out	Outlet
p	Particles
ref	Reference
s	Saturation
se	Separation
st	Stripping
sur	Surface
th	Throat
tu	Turbulent
v	Vapor
w	Wilson point
0	Stagnation condition

Chapter 1
Offshore Processing of Natural Gas

1.1 Overview of Natural Gas Processing

Natural gas, a hydrocarbon-rich gaseous fossil fuel, is commonly found in oil fields, natural gas fields, and coal beds [7, 55]. It plays a crucial role in the global energy landscape due to its versatility and cleaner-burning properties compared to other fossil fuels. The history of natural gas utilization can be traced back to 1821 when William Hart drilled the first documented natural gas well in Fredonia, New York, marking a significant milestone in energy development. This achievement earned Hart the recognition as the "Father of Natural Gas" in America. Initially, natural gas was primarily discovered as a byproduct during crude oil exploration, and its use was limited due to the lack of infrastructure to safely transport it over long distances. During the nineteenth century, natural gas was predominantly utilized locally, primarily for lighting purposes, because the technology to transport it safely over long distances had not yet been developed. However, the landscape changed significantly after World War II. Advances in engineering and technology enabled the construction of safe and reliable long-distance gas pipelines, facilitating the widespread adoption of natural gas as a major energy source [30, 52, 55]. This development was pivotal in expanding the use of natural gas beyond local applications to broader, more industrial uses, contributing to its rise as a vital component of the global energy mix.

In its pure state, natural gas is a colorless, shapeless, and odorless substance, making it difficult to detect without specialized equipment. Despite these characteristics, natural gas is highly combustible and releases a significant amount of energy when burned, making it an efficient and powerful energy source [54]. What distinguishes natural gas from other fossil fuels like coal and crude oil is its environmental profile. It is widely regarded as a cleaner alternative due to its lower emission levels when burned. Unlike other fossil fuels, which release large quantities of harmful compounds and particulate matter that can negatively impact human health, natural gas combustion results in minimal emissions of sulfur dioxide, nitrogen oxides,

H. Ding et al., *Supersonic Separators for Sustainable Gas Purification*, Green Energy and Technology, https://doi.org/10.1007/978-3-031-85213-8_1

and carbon dioxide. These lower emission levels contribute to mitigating environmental issues such as acid rain, ozone depletion, and the accumulation of greenhouse gases [46]. The shift from higher-emission fossil fuels to natural gas for power generation represents a significant step toward improving energy efficiency and reducing carbon emissions on a global scale. This transition is not only crucial for meeting current energy demands but also for addressing future environmental challenges. Furthermore, the development and implementation of carbon capture and storage (CCS) technologies offer additional opportunities to reduce carbon dioxide emissions, making natural gas a cornerstone of future low-carbon energy strategies [55].

Global natural gas reserves continue to grow, underscoring the importance of this resource in meeting future energy needs. According to the BP Statistical Review of World Energy [14], the world's proven natural gas reserves reached 187.1 trillion m^3 by the end of 2014, marking a slight increase from 186.5 trillion m^3 at the end of 2013. This growth, although modest, reflects the ongoing exploration and development efforts aimed at ensuring a stable supply of natural gas for the future. In 2014, global natural gas production increased by 1.6%, a rate that was approximately four times the growth rate of global consumption, which stood at 0.4%. This discrepancy between production and consumption highlights the varying dynamics in different regions, with North America experiencing more robust growth compared to other areas where production and consumption growth rates were below average. The safety of natural gas during transportation, storage, and use is another critical factor contributing to its widespread adoption. It is a highly reliable energy source with extensive applications across various sectors. In residential, commercial, and industrial settings, natural gas is commonly used for heating. It is also a primary fuel for thermal and electrical power generation, providing a stable and efficient source of energy for numerous applications. In the petrochemical industry, natural gas serves as a fundamental feedstock or raw material, particularly in the production of ethylene, which is a key building block for a wide range of chemical products. Additionally, the fertilizer industry relies heavily on natural gas for the production of ammonia, an essential component of agricultural fertilizers. Beyond these applications, natural gas is also utilized in the production of hydrogen, sulfur, and carbon black, all of which are important industrial materials [41, 51, 58].

On a global scale, natural gas accounts for 23.7% of primary energy consumption, highlighting its significant role in the global energy portfolio. The demand for natural gas is projected to continue growing, driven by its increasing use in the power generation and industrial sectors. According to the BP Energy Outlook [13], global natural gas demand is expected to grow at an annual rate of 1.9%, with over 80% of this growth attributed to the power and industrial sectors. This projected increase underscores the critical role that natural gas will play in meeting future energy demands while supporting efforts to reduce carbon emissions and promote environmental sustainability.

Because natural gas extracted directly from production wells often contains various impurities such as hydrocarbons, carbon dioxide, hydrogen sulfide, and water, it typically cannot be used in its raw form. It must undergo multiple treatment stages

Table 1.1 Typical pipeline gas specifications

Characteristics	Specification
Water content	4–7 lbm H_2O/MMscf of gas
Hydrogen sulfide content	0.25–1.0 grain/100 scf
Gross heating value	950–1200 Btu/scf
Hydrocarbon dew point	14–40 °F at specified pressure
Mercaptan content	0.25–1.0 grain/100 scf
Total sulfur content	0.5–20 grain/100 scf
Carbon dioxide content	2–4 mol%
Oxygen content	0.01 mol% (max)
Nitrogen content	4–5 mol%
Total inert content ($N_2 + CO_2$)	4–5 mol%
Sand, dust, gums, and free liquid	None
Typical delivery temperature	Ambient
Typical delivery pressure	400–1200 psig

before it can be utilized. The goal of a natural gas processing plant is to remove these impurities and purify the gas to meet the specifications required for pipeline transport and practical applications. Substandard natural gas can lead to issues such as pipeline corrosion and blockages, which in turn may result in unsafe operations. Table 1.1 presents the typical pipeline gas specifications, which provide standards to ensure gas quality, and to maintain the cleanliness and safety of fuel gas. The product gas must meet the heating value or Wobbe Index specifications to reduce emissions of nitrogen oxides, carbon monoxide, and particulates, which is essential for the optimal operation of gas turbines and combustion equipment. Additionally, the interchangeability and similarity of the product gas with existing pipeline gases in terms of properties must be considered to enhance transportation efficiency.

1.1.1 Gas Plant with Hydrocarbon Dew-Point Controlling

If the heavy hydrocarbon content in natural gas is too high, it may exceed the maximum allowable British thermal unit (Btu) specifications. In such cases, it is necessary to remove some of the heavier components to meet the required standards. The removed liquid hydrocarbons typically have a higher value than natural gas of the same heating value. For example, ethane can be used as a feedstock for petrochemical plants, propane and butane can be sold as liquefied petroleum gas, and C_{5+} components (condensates) can be exported to refineries as gasoline blending components. The characteristics of various natural gas liquids (NGLs) products can be found in the Gas Processors Suppliers Association Engineering Data Book [36].

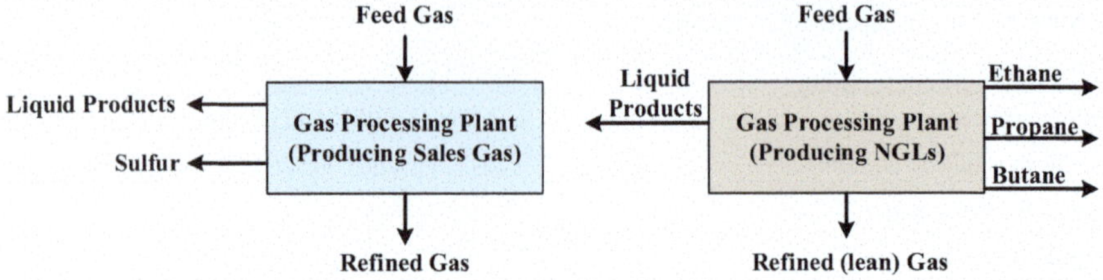

Fig. 1.1 Two different schemes of natural gas processing plants

Even slight variations in product specifications can significantly impact processing options, thereby affecting the cost and complexity of gas processing plants.

The composition of the feed gas, product specifications, and emission restrictions all influence the configuration and complexity of a natural gas processing plant. The value of liquid products is also affected by the level of recovery of NGLs components, which in turn impacts the process complexity. Figure 1.1 presents simplified diagrams of two natural gas processing plant configurations. The first approach involves directly removing heavier components such as condensates and sulfur to meet the required specifications. The second approach involves deep processing of the feed gas to recover individual NGLs components. The residual gas must be compressed to meet usage requirements, which increases both cost and complexity. A natural gas processing plant must be designed "fit for purpose," balancing economic and environmental requirements. While both approaches require the removal of sulfur compounds and other pollutants to meet emission standards, the level of processing in the second approach is specific and depends on the commercial agreements between producers, product distributors, and buyers.

Raw gas entering a processing plant may be relatively low in C_{2+} hydrocarbons, meaning it is lean in composition. Such lean gas can be processed using the units depicted in Fig. 1.2. The primary processing steps include the removal of acid gases, dehydration of the gas, and control of the hydrocarbon dew point. Additional systems, such as sulfur recovery, tail gas treatment, and sulfur formation, are necessary to comply with environmental regulations. In cases where the gas contains liquid condensate, a stabilization unit for the condensate will be needed. If the gas has a high-nitrogen content (greater than approximately 5 mol%), nitrogen rejection is required. Other units, such as gas compression, may also be necessary depending on the specific characteristics of the gas.

The raw gas extracted from production wells first undergoes separation at the inlet facilities, which typically include slug catchers, separation equipment, and pressure protection systems to prevent issues like blockages from affecting the gas processing plant. Slug catchers provide sufficient buffer capacity to accommodate the maximum liquid slug during pigging operations and ensure a steady liquid flow to the stabilization unit. Generally located at a safe distance from the gas processing plant, slug catchers have two main designs: "vessel-type" and "finger-type." The vessel-type slug catcher consists of a phase separator for vapor, liquid, and water, along with a

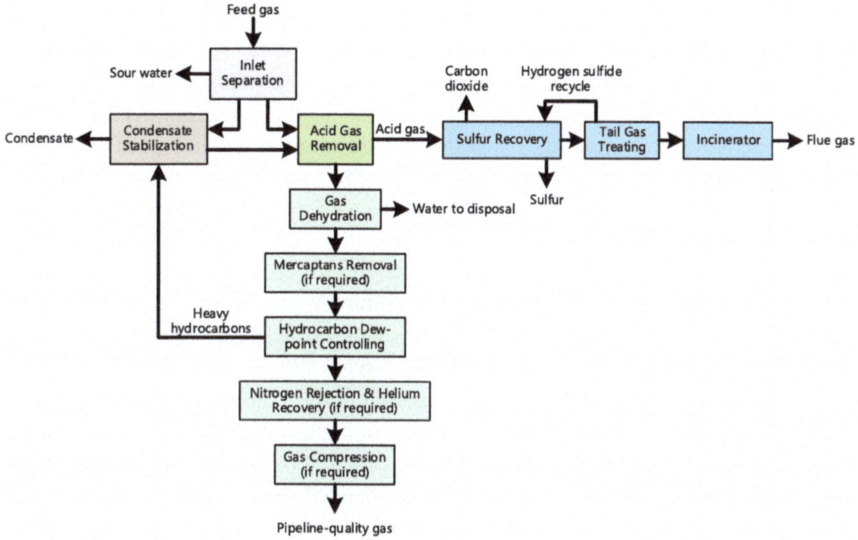

Fig. 1.2 Process units in a gas processing plant with hydrocarbon dew-point controlling

storage vessel to hold the slug during pigging, making it suitable for handling raw gas with a high gas-to-liquid ratio and stable flow, typically used in space-constrained locations like offshore platforms. In contrast, the finger-type slug catcher, which is more common in onshore plants where space is not limited, is more cost-effective in high-pressure services and consists of multiple long pipes that provide buffer capacity. Gas from the slug catcher is directed to a high-pressure separator to protect the gas processing plant from upstream disturbances. The water separated from the liquid in the slug catcher is sent to the sour water stripper for treatment, while the hydrocarbon liquids are processed in the condensate stabilization unit to reduce vapor pressure for easier storage and transportation. The produced water is typically mixed with monoethylene glycol (MEG) used for hydrate control, and the MEG is recovered through a regenerator. If salts are present, some MEG is treated in the recovery unit, where other contaminants, such as corrosion inhibitors, are also removed.

The condensate oil extracted contains dissolved light hydrocarbons and H_2S, which need to be removed. The primary objective of the condensate stabilization unit is to produce condensate oil with H_2S concentrations that comply with a maximum of 4 ppm and a Reid Vapor Pressure within the range of 8–12 psi. Vapor generated at the top of the stabilization tower is compressed and recycled back to the high-pressure separator. Typically, the H_2S levels in the condensate oil from this stabilization process meet the specified criteria. However, the oil may also contain elevated concentrations of organic sulfur compounds, including carbonyl sulfide (COS) and mercaptans. If the condensate is intended for export, it must adhere to total sulfur content standards, which may require an additional unit to eliminate mercaptans.

Conversely, if the condensate oil is directed to a refinery, it can be blended with the feedstock for further processing.

The Acid Gas Removal Unit (AGRU) is capable of removing acidic components to meet specifications for sulfur and carbon dioxide. H_2S must be removed, adhering to a standard limit of 4 ppmv or 1/4 grain per 100 standard cubic feet of gas. If the levels of organic sulfur compounds, such as COS and mercaptans, exceed total sulfur specifications, they must also be eliminated. The acid gas generated from the amine regenerator contains high concentrations of H_2S, which cannot be safely vented or flared due to pollution concerns. If reinjection wells are accessible, the acid gas can be reinjected into reservoirs for sequestration, eliminating the need for a sulfur recovery unit (SRU). In situations where reinjection is not feasible, H_2S is typically processed in an SRU. This unit can be integrated with a tail gas treatment system to achieve a sulfur removal efficiency of 99.9%, thus meeting current emission standards.

The gas processed by the AGRU is directed to the gas dehydration unit to ensure it meets the water dew-point requirements for pipeline transportation, usually set at 7 pounds of water per Million Standard Cubic Feet (MMscf). In colder regions, this specification may be reduced to as low as – 40 °F to prevent hydrate formation within the pipeline. Depending on the plant's capacity and the degree of drying needed, various dehydration techniques can be utilized, including glycol dehydration and solid desiccant methods, such as molecular sieves, silica gel, or activated alumina.

Natural gas extracted from the wellhead may contain mercaptans and other sulfur compounds. To meet the total sulfur content limits in the processed gas or natural gas liquids and to adhere to environmental regulations, it may be necessary to remove these mercaptans. This removal is also crucial to prevent corrosion and damage to downstream equipment. Various techniques exist for eliminating mercaptans from natural gas streams, but the concentration of mercaptans in the feed gas will likely be the determining factor in selecting the most suitable removal method. The hydrocarbon dew-point temperature must be lowered to below the coldest ambient temperature encountered during transmission to prevent the condensation of hydrocarbon liquids within the pipeline, which poses a significant safety risk. It is important to note that, depending on the phase behavior of the pipeline gas, the hydrocarbon dew point can actually rise when the pressure drops. This factor needs to be carefully considered in the unit's design.

The nitrogen content in natural gas varies depending on the source. In some regions, such as the South China Sea, natural gas can contain 30–50% nitrogen. Onshore facilities where nitrogen injection is used for enhanced oil recovery can also have high-nitrogen levels. The purpose of the nitrogen rejection unit (NRU) is to remove nitrogen from the residue gas, thereby improving the lower heating value of the treated gas [40]. Although nitrogen rejection is less common in the gas processing industry, it will become increasingly important as we move toward lower-quality gas feedstocks. The optimal nitrogen rejection process depends largely on the nitrogen content in the feed gas. High concentrations of nitrogen should be removed downstream of the NGLs recovery unit, where heavy hydrocarbons must also be recovered to prevent them from solidifying in the NRU's cryogenic section. For liquefied natural gas (LNG) production, nitrogen must be reduced to below

1 mol% to meet LNG specifications, as higher nitrogen content increases the power consumption by lowering the liquefaction temperature [40]. Although cryogenic separation is the most efficient method for nitrogen removal [18, 20], alternatives like membrane separators and molecular sieves can be used, albeit with limited capacity and efficiency. The rejected nitrogen often contains significant hydrocarbons, which could pose environmental concerns and may require flaring. Additionally, natural gas may contain helium, a valuable noble gas with medical and industrial applications. Helium recovery is typically uneconomical unless its content exceeds 0.5 vol%, but it can be integrated into the NRU if recoverable quantities are present [30].

The pressure of feed gas entering the plant is usually lowered to enable phase separation. Subsequently, the residual gas often requires recompression to match pipeline pressure. This processed gas, known as sales gas, is then delivered through a pipeline network to various industrial and residential users. For those involved in natural gas trading, precise measurement of the sales gas is crucial, as it affects the determination of energy content, identification of gas additives, water content, and flow rate.

1.1.2 Gas Plant for Natural Gas Liquid Production

When the feed gas contains a substantial quantity of liquids, there are financial benefits to extracting liquefied petroleum gas (and occasionally liquid ethane) as byproducts. A liquid recovery facility generally consists of storage systems, pipelines, metering devices, and custody transfer arrangements, all of which must incorporate safety measures to prevent liquid leaks or spills. This type of facility is more complex and costly than a basic hydrocarbon dew-point control plant. Figure 1.3 presents a block flow diagram of a gas plant designed for NGLs production, with its overall structure resembling that of a hydrocarbon dew-pointing facility.

The removal of CO_2 is essential to comply with the sales gas specifications, which usually limit CO_2 concentrations to 2–3 mol%. In some instances, it may be necessary to lower CO_2 levels further to prevent freezing in the cold section of the NGLs recovery unit. During a propane recovery process, a CO_2 concentration of around 2 mol% is generally acceptable due to the relatively higher operating temperature of the NGLs column. However, if the gas is directed toward a liquefaction unit, CO_2 levels must be reduced to 50 ppmv. When the NGLs recovery unit is designed for ethane recovery, the demethanizer column operates at much lower temperatures, increasing the risk of CO_2 freezing. Although freezing may not pose a significant challenge if a considerable amount of CO_2 condenses with the ethane product, the resulting liquid may not comply with the CO_2 specifications, as typical Y-Grade NGLs limit the CO_2 content in ethane to 500 ppmv. Therefore, when ethane recovery is necessary, the design must ensure adequate CO_2 removal from the feed gas to prevent freezing and meet the CO_2 specification for the ethane product.

For NGLs recovery, it is essential for the deethanizer or demethanizer to function at low temperatures. This necessitates the removal of adequate water to prevent

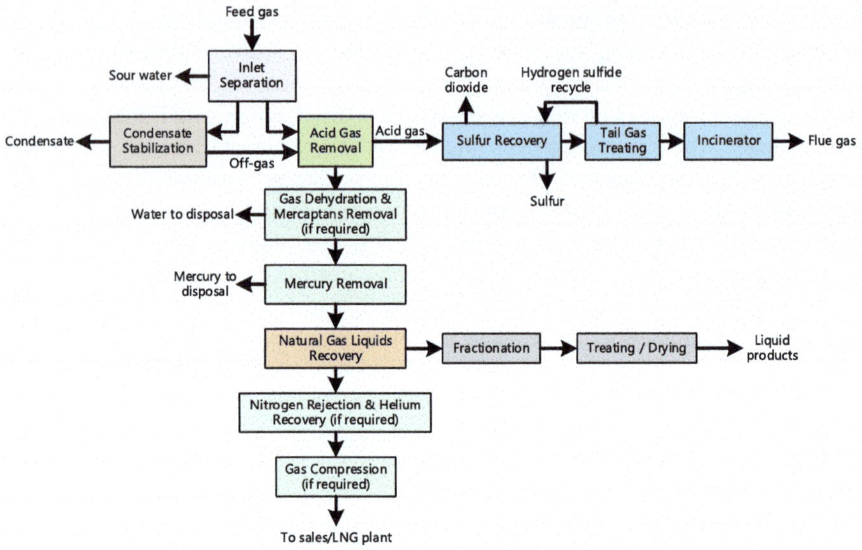

Fig. 1.3 Process units in a gas processing plant for natural gas liquid production

hydrate formation within the columns. When only propane recovery is planned, the column can operate at a warmer temperature, approximately – 60 °F. In such cases, utilizing Dehydration using Reboiled Triethylene Glycol with Zero Emission (DRIZO) triethylene glycol (TEG) dehydration may suffice. The TEG unit is typically more compact and cost-effective than a molecular sieve unit, making it well-suited for offshore applications. However, if ethane recovery is needed, molecular sieve dehydration becomes essential.

Mercury is commonly found in the feed gas at parts per trillion by weight or higher concentrations. To mitigate the risks associated with mercury corrosion of brazed aluminum heat exchangers, it is necessary to remove mercury using sulfur-impregnated carbon beds or metal sulfide beds. Brazed aluminum exchangers are frequently utilized in the NGLs recovery process due to their effectiveness in maximizing NGLs recovery. However, aluminum is susceptible to corrosion from elemental mercury, which can lead to rapid degradation and failure of the exchangers. Typically, the mercury removal system is designed to reduce mercury levels to below 0.01 mg/Nm^3. This removal process can be positioned either before the AGRU or after the dehydration unit. By installing the mercury removal unit upstream, it is possible to prevent the migration of mercury to downstream equipment, ensuring that both the sales gas and liquid products remain free of mercury contamination.

There are various patents and proprietary technologies related to NGLs recovery. In contrast to an AGRU, selecting the appropriate NGLs recovery process requires careful consideration of factors such as desired recovery levels, feed gas pressure, temperature, gas composition, product specifications, and the flexibility of NGLs recovery. The design must accommodate both rich and lean gas scenarios. If nitrogen

Table 1.2 Maximum levels of major contaminants of common liquid products

	H_2S (ppmw)	Total sulfur (ppmw)	CO_2 (ppmw)	H_2O (ppmw)
High-ethane raw streams	50	200	3500	No free
Ethane-propane mixes	Pass copper strip test	143	3000	No free
High-purity ethane	10	70	5000	No free
Commercial propane	Pass copper strip test	185	–	Pass ASTM D2713-91 (2001) standard test
Commercial butane	Pass copper strip test	140	–	–
Commercial butane-propane mixes	Pass copper strip test	140	–	–
Propane HD-5	Pass copper strip test	123	–	Pass ASTM D2713-91 (2001) standard test

injection is planned for enhanced recovery within the same hydrocarbon reservoir, the design should account for a higher nitrogen concentration in the feed gas. Typically, the initial design of the NGLs recovery unit's front section is influenced by the rich gas case during summer, as this scenario presents the highest refrigeration demand. Conversely, the lean gas and high-nitrogen gas scenarios dictate the design of the demethanizer and the residue gas compressor due to increased gas flow. The NGLs recovery unit can be configured for either propane or ethane recovery. To enhance operational flexibility, the process may be set up for ethane recovery while allowing for ethane rejection during periods of low margins. Alternatively, the unit can be designed for propane recovery but operated for ethane recovery when needed. The composition of NGLs products is determined by the targeted recovery of propane and ethane. The NGLs product pump and pipeline must be designed for the ethane recovery scenario, as this will entail higher flow rates and pipeline pressures. Typically, NGLs is transported to a fractionation center, where final liquid products are produced. Additional treatment and processing may be required to ensure that the products meet the specified standards (see Table 1.2).

1.1.3 Finding the Best Gas Processing Route

When establishing the optimal processing scheme for a gas processing project that meets economic and environmental standards, it is crucial to not only choose the appropriate technology for each processing unit but also to consider the interactions among different gas processing units. Adjusting the sequence of processing steps can enhance overall operational efficiency, leading to greater flexibility in the gas processing plant. A range of technologies, both conventional and proprietary, can be employed in the design of a gas processing facility [41]. The design should be "fit-for-purpose," addressing current requirements while also accommodating potential future modifications for different gas compositions and NGLs recovery needs. The design must be reviewed by plant operators to secure their approval. The design of components in the front section of the gas plant, such as the slug catcher and condensate stabilization unit, may follow conventional methods. In contrast, the designs of the gas treating unit and sulfur recovery/handling system are more complex and depend on the acid gas content of the feed gas. The NGLs recovery unit's design is influenced by the desired levels of NGLs recovery, energy efficiency, and operational flexibility. Historically, most NGLs recovery units were engineered for extremely high propane recovery (typically 99%), as NGLs was valued significantly more than natural gas. Such high-NGLs recovery units require the integration of turboexpanders, feed gas and residue gas compressors, and propane refrigeration, which are all more expensive than simpler hydrocarbon dew-point controlling plants. However, with the increasing importance of shale gas, remote gas, and distributed gas in the global energy landscape, NGLs availability is no longer an issue. Consequently, the economic incentive to maximize NGLs production has diminished. Current trends in NGLs design emphasize reducing capital expenditures. Ethane demand is primarily driven by new petrochemical plants. The use of prefabricated modular equipment is becoming common, as it shortens field construction time and lowers capital and operational costs. These modular units can be packaged in various capacities, typically ranging from 60 to 100 million standard cubic feet per day (MMscfd), with some reaching up to 200 MMscfd. When implemented effectively, this modular design can create substantial value for gas producers and processors. In conclusion, there is no universally optimal design for a gas processing plant. The design of NGLs recovery units continues to evolve based on market economics. Designs that are considered optimal today may not remain effective in future as feed gas compositions change. Therefore, a well-optimized gas plant design must be adaptable and capable of being revamped to meet future requirements and economic considerations while maintaining essential equipment.

1.2 Classification of Natural Gas

Natural gas is a combustible gas that is colorless, shapeless, and odorless in its pure state, releasing a significant amount of energy when burned. Compared to other fossil fuels, such as coal and crude oil, it is considered an environmentally friendly clean fuel. The formation of natural gas can occur through one or more of three primary processes: thermogenic, biogenic, and abiogenic. Natural gas can be categorized into conventional and unconventional natural gas based on its source.

1.2.1 Conventional Gas

Conventional natural gas [55] primarily exists in deep reservoirs, where it is either associated with crude oil (associated gas) or contains little to no crude oil (non-associated gas). Associated gas coexists with oil in the reservoir rock and may be present in various forms, such as dissolved gas within the oil or as gas-cap gas located above the oil reservoir (casing head gas). This associated gas, also known as wet gas, generally has a lower methane content and is richer in higher molecular weight paraffinic components. Traditionally, gas streams rich in NGLs (rich gas) contain a high proportion of (C_{2+}) products, while (C_{5+}) components are commonly referred to as natural gasoline or gas condensate. Non-associated gas, also referred to as gas well gas or dry gas, is produced from geological formations that typically contain little to no crude oil or higher-boiling hydrocarbons (gas liquids). This type of gas is usually rich in methane and significantly lower in higher molecular weight hydrocarbons and condensate materials. It may also contain non-hydrocarbon gases, such as carbon dioxide and hydrogen sulfide [54]. It is important to note that after gas processing, methane is produced in a pure form, regardless of whether the raw gas is associated or non-associated.

1.2.2 Unconventional Gas

Unconventional natural gas encompasses several types, including shale gas, coalbed methane, deep basin gas, and natural gas hydrates. Among these, shale gas is particularly noteworthy due to its occurrence in low-permeability geological formations such as shale, impermeable (tight) sandstones, siltstones, and carbonates, including limestones, dolomites, and chalk reservoirs. The methane extracted from these reservoirs is distinct from that associated with petroleum, as it is not linked to crude oil deposits [55]. The origins of shale gas can be traced to a combination of processes, which may include thermal generation, biological activity, or a mixture of both [45]. For instance, the Upper Devonian New Albany shale located in the Illinois Basin has been identified as a significant source of shale gas, as well as shallow aquifers found in southwestern Ontario, Canada [25, 35, 38, 48, 49]. Research has employed stable isotope analysis alongside hydrogeochemical characteristics to establish the source of natural gas within these shale basins. Understanding the origin of natural gas is crucial for enhancing the development and exploration of methane production. This knowledge is particularly valuable in certain contexts where stimulation techniques may be necessary to boost production [35, 48, 49]. In regions characterized by low concentrations of biodegradable organic matter, methods such as stimulation techniques or enhanced microbial methane production can be utilized effectively [37, 48–50, 57]. Additionally, stimulation techniques designed for methane production apply to various challenging environments, including deep or thin unmineable coal seams, gas-depleted shale formations, and coal waste [47]. Notably, research conducted by Meslé et al. [39] has indicated that biostimulation techniques serve as reliable methods for identifying methanogenic communities within organic-rich rocks. Their findings revealed the presence of viable methanogenic populations in shale formations, demonstrating their capability to produce methane efficiently. Overall, the exploration and understanding of unconventional natural gas sources, particularly shale gas, represent a vital aspect of the energy landscape, influencing both production techniques and future research directions in the field.

Coalbed methane, also referred to as coal mine methane, is a natural gas that exists within coal seams and is formed through the complex processes of coalification. This process involves the gradual transformation of organic material into coal under varying geological conditions. Thermogenic coalbed methane is generated when buried organic matter undergoes thermal degradation due to elevated temperatures and pressures associated with deep burial. In contrast, biogenic coalbed methane is produced by microbial activity and is typically found in coal that exhibits lower thermal maturity; see Fig. 1.4 for a visual representation of these processes. The primary constituent of the gas produced from coal seams is methane, which is often accompanied by smaller quantities of other gases, including ethane, carbon dioxide, nitrogen, hydrogen sulfide, and sulfur dioxide [19, 42, 52]. For instance, a detailed study conducted in the Velenje Basin [26, 27, 31] utilized stable isotopes and the carbon dioxide-methane index (CMDI) to effectively trace the origins of gas within that specific geological setting. In this dynamic coalbed methane system, several critical factors influence methane behavior and production, including the distance to the longwall, the rate at which the longwall advances, the depth of the mining site, and the pre-mining regional geological and stress conditions [31]. Each mining section within the Velenje coal mine exhibits its own distinct gas characteristics, which encompass not only the origins of methane but also the concentrations of both methane and carbon dioxide. Methane's relatively rapid migration allows it to accumulate as free gas within water-saturated strata, while carbon dioxide tends to be either retained within the coal structure or dissolved in the surrounding water. Furthermore, both methane and carbon dioxide can be adsorbed onto the microporous surfaces of coal [26]. Given their potent greenhouse gas properties, it is crucial to monitor methane and carbon dioxide concentrations at the working face to mitigate potential explosive hazards. However, effective gas monitoring can also play a significant role in reducing methane emissions, as methane is a valuable energy resource that can be utilized for various applications [27]. In conclusion, the exploration and production of coalbed methane are fundamentally dependent on accurately assessing reservoir characteristics. Key factors include determining gas saturation, evaluating coal permeability, and understanding the continuity of the main coal seams [42]. This knowledge is vital for optimizing extraction techniques and ensuring safe and efficient gas production.

Deep saline gas is formed as gas migrates through aquifers to reservoir rocks, where these aquifers are predominantly saturated with methane. The solubility of methane in water is relatively low, which means that the concentration of gas in these aquifers is largely influenced by various factors such as pressure, salinity, and temperature. Natural gas hydrates, on the other hand, are crystalline structures that

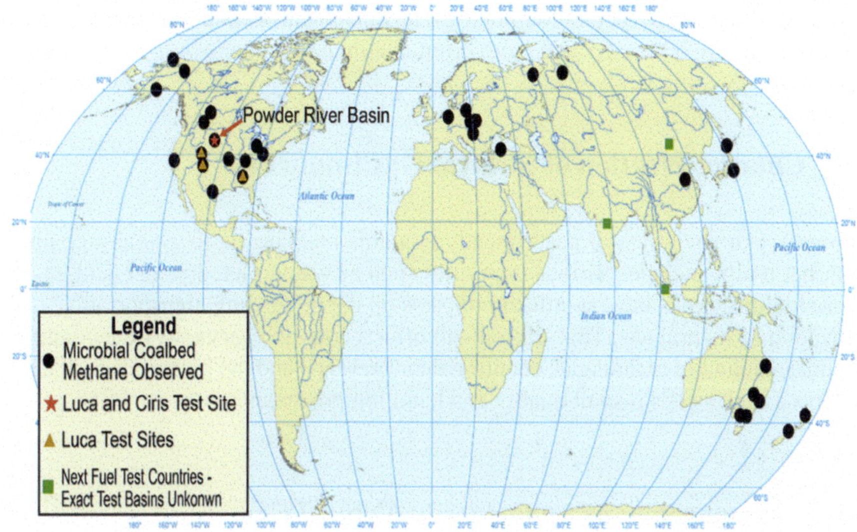

Fig. 1.4 Map showing global microbial coal bed methane production, and pilot test locations [16]

consist of hydrogen-bonded water molecules, creating a lattice-like formation that resembles ice. In this structure, gas molecules become trapped within cages formed by the water molecules. Under different temperature and pressure conditions, nearly all gases can potentially form hydrates. The process of hydrate crystal formation involves a nucleation step followed by the growth of crystals from these nucleation sites. Initially, nucleation seeds are generated, but they remain in an unstable state until they reach a critical size. These nuclei are represented by seeds of specific dimensions, and the time required for nucleation to occur is known as the incubation period. Once a stable nucleus is formed, the subsequent growth of natural gas hydrate crystals can commence. Multiple factors can influence the formation and dissociation of natural gas hydrates, including the composition of the gas, temperature, pressure, the salinity of the reservoir, and the characteristics of the porous media in which the hydrates form. Methane hydrates are primarily generated through the microbial reduction of carbon dioxide present in organic matter, typically within shallow sedimentary basins. In certain cases, mixed origins, combining both thermogenic and biogenic processes, can be identified. Commonly referred to as "flammable ice," methane hydrates are generally found in marine environments, such as deep-sea sediments, and terrestrial sediments located in polar regions. They may also be discovered in deep lake sediments, often situated in or beneath permafrost layers, which help to stabilize their structure. Various resources for methane hydrate production are described in the "Methane Hydrate Resource Pyramid," which outlines the different types of hydrates and their potential for extraction [55]. According to BP's energy Outlook (2018), by the year 2050, the extraction of methane hydrates could potentially account for 5% or more of the total global natural gas production. This ambitious goal is achievable if both private and public sectors commit to the advancement of extraction technologies, ensuring they are both safer and more profitable in the long term.

1.3 Characterization of Natural Gas Impurities

Naturally extracted natural gas typically contains methane as its primary component, but it also includes various impurities such as water, mercury, and acid gases. These impurities can have significant impacts on the processing, transportation, and utilization of natural gas. This section will offer a concise overview of the common impurities found in natural gas, detailing their presence and the potential challenges they pose to the gas's overall quality and handling processes.

1.3.1 Composition of Impurities in Natural Gas

The escalating global demand for natural gas has not only driven exploration but also the successful discovery of vast reserves situated in increasingly remote and previously inaccessible regions. These advancements have been largely attributed to significant strides in modern technology, which have enabled the extraction of natural gas from challenging environments. However, it's important to recognize that when natural gas is extracted from these subterranean reservoirs, it is not in a pure form. Instead, it often contains a mixture of various pollutants and heavy hydrocarbons that must be addressed before the gas can be considered suitable for commercial use [28]. The primary constituents of the extracted gas include propane, methane, and ethane, which together form the bulk of its overall composition. As detailed in Table 1.3, the composition of natural gas is predominantly made up of a range of combustible hydrocarbons, which are essential to its function as a fuel source. In addition to these hydrocarbons, natural gas can also include non-combustible gases such as nitrogen (N_2), carbon dioxide (CO_2), and hydrogen sulfide (H_2S), each of which has implications for the processing and refining stages. Furthermore, trace quantities of other elements like argon, hydrogen, and helium [53] may also be present, adding to the complexity of the gas mixture and requiring specialized treatment to ensure the gas meets industry standards for purity and performance.

Before natural gas can be sent for further processing, it undergoes an initial conditioning phase to address the presence of impurities. This preliminary step is crucial to ensure that the gas is in a suitable state for more advanced treatment. The conditioning process typically involves two key procedures: drying and sweetening. Drying is performed to remove any water content from the natural gas, which is essential to prevent corrosion and other issues during subsequent processing stages. The sweetening step is specifically aimed at removing sulfur compounds and acidic gases, primarily hydrogen sulfide (H_2S) and carbon dioxide (CO_2), from the gas stream. This process is critical because both hydrogen sulfide and carbon dioxide can have

Table 1.3 Components of natural gas

Component	Vol. %
Methane (CH_4)	> 85
Ethane (C_2H_6)	3–8
Propane (C_3H_8)	1–5
Butane (C_4H_{10})	1–2
Pentane (C_5H_{12})	1–5
Carbon dioxide (CO_2)	1–2
Hydrogen sulfide (H_2S)	1–2
Nitrogen (N_2)	1–5
Helium (He)	< 0.5

detrimental effects on the quality of the natural gas and pose environmental and safety hazards. By effectively eliminating these contaminants, the sweetening process helps to ensure that the natural gas meets industry standards and is suitable for further refinement and commercial use.

1.3.2 *Influence of Heavy Hydrocarbons, Carbon Dioxide, and Other Impurities on the Quality and Utilization of Natural Gas*

The presence of heavy hydrocarbons, carbon dioxide (CO_2), and other impurities in natural gas can have a significant impact on its quality and utility. Heavy hydrocarbons, such as ethane, propane, and butane, contain higher energy content per unit volume compared to methane, the primary constituent of natural gas. This higher energy content can be advantageous in applications that demand higher heating values, as it boosts the overall energy efficiency of the gas. However, the inclusion of these heavier hydrocarbons, along with impurities like CO_2, can also present challenges.

While the enhanced energy content is beneficial, the presence of heavy hydrocarbons can complicate the combustion process. These hydrocarbons may require more sophisticated combustion management to achieve complete and efficient burning, as incomplete combustion could lead to the formation of harmful byproducts such as carbon monoxide and unburned hydrocarbons. Additionally, impurities like CO_2 do not contribute to the energy value of the gas and can dilute its overall heating potential, which might necessitate further processing to optimize the gas for specific applications.

This reduction in heating value could impact the performance of appliances and industrial equipment designed to operate with a certain energy output, potentially leading to inefficiencies or operational issues. Therefore, while heavy hydrocarbons can enhance the energy content of natural gas, their presence, along with other impurities, often requires additional processing or treatment to ensure that the gas burns efficiently and meets the desired standards for various uses. This balance between maximizing energy content and ensuring clean, efficient combustion underscores the need for careful management of natural gas composition before it reaches the end user.

The presence of carbon dioxide (CO_2) and other impurities in natural gas poses serious challenges, particularly when it comes to the transportation of the gas through pipeline systems. Carbon dioxide in contact with water is corrosive, and its presence in the natural gas stream can lead to the deterioration of pipeline materials, potentially causing leaks, reducing the lifespan of the infrastructure, and increasing maintenance costs. This necessitates the implementation of additional treatments or measures to remove or neutralize these contaminants before transportation, ensuring the safety and durability of the pipelines.

Moreover, the environmental implications of CO_2 in natural gas are significant. CO_2 is a well-known greenhouse gas, and its release into the atmosphere contributes directly to global warming. The accumulation of CO_2 from various human activities, including the extraction and use of natural gas, is a major factor in the rising levels of greenhouse gases in the atmosphere [1]. This, in turn, accelerates climate change, manifesting in higher global temperatures, more extreme weather events, and other ecological disruptions [8].

Furthermore, the inclusion of CO_2 in natural gas increases the overall carbon footprint of the fuel. As the world becomes more conscious of the need to reduce greenhouse gas emissions, the carbon footprint of energy sources is a critical factor in their sustainability and long-term viability. To mitigate the environmental impact of natural gas, especially in terms of CO_2 emissions, it is crucial to employ carbon capture and storage (CCS) technologies or other innovative mitigation strategies [62]. These technologies can effectively capture CO_2 before it is released into the atmosphere, storing it in a way that prevents it from contributing to climate change.

Addressing the issue of CO_2 and other impurities in natural gas is not just a matter of maintaining infrastructure; it is also about fulfilling global commitments to reduce greenhouse gas emissions and combat climate change. By investing in and deploying these mitigation technologies, the natural gas industry can play a vital role in the transition to a more sustainable energy future, minimizing its environmental impact while continuing to meet the world's energy needs.

The presence of heavy hydrocarbons and impurities in natural gas necessitates additional processing steps to ensure that the gas meets required quality standards and is suitable for various applications. These steps, which involve separating and purifying the gas, introduce greater complexity into the processing workflow. As a result, they increase the operational costs for upstream processing facilities, potentially impacting both efficiency and the overall economic viability of natural gas production.

The influence of heavy hydrocarbons, carbon dioxide (CO_2), and other contaminants on the quality and usability of natural gas can vary widely depending on the specific application and the regulatory standards in place. For instance, while heavy hydrocarbons can enhance the energy content of natural gas, they may also necessitate further treatment to ensure safe and efficient combustion. CO_2, on the other hand, can lead to pipeline corrosion and must be removed to preserve the integrity of transport infrastructure.

Given these challenges, there is a strong emphasis on improving gas processing techniques and developing innovative technologies. These efforts aim not only to address the issues posed by impurities but also to optimize the overall efficiency of natural gas utilization. By enhancing processing methods, the industry seeks to maximize the utility of natural gas while minimizing environmental impacts and adhering to increasingly stringent regulatory requirements.

In emerging economies, the transition to renewable energy sources, while increasingly prioritized for future power generation, is accompanied by significant challenges. These include issues such as unreliable power distribution and high costs associated with implementing and maintaining new energy technologies. Despite these challenges, natural gas (NG) continues to play a crucial role in the energy landscape [34, 56]. This is largely due to its relatively favorable environmental profile compared to other fossil fuels. Natural gas emits less carbon dioxide (CO_2) than oil and coal, primarily because of its higher hydrogen-to-carbon (H/C) ratio. This characteristic makes natural gas a more environmentally friendly option, as it produces fewer greenhouse gases per unit of energy generated.

The energy industry is actively pursuing the replacement of more carbon-intensive systems with modern, eco-friendly alternatives that offer a lower carbon footprint. In this context, natural gas is seen as a dependable energy source in the medium term, bridging the gap as renewable energy technologies are further developed and integrated. However, a significant challenge in utilizing natural gas effectively is the fact that a substantial portion of confirmed natural gas reserves—ranging from 15 to 80%—contains CO_2 [5]. This presence of CO_2 in natural gas reserves necessitates the development of innovative approaches to exploration and extraction. Addressing this challenge is crucial for maximizing the benefits of natural gas while minimizing its environmental impact, particularly in the context of reducing CO_2 emissions and ensuring sustainable energy practices.

1.3.3 Significance of Purifying Natural Gas for Efficient Combustion and Energy Conversion

Fuel emissions represent a major environmental concern in the offshore oil and gas sector, largely due to the extensive hydrocarbon processing and transportation activities that occur on oil and gas platforms [59]. These emissions primarily stem from the operation of gas-fired power generation systems, which are common on such platforms [10]. The combustion of hydrocarbons in these systems releases a variety of pollutants, including carbon dioxide (CO_2), which significantly impacts air quality and contributes to climate change.

Research into the emissions associated with offshore oil and gas operations has highlighted the need for effective mitigation strategies to address these environmental challenges. For instance, studies by Kheshgi and Prince [29] have examined the CO_2 emissions associated with ethanol fermentation, providing insights into the potential for reducing greenhouse gas emissions through alternative processes. Additionally, Xu et al. [61] have contributed to the understanding of CO_2 emissions, further emphasizing the importance of finding and implementing methods to minimize the carbon footprint of hydrocarbon extraction and processing activities.

Addressing these emission concerns involves a multifaceted approach, including improving the efficiency of power generation systems, adopting cleaner technologies, and implementing carbon capture and storage (CCS) solutions. By focusing on these strategies, the offshore oil and gas sector can reduce its environmental impact while continuing to meet energy demands.

Water vapor is a common component of natural gas, often present due to variations in temperature and pressure during extraction and processing. The interaction of water vapor with other impurities in natural gas, such as carbon dioxide (CO_2) and hydrogen sulfide (H_2S), can significantly impact the integrity and functionality of pipeline systems. Specifically, when moisture is present, these impurities can accelerate the corrosion of pipeline materials, leading to potential structural failures and increased maintenance requirements.

Moreover, the presence of water vapor can lead to the formation of hydrates-solid, ice-like compounds that can precipitate under certain conditions. These hydrates are problematic as they can drastically reduce the flow capacity of pipelines and cause blockages, thereby interrupting the continuous transport of natural gas and potentially leading to operational downtime.

Water in a gas pipeline can also contribute to slug flow, a phenomenon where large, irregular slugs of liquid form and travel through the pipeline. This can diminish the overall efficiency of gas transportation by causing pressure fluctuations and reducing the effective flow rate. Additionally, the accumulation of water and other contaminants can result in the deposition of particles and heavy hydrocarbons on heat transfer surfaces, such as those found in heat exchangers and compressors.

The buildup of these deposits can impair the effectiveness of heat transfer surfaces, leading to reduced heat transfer rates and compromised system performance. This reduction in efficiency not only impacts the operational costs but also affects the overall energy conversion process.

To mitigate these issues, it is essential to implement effective cleaning and processing techniques for natural gas. Removing impurities and controlling moisture content are crucial steps in maintaining clean heat transfer surfaces, optimizing energy conversion, and enhancing the overall efficiency of natural gas systems. This ensures the reliable and efficient operation of pipelines and processing equipment while minimizing the risk of corrosion, blockages, and performance degradation.

Eliminating impurities from natural gas while maintaining affordable electricity prices presents a substantial challenge, particularly as the urgency to combat climate change intensifies due to increasing emissions like carbon dioxide (CO_2). The process of preparing natural gas for pipeline transport and delivery involves several critical steps to ensure that the gas is dry and fully gaseous, meeting the stringent quality standards required for safe and efficient distribution.

However, natural gas often contains various impurities, including solids, moisture, and corrosive substances, that can severely impact the performance and longevity of combustion equipment such as burners, valves, and heat exchangers. These impurities, if not properly removed, can lead to significant wear and tear on this equipment, causing corrosion, clogging, and other forms of damage that can impair functionality. The presence of moisture, for example, can promote the formation of corrosive acids when combined with CO_2 and hydrogen sulfide (H_2S), accelerating the degradation of metal components. Solids and other particulates can accumulate within the equipment, leading to blockages that reduce efficiency and increase the risk of malfunction. Corrosive substances can further exacerbate these issues, leading to costly repairs and replacements.

By effectively removing these impurities through advanced processing techniques, not only can the operational lifespan of combustion equipment be extended, but the overall reliability and efficiency of the system can be significantly improved. This reduction in impurities minimizes the need for frequent maintenance and lowers the likelihood of operational downtime, which is crucial for maintaining a consistent and affordable electricity supply.

In a typical natural gas processing sequence, several critical steps are meticulously executed to ensure the gas meets the necessary quality standards for safe and efficient use. These steps include the removal of hydrogen sulfide (H_2S), which is essential due to its highly corrosive and toxic nature. Dehydration is performed to eliminate water vapor, which can lead to the formation of hydrates and corrosion in pipelines. Use propane for water dew-point adjustment and heavy hydrocarbons (specifically C_3^+) for hydrocarbon dew-point adjustment. Additionally, carbon dioxide (CO_2) is extracted to reduce its corrosive potential and environmental impact [11]. Impurities such as sulfur compounds are of particular concern, as they can disrupt the combustion process by forming sulfur oxides (SO_x), which not only contribute to air pollution but also pose a significant risk of corrosion to combustion equipment. By effectively removing these sulfur compounds and other contaminants, the overall combustion efficiency is improved, leading to reduced harmful emissions and enhanced performance of the natural gas processing system.

Certain impurities, such as volatile organic compounds (VOCs) and hazardous air pollutants (HAPs), present serious risks to both air quality and public health. VOCs are known to contribute to the formation of ground-level ozone and smog, while HAPs can have various harmful effects, including respiratory issues and increased cancer risks. Removing these contaminants from natural gas is crucial for mitigating the release of these harmful emissions into the atmosphere. This process not only helps in fostering more sustainable and environmentally friendly energy conversion practices but also aligns with stricter regulatory standards aimed at protecting public health and the environment.

In addition to the environmental and health considerations, impurities in natural gas can significantly impact its use as a feedstock in industrial processes, such as those in refineries or petrochemical plants. For instance, the presence of impurities can poison or deactivate catalysts used in these catalytic processes, leading to reduced catalyst effectiveness and diminished overall process efficiency. Catalysts are essential for accelerating chemical reactions in industrial processes, and their performance is critical for achieving high yields and economic viability. By purifying natural gas, the lifespan and effectiveness of these catalysts are preserved, which helps optimize the efficiency of industrial conversion processes. This purification process ensures that natural gas can be used effectively and efficiently, contributing to improved operational performance and reduced costs in industrial applications.

Injecting carbon dioxide (CO_2) into oil fields provides dual benefits by addressing two significant concerns: it aids in the removal of CO_2 from natural gas (NG) and simultaneously enhances oil recovery through a process known as enhanced oil recovery (EOR) [3]. This technique not only helps in reducing the CO_2 content in natural gas but also improves the extraction of oil from reservoirs, making it a valuable tool in both environmental management and resource optimization.

de Carvalho Reis et al. [9] explored optimization techniques for membrane permeation systems designed to tackle both CO_2 removal and EOR requirements. Their research focused on improving the efficiency of final polishing chemical absorption by incorporating time-dependent CO_2 levels. This approach aimed to refine the membrane permeation process, thereby enhancing its ability to manage varying CO_2 concentrations and improve the overall effectiveness of CO_2 removal and oil recovery.

Building on this work, de Carvalho Reis et al. [10] further advanced membrane permeation technology by integrating time-varying CO_2 concentrations into the system. This advancement aimed to better address the challenges associated with bulk CO_2 removal and to enhance the absorption efficiency of polishing chemicals. The incorporation of dynamic CO_2 levels into the membrane permeation process represents a significant step forward in optimizing both CO_2 management and oil recovery.

Additionally, Anwar et al. [2] highlighted several innovative CO_2 mitigation strategies, emphasizing the development and application of advanced technologies for capturing and managing CO_2 emissions. Their research included a variety of approaches such as algal CO_2 capture, which utilizes algae to absorb CO_2, nanotechnology-based capture methods that employ nanomaterials for more efficient CO_2 capture, and biochar CO_2 capture, which uses biochar to sequester CO_2. These strategies represent a broad spectrum of solutions aimed at reducing CO_2 emissions and addressing the broader issue of climate change.

To summarize, the purification of natural gas is crucial for several reasons: it ensures efficient combustion, protects equipment, improves heat transfer efficiency, manages emissions, and maintains catalyst effectiveness. These processes are essential for optimizing energy conversion, minimizing environmental impact, and enhancing overall system performance.

1.4 General Overview of Supersonic Separator for Natural Gas Processing

In 1893, at the World's Columbian Exposition in Chicago, the largest engineering and liberal arts exhibition showcased the most advanced and promising technologies from engineers around the world. In an unnoticed corner, an important machine that would influence the future of aerodynamics was displayed. This was a single-stage steam turbine for ship propulsion, invented by Swedish engineer Carl G. P. de Laval. While there was nothing particularly special about the steam supplied to the counter-rotating and driving turbines, the novelty lay in the way high-temperature, high-pressure steam was channeled through a series of converging and diverging pipes. The effectiveness of the invention was evident, with a rotational speed of 3000 r/min breaking the records of that time. Little did Laval know that his device would open the door to research in the field of supersonic separation in the twentieth century. The development of supersonic separators has a history of over a century and has proven highly effective in the field of natural gas processing. This section will introduce the development history of supersonic separators.

1.4.1 U-shaped Supersonic Separator

In 1970, Garrett [21] developed a device for separating condensable components, which consisted of a Laval nozzle connected to a curved channel with a rectangular cross-section. The feed gas achieved supersonic speeds and low temperatures as it passed through the throat, and the diverging section had a curved concave shape. Under the action of centrifugal force, the condensed droplets collided with the wall, forming a liquid film that was separated through the permeable wall. Because the liquid film filled the pores of the permeable wall, it prevented gas from returning to the diverging section. The viscous liquid film acted as a liquid seal for the permeable wall, which could be made of permeable metal, such as sintered or particulate metal. The device could reinject some of the separated liquid into the diverging section near the throat to form a liquid film on the permeable wall surface.

In the same year, Garrett and McDonald [23] introduced several enhancements to the separator. The first modification involved injecting a specific substance either before or after the throat, which would mix with the gas stream to improve the recovery efficiency of certain components. The method of enhancing separation efficiency depended on the nature of the injected substance. Additionally, the device was designed to be heated, allowing for the removal of ice, hydrates, and other solid materials that could accumulate on the permeable wall at low temperatures.

Garrett's second improvement was to change the supersonic separation channel from a rectangular cross-section to shapes such as triangular sections with gradually decreasing cross-sectional areas [22]. This structure was more favorable for droplet condensation on the wall and corrosion prevention. Additionally, it allowed for the use of cheaper, rougher, and more permeable materials. The permeable wall could also be automatically replaced through mechanical means, and since the permeable wall did not completely define the fluid flow boundary, its position requirements were not stringent.

Garrett's third improvement allowed the supersonic fluid to be in the optimal position for practical operation by moving the shock wave as it began to pass through the supersonic separation section. To achieve this, the throat area and the shape of the expansion section needed to be adjustable. Despite Garrett's numerous improvements to the supersonic separator, its complexity, along with the need for electromagnetic fields or inhibitor injections to solve hydrate freezing problems, hindered its widespread adoption.

In 1978, Linhardt and Beveridge [32] introduced a supersonic separator that employed centrifugal force and oblique shock waves to remove micron and submicron-sized particles from gases. This design addressed the limitations of earlier devices, such as significant pressure drops and the reliance on electrostatic generators. It was especially efficient in separating sulfur-containing and water-soluble components from flue gases. However, the separation process occurred downstream of the shock wave, resulting in the formation of intense vortices, which hindered the device's effectiveness across all Mach numbers. Consequently, it did not fully exploit the supersonic potential of the nozzle. Moreover, the shock waves induced secondary evaporation of the droplets, leading to a loss of valuable pressure energy.

In 1994, Nasikas [43] developed a nozzle-based separator that relied on normal shock waves for separation. The nozzle generated a normal shock wave at the throat, causing the fluid velocity to drop from supersonic to subsonic after passing through the shock wave, while particles or droplets retained their higher velocity due to inertia. As the gas changed direction following the curved shape of the diverging section, the particles or droplets maintained their original direction. This allowed the gas and particles or droplets to enter two separate regions divided by separation blades, achieving separation. However, this method made it difficult to control the generation and positioning of the normal shock wave, and it did not address the issues present in Linhardt's device.

Haghighi [24] proposed a novel supersonic separator. The device features a U-shaped channel downstream of the nozzle, where droplet separation is achieved by a separation plate and an outlet located at the outer wall of the U-shaped channel. This design avoids the issue of hydrate formation that can clog the outer wall, a defect found in Garrett's design. By adjusting the bending radius of the U-shaped section, the centrifugal force and separation efficiency can be maximized, while preventing shock wave formation before separation. The key significance of this design lies in the elimination of the cyclone, as the generation and dissipation of cyclonic flow cause significant pressure loss. The absence of internal intermediaries or annular cross-sections makes the design simpler and more flexible. The device fully utilizes the pure axial flow inside the nozzle, preventing turbulence and flow separation, and reducing liquid film and gas shear force by limiting axial fluid slippage. However, the design does not provide detailed information on the placement of structures such as the liquid outlet, nor does it consider their impact on separation efficiency.

1.4.2 Twister BV Supersonic Separator

The first group to research supersonic separators was a Dutch engineering team, now part of Twister BV. The company developed two types of supersonic separators, namely Twister™ Mark I and Twister™ Mark II [6]. The design concept of Twister™ Mark I is illustrated in Fig. 1.5. The nozzle first expands the saturated gas to reach supersonic speeds, resulting in a drop in pressure and temperature, which causes water or condensate to precipitate. Fins located in the supersonic region then induce a strong swirl in the airflow, using centrifugal force to fling the precipitated droplets to the wall. Gas–liquid separation is achieved in the cyclone separator section, with the separated gas entering the diffuser section, where its velocity decreases and pressure is restored. Any small amount of gas still entrained in the separated liquid can be filtered and separated before being mixed with dry gas. However, the vortex model established within the nozzle becomes asymmetric as the guide vane angle of attack increases, affecting the interception efficiency of the guide vanes and the vortex within the nozzle. Additionally, when the supersonic fluid collides with the swirl vanes inside the nozzle, strong shock waves are generated, leading to significant energy loss. This also has a substantial impact on the flow field within the nozzle, disrupting the low-temperature, low-pressure environment, and causing the already condensed droplets to re-evaporate, ultimately reducing the separation efficiency of the supersonic separator.

To overcome the aforementioned issues, Twister BV developed a new supersonic separator, Twister™ Mark II. The design concept diagram is shown in Fig. 1.6. Compared to its predecessor, the main improvement in Twister™ Mark II is the repositioning of the swirling device to the inlet. The swirl generator operates based on the principle of conservation of angular momentum, where guide vanes on the internal central body induce gas rotation. During the axial acceleration of the fluid flow, the contraction of the structure causes an increase in tangential velocity. The strong swirl

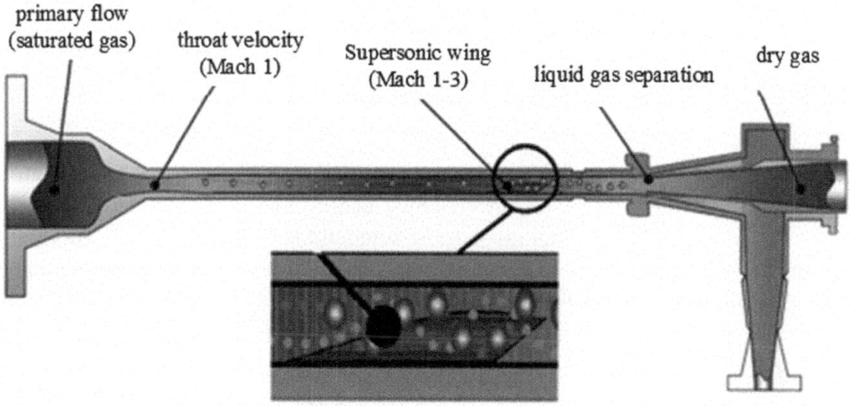

Fig. 1.5 Design concept of Twister™ Mark I [15]

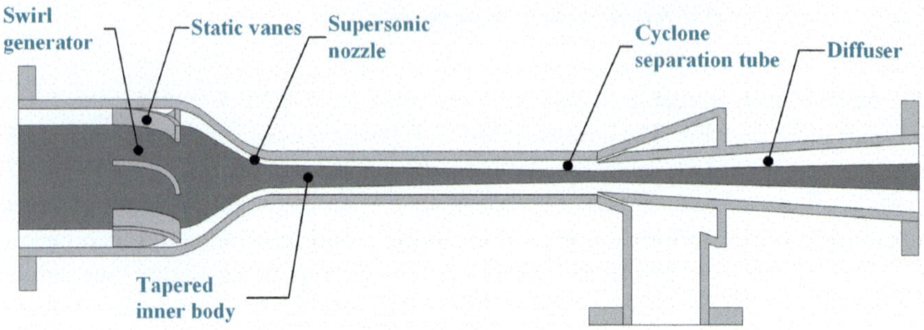

Fig. 1.6 Design concept of Twister™ Mark II

of the fluid during condensation improves separation efficiency. Twister™ Mark II effectively addresses the shortcomings of Twister™ Mark I; however, in Twister™ Mark II, the fluid first passes through the cyclone, generating high-speed rotation before entering the nozzle's converging section, throat, diverging section, and straight section. The viscous resistance leads to a gradual attenuation of the fluid's rotational strength, ultimately reducing the separation efficiency of the supersonic separator. Furthermore, as the operating parameters change, the location and intensity of the shock wave within the nozzle also vary, thereby affecting the separation performance of the supersonic separator.

The third-generation supersonic separator, Twister Mark III, known as the Reflux Twister, has been in development since 2008. The prototype replaced the internal central body with an intermediate body, resulting in reduced pressure drop and shorter equipment length. In October 2009, the new device underwent performance testing in a test loop in the Netherlands. The separation efficiency of this device is reportedly up to 98%, with a dew-point reduction of up to 35 K and a reduction in pressure drop by 20%. Despite the promising separation performance of Twister Mark III, there has been no news of its commercialization [44].

1.4.3 TransLang Supersonic Separator

Russia's ENGO subsidiary, TransLang, successfully developed the 3S separator (supersonic separator) based on aerodynamic achievements from aerospace technology. The design concept diagram of the 3S separator is shown in Fig. 1.7. This low-investment, low-operating-cost separator has been applied in natural gas processing and treatment in oil and gas fields. Since 1996, the company has conducted research and testing on the 3S, developing an optical method for measuring droplet size and establishing a computational model for the supersonic separator [17].

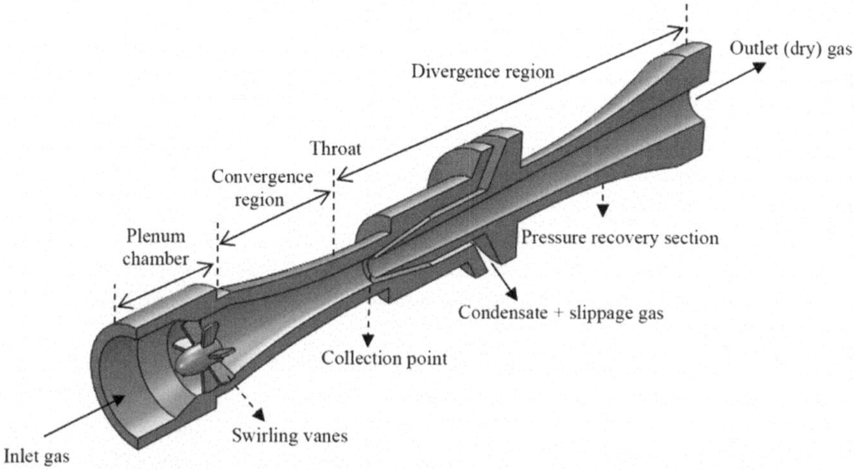

Fig. 1.7 Design concept of 3S separator [60]

Depending on the application, the 3S can be equipped with either subsonic or supersonic connections and can utilize different cyclones and diffusers. However, the structure of the 3S separator is not significantly different from that of the Twister™ Mark II, as it also features a swirl generator at the subsonic inlet before the nozzle, without an intermediate body. Nonetheless, the device still faces challenges in adapting to the constantly changing inlet and outlet pressure conditions in practical applications.

1.4.4 Supersonic Separator Developed by Chinese Scholars

Since the beginning of the twenty-first century, supersonic separators, which hold great potential in the field of natural gas processing, have gradually entered the focus of Chinese scholars and have garnered widespread attention. Several Chinese universities have designed novel structures for supersonic separators.

In 2006, Beijing University of Technology in China developed a supersonic separator with a rear-mounted cyclone, similar to the Twister™ Mark I. In 2009, they proposed a nozzle with an annular internal cross-section, where changes in the shape of the internal body enabled variations in the cross-sectional area of the nozzle's convergent section, throat, and divergent section. Building on this, in 2011, they developed a natural gas supersonic expansion refrigeration and cyclone separator, which had a similar internal structure to the previous designs but differed in that the swirl vanes were located at the nozzle's convergent position, which, being in the fluid acceleration zone, was not conducive to reducing flow resistance. In 2011,

Beijing University of Technology also developed and fabricated a wet gas recirculation supersonic separation tube [4]. The fluid could recirculate through the device's recirculation port into the straight tube section, causing the wet gas to repeatedly cycle through the built-in cyclone generator, thereby enhancing the separation efficiency of the supersonic separator. In 2015, the university developed a new supersonic separator in which the cyclone was fixed at the entrance of the straight tube section. The cyclone used a single-piece helical structure, which not only generated sufficient swirl in the fluid but also overcame the problem of excessive resistance. Despite the structural improvements in these new devices, the fact that the cyclone generator is located in the high-speed region of the nozzle does not fundamentally solve the problem of high flow losses.

Beihang University in China developed a dual-throat, self-starting supersonic swirl separator by modifying part of the cavity wall to be porous, as illustrated in Fig. 1.8 (Jing et al. 2014). The main feature of this device is the use of a porous wall structure in the convergent and divergent sections of the diffuser. This design forms a normal shock wave under conditions far below the nozzle's Mach number in the diffuser, thus minimizing shock losses. At the same time, the device ensures that the entire nozzle remains within the supersonic region, which helps improve separation efficiency. Additionally, due to the use of a porous structure in the diffuser's convergent and divergent sections, some of the fluid can enter the overflow cavity through the porous wall and then re-enter the diffuser's expansion section through the porous wall. The fluid then enters the diffuser's throat and expands, further decelerating and increasing in pressure, before the dry gas enters the gas pipeline. The outer layer of wet gas slows down and increases in pressure within the separator, flowing through the wet gas outlet into the wet gas processing system. This setup allows for adaptive control of the diffuser throat's flow area. However, the liquid discharge gap is an annular structure formed by concentric inner and outer sleeves. In cases of excessive condensate production, increasing the size of this gap can cause a shock wave to form in the nozzle's divergent section, leading to a rise in static temperature within the nozzle, condensate evaporation, and consequently a decrease in separation efficiency and significant loss of pressure energy. Therefore, the liquid discharge gap structure is a drawback of the supersonic swirl separation device.

Dalian University of Technology has been dedicated to the structural optimization and development of new supersonic separators for many years. In 2009, they proposed a conical-core supersonic swirl separator, as shown in Fig. 1.9, which is easy to manufacture, has a short droplet settling distance, low-pressure loss, and features a supersonic nozzle with an annular cross-section [33]. The working region of the conical-core supersonic separator is an annular cross-sectional channel formed by the sleeve and conical core, which first converges and then diverges. This device can operate at a low-pressure ratio, giving it significant industrial application value. However, due to the presence of the conical core and the radially arranged swirl

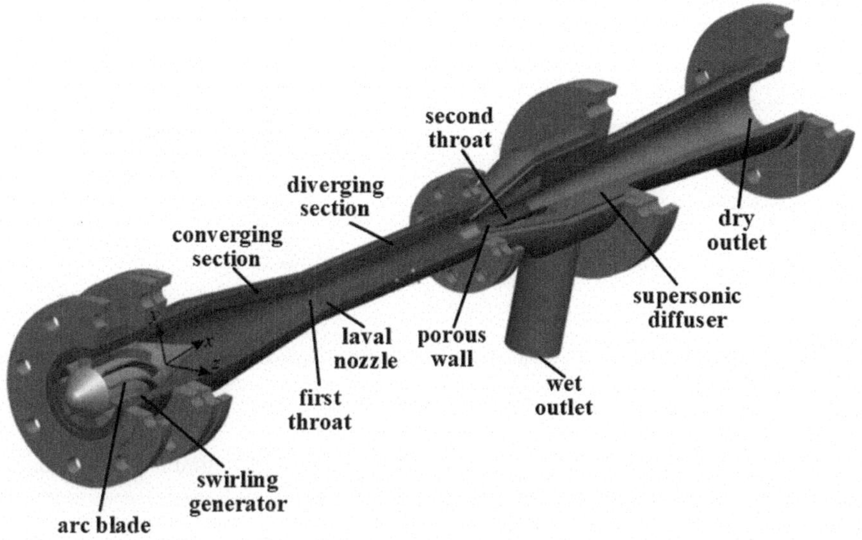

Fig. 1.8 Schematic graph of constructed dual-throat supersonic separation with porous wall [15]

blades, the structure experiences increased resistance. Additionally, in high-pressure systems, the conical core is prone to damage due to disturbances in the internal flow field of the equipment.

Ding et al. [12] from Tianjin University conducted a study on the performance of a supersonic separator with inner body, using a swirl multi-fluid slip model with homogenous/heterogeneous condensation and evaporation processes, as shown in

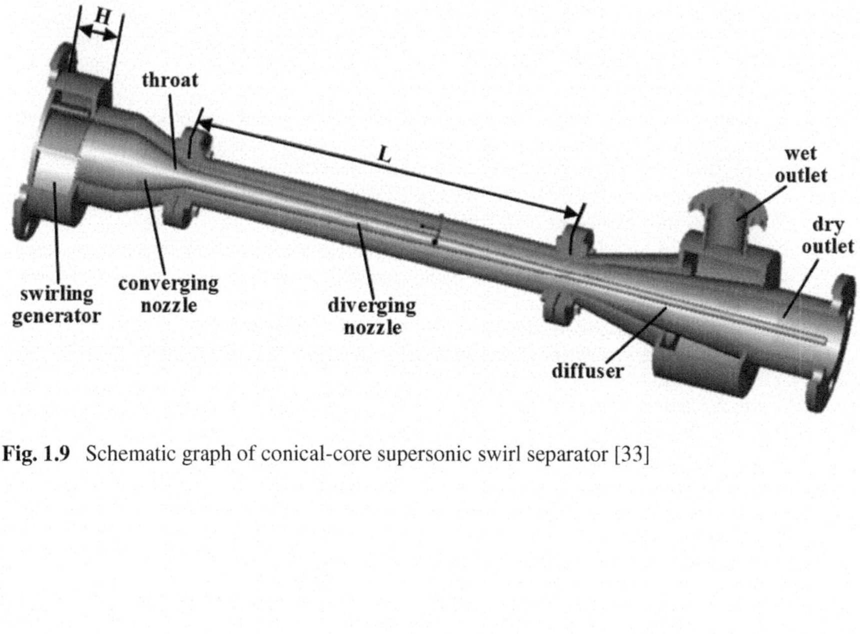

Fig. 1.9 Schematic graph of conical-core supersonic swirl separator [33]

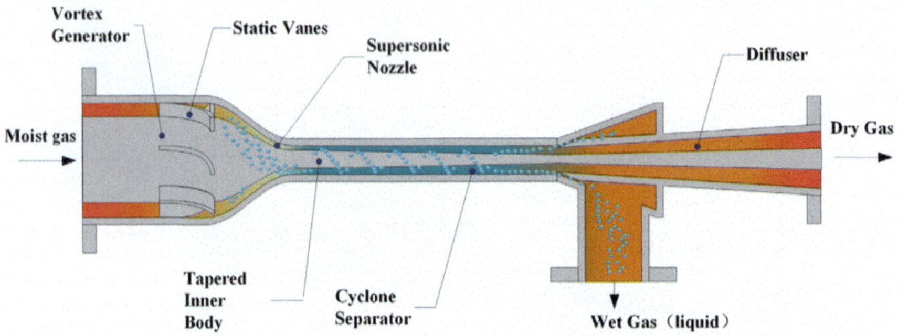

Fig. 1.10 Schematic diagram of the internal structure of the supersonic separator with inner body [12]

Fig. 1.10. The results indicated that it exhibited excellent dehydration performance and holds great potential for applications in natural gas dehydration.

References

1. Anvari S, Mahian O, Solomin E, Wongwises S, Desideri U (2021) Multi-objective optimization of a proposed multi-generation cycle based on Pareto diagrams: performance improvement, cost reduction, and CO_2 emissions. Sustain Energy Technol Assess 45:101197
2. Anwar MN, Fayyaz A, Sohail NF, Khokhar MF, Baqar M, Khan WD et al (2018) CO_2 capture and storage: a way forward for sustainable environment. J Environ Manage 226:131–144
3. Araújo ODQF, de Carvalho Reis A, de Medeiros JL, do Nascimento JF, Grava WM, Musse APS (2017) Comparative analysis of separation technologies for processing carbon dioxide rich natural gas in ultra-deepwater oil fields. J Clean Prod 155:12–22
4. Bao LL, Liu ZL, Jiang WM (2011) The research of dehydration performance and flow characteristic of the recycling supersonic separator. J Eng Thermophys 32:1031–1034
5. Burgers WFJ, Northrop PS, Kheshgi HS, Valencia JA (2011) Worldwide development potential for sour gas. Energy Procedia 4:2178–2184
6. Cao X, Bian J (2019) Supersonic separation technology for natural gas processing: a review. Chem Eng Process Process Intensif 136:138–151
7. Carroll JJ (2010) Acid gas injection and carbon dioxide sequestration. Wiley
8. Chai X, Tonjes DJ, Mahajan D (2016) Methane emissions as energy reservoir: context, scope, causes and mitigation strategies. Prog Energy Combust Sci 56:33–70
9. de Carvalho Reis A, de Medeiros JL, Nunes GC, Araújo ODQF (2017) Upgrading of natural gas ultra-rich in carbon dioxide: optimal arrangement of membrane skids and polishing with chemical absorption. J Clean Prod 165:1013–1024
10. de Carvalho Reis A, de Medeiros JL, Nunes GC, Araújo ODQF (2018) Lifetime oriented design of natural gas offshore processing for cleaner production and sustainability: high carbon dioxide content. J Clean Prod 200:269–281
11. de Oliveira Arinelli L, Teixeira AM, de Medeiros JL, Ofélia de Queiroz FA (2019) Supersonic separator for cleaner offshore processing of natural gas with high carbon dioxide content: environmental and economic assessments. J Clean Prod 233:510–521
12. Ding H, Sun C, Wang C, Wen C, Tian Y (2020) Prediction of dehydration performance of supersonic separator based on a multi-fluid model with heterogeneous condensation. Appl Therm Eng 171:115074

13. Dudley B (2018) BP energy outlook. Report. BP Energy Economics, London, UK
14. Dudley B (2018) BP statistical review of world energy 2018. Energy Economic, Centre for Energy Economics Research and Policy. British Petroleum
15. Eriqitai, Duan R, Wu M (2014) Performance of dual-throat supersonic separation device with porous wall structure. Chin J Chem Eng 22(4):370–382
16. Faramawy S, Zaki T, Sakr AE (2016) Natural gas origin, composition, and processing: a review. J Nat Gas Sci Eng 34:34–54
17. Feygin V, Imayev S, Alfyorov V, Bagirov L, Dmitriev L, Lacey J (2006) Supersonic gas technologies. TransLang Technologies Ltd., Calgary, Canada
18. Finn AJ (2007) Nitrogen rejection strategies. Hydrocarb Eng 12(10):49–52
19. Flores RM (1998) Coalbed methane: from hazard to resource. Int J Coal Geol 35(1–4):3–26
20. Garcel JC, Perisse B, Beresford R, Finn A, Healy M (2007) Liquefaction of non conventional gas. In: Offshore Mediterranean conference and exhibition (OMC-2007), Mar 2007. OMC
21. Garrett RL (1970) Supersonic flow separator with film flow collector (No. US 3528217)
22. Garrett RL, McDonald Jr WJ (1970) Supersonic flow separator with admixing (No. US 3528218)
23. Garrett RL, McDonald Jr WJ (1970) Triangular supersonic flow separator (No. US 3528221)
24. Haghighi M (2010) Supersonic separators: a gas dehydration device. Doctoral dissertation, Memorial University of Newfoundland
25. Hamilton SM, Grasby SE, McIntosh JC, Osborn SG (2015) The effect of long-term regional pumping on hydrochemistry and dissolved gas content in an undeveloped shale-gas-bearing aquifer in southwestern Ontario, Canada. Hydrogeol J 23(4):719
26. Kanduć T, Pezdic J (2005) Origin and distribution of coalbed gases from the Velenje basin, Slovenia. Geochem J 39(5):397–409
27. Kanduč T, Markič M, Zavšek S, McIntosh J (2012) Carbon cycling in the Pliocene Velenje Coal Basin, Slovenia, inferred from stable carbon isotopes. Int J Coal Geol 89:70–83
28. Karimi A, Abdi MA (2009) Selective dehydration of high-pressure natural gas using supersonic nozzles. Chem Eng Process Process Intensif 48(1):560–568
29. Kheshgi HS, Prince RC (2005) Sequestration of fermentation CO_2 from ethanol production. Energy 30(10):1865–1871
30. Kidnay AJ, Parrish WR, McCartney DG (2019) Fundamentals of natural gas processing. CRC Press
31. Lazar J, Kanduč T, Jamnikar S, Grassa F, Zavšek S (2014) Distribution, composition and origin of coalbed gases in excavation fields from the Preloge and Pesje mining areas, Velenje Basin, Slovenia. Int J Coal Geol 131:363–377
32. Linhardt HD, Beveridge JH, Boone CG (1978) Investigation of high velocity wedge separator for particle removal in coal gasification plants. Department of Energy
33. Ma Q, Hu D, He G, Hu S, Liu W, Xu Q, Wang Y (2009) Performance of inner-core supersonic gas separation device with droplet enlargement method. Chin J Chem Eng 17(6):925–933
34. Mahian O, Javidmehr M, Kasaeian A, Mohasseb S, Panahi M (2020) Optimal sizing and performance assessment of a hybrid combined heat and power system with energy storage for residential buildings. Energy Convers Manage 211:112751
35. Martini AM, Walter LM, McIntosh JC (2008) Identification of microbial and thermogenic gas components from Upper Devonian black shale cores, Illinois and Michigan basins. AAPG Bull 92(3):327–339
36. Matthews C (2011) Engineers' data book. Wiley
37. McIntosh JC, Walter LM, Martini AM (2002) Pleistocene recharge to midcontinent basins: effects on salinity structure and microbial gas generation. Geochim Cosmochim Acta 66(10):1681–1700

38. McIntosh JC, Grasby SE, Hamilton SM, Osborn SG (2014) Origin, distribution and hydrogeochemical controls on methane occurrences in shallow aquifers, southwestern Ontario, Canada. Appl Geochem 50:37–52
39. Meslé M, Périot C, Dromart G, Oger P (2013) Biostimulation to identify microbial communities involved in methane generation in shallow, kerogen-rich shales. J Appl Microbiol 114(1):55–70
40. Mokhatab S, Mak JY, Valappil JV, Wood D (2013) Handbook of liquefied natural gas. Gulf Professional Publishing
41. Mokhatab S, Poe WA, Mak JY (2018) Handbook of natural gas transmission and processing: principles and practices. Gulf Professional Publishing
42. Moore TA (2012) Coalbed methane: a review. Int J Coal Geol 101:36–81
43. Nasikas A (1994) Method and mechanism for the supersonic separation of droplets from a gas stream (No. US 5306330)
44. Pallanich J (2009) Annular twister negotiates subsea turn. Offshore Engineer (SEP), 29–30
45. Prinzhofer A, Battani A (2003) Gas isotopes tracing: an important tool for hydrocarbons exploration. Oil Gas Sci Technol 58(2):299–311
46. Ramanathan V, Feng Y (2009) Air pollution, greenhouse gases and climate change: global and regional perspectives. Atmos Environ 43(1):37–50
47. Ritter D, Vinson D, Barnhart E, Akob DM, Fields MW, Cunningham AB et al (2015) Enhanced microbial coalbed methane generation: a review of research, commercial activity, and remaining challenges. Int J Coal Geol 146:28–41
48. Schlegel ME, McIntosh JC, Bates BL, Kirk MF, Martini AM (2011) Comparison of fluid geochemistry and microbiology of multiple organic-rich reservoirs in the Illinois Basin, USA: evidence for controls on methanogenesis and microbial transport. Geochim Cosmochim Acta 75(7):1903–1919
49. Schlegel ME, Zhou Z, McIntosh JC, Ballentine CJ, Person MA (2011) Constraining the timing of microbial methane generation in an organic-rich shale using noble gases, Illinois Basin, USA. Chem Geol 287(1–2):27–40
50. Schlegel ME, McIntosh JC, Petsch ST, Orem WH, Jones EJ, Martini AM (2013) Extent and limits of biodegradation by in situ methanogenic consortia in shale and formation fluids. Appl Geochem 28:172–184
51. Schoell M (1983) Isotope techniques for tracing migration of gases in sedimentary basins. J Geol Soc 140(3):415–422
52. Speight JG (1993) Gas processing: environmental aspects and methods. Butterworth-Heinemann
53. Speight JG (2011) Production of hydrocarbons from natural gas. In: Handbook of industrial hydrocarbon processes, 127–162
54. Speight JG (2015) Liquid fuels from natural gas. In: Handbook of alternative fuel technologies, 1:153–170
55. Speight JG (2018) Natural gas: a basic handbook. Gulf Professional Publishing
56. Stram BN (2016) Key challenges to expanding renewable energy. Energy Policy 96:728–734
57. Strapoć D, Mastalerz M, Schimmelmann A, Drobniak A, Hasenmueller NR (2010) Geochemical constraints on the origin and volume of gas in the New Albany Shale (Devonian–Mississippian), eastern Illinois Basin. AAPG Bull 94(11):1713–1740
58. Tabak J (2009) Natural gas and hydrogen. Facts on File
59. Teixeira AM, de Oliveira Arinelli L, de Medeiros JL, Ofélia de Queiroz FA (2019) Economic leverage affords post-combustion capture of 43% of carbon emissions: supersonic separators for methanol hydrate inhibitor recovery from raw natural gas and CO_2 drying. J Environ Manage 236:534–550
60. Vaziri BM, Shahsavand A (2013) Analysis of supersonic separators geometry using generalized radial basis function (GRBF) artificial neural networks. J Nat Gas Sci Eng 13:30–41

61. Xu Y, Isom L, Hanna MA (2010) Adding value to carbon dioxide from ethanol fermentations. Bioresour Technol 101(10):3311–3319
62. Yazdani S, Salimipour E, Moghaddam MS (2020) A comparison between a natural gas power plant and a municipal solid waste incineration power plant based on an emergy analysis. J Clean Prod 274:123158

Chapter 2
Conventional Natural Gas Processing

2.1 Overview of Conventional Separation Technologies

Many methodologies address the imperative task of natural gas separation and purification, each endowed with distinct benefits and operational nuances. To effectively navigate this selection process, several pivotal factors demand meticulous consideration. Primarily, the nature and concentration of impurities, notably hydrogen sulfide (H_2S) and carbon dioxide (CO_2), hold significant sway over the choice of technique. It is incumbent upon engineers to assess these levels comprehensively, as they directly impact the separation process. Moreover, the operational parameters, encompassing gas throughput, ambient temperature, and prevailing pressure, are indispensable factors in determining the selected method's optimal performance and economic viability. These elements necessitate careful calibration to align with the overarching objectives of efficiency and sustainability. Environmental stewardship and regulatory adherence emerge as critical facets in this decision-making matrix. The stringent guidelines governing emissions control and waste management dictate that the chosen approach must be technically proficient and environmentally conscientious, thereby safeguarding compliance with legislative mandates. Economic prudence further complicates the selection process, requiring a thorough analysis of both initial capital outlays and recurring operational expenses. This financial scrutiny is essential in identifying the most cost-efficient alternative that ensures long-term fiscal responsibility without compromising environmental or technical standards.

As shown in Fig. 2.1, a comprehensive array of strategies is delineated for natural gas processing, encompassing chemical, physical, or hybrid absorption, adsorption, membrane separation, or a combination thereof.

Chemical absorption is distinguished by the utilization of reactive solvents, predominantly amines, which directly interact with impurities, notably high concentrations of hydrogen sulfide (H_2S) and carbon dioxide (CO_2), facilitating their

H. Ding et al., *Supersonic Separators for Sustainable Gas Purification*, Green Energy and Technology, https://doi.org/10.1007/978-3-031-85213-8_2

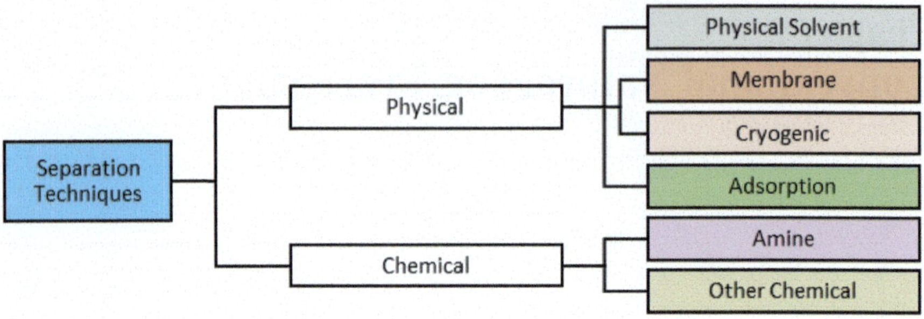

Fig. 2.1 Most commonly used natural gas processes

removal from the gaseous mixture. This process is particularly effective in situations where the impurities are present in high concentrations. The amines, such as monoethanolamine (MEA), diethanolamine (DEA), and methyldiethanolamine (MDEA), react with the impurities, forming stable salts or complexes that can be easily separated. The choice of amine depends on the specific impurity concentration, temperature, and pressure conditions, as well as the desired purity of the final gas product. Contrastingly, physical absorption operates under a principle devoid of chemical alteration, wherein inert solvents are employed to sequester impurities by means of dissolution. This method is optimally suited to scenarios characterized by lower concentrations of impurities and a fluctuating range of process conditions, demonstrating notable flexibility and adaptability. Solvents in physical absorption typically include glycols (triethylene glycol, TEG) and alcohols. These solvents absorb impurities without undergoing a chemical reaction, allowing for easier regeneration and reuse. Physical absorption is often used with other methods to achieve the desired purity levels. A noteworthy innovation is the concept of hybrid absorption, which artfully merges the attributes of chemical and physical absorption techniques. This amalgamation amplifies the process's performance capabilities, catering specifically to applications that require the precision and efficacy of both chemical reactivity and physical solvency.

Hybrid absorption systems can offer the best of both worlds, combining the selectivity and reactivity of chemical absorption with the stability and renewability of physical absorption. This approach is instrumental in complex gas mixtures where multiple impurities need to be removed efficiently. Adsorption is yet another formidable approach, harnessing the surface properties of solid adsorbents to capture minute quantities of contaminants selectively. This method excels in eradicating trace-level impurities, underscoring its utility in achieving ultra-pure gas streams. Adsorbents commonly used include activated carbons, zeolites, and metal–organic frameworks (MOFs). These materials provide a large surface area for the adsorption of impurities, making them ideal for removing residual contaminants after primary treatment stages. Membrane separation presents an alternative solution, deploying selectively permeable barriers that permit the passage of desired gases while concurrently retaining unwanted components. This technique is lauded for its compact design and energy-efficient profile, making it a preferred choice for installations where spatial constraints and operational economy are paramount. Membranes used in natural gas processing are typically polymeric or inorganic, with pore sizes tailored to separate specific molecules. They offer a continuous operation mode and require minimal maintenance compared to traditional separation methods.

The strategic integration of these methodologies may be deemed necessary to attain the requisite purity levels and operational efficiency. By judicious consideration of these factors and selecting an appropriate process, raw natural gas is transformed to eliminate impurities from gas streams, thereby securing adherence to environmental statutes and optimizing cost structures. This balanced approach ensures that the purification process is effective, economically viable, and ecologically responsible. The strategic integration of purification methodologies is essential to achieve the desired purity levels and operational efficiency. The selection of an apt process depends on several factors, including the specific impurities present in the raw natural gas, the required purity level, and the economic constraints faced by the operators. Through careful analysis of these variables, the purification process is tailored to meet stringent environmental regulations while maintaining a cost-effective approach. By judiciously considering these factors, the purification process is optimized to ensure that the natural gas meets the requisite standards for use. This process eliminates contaminants and hydrocarbons from the gas stream and complies with environmental statutes, thereby mitigating the environmental impact of natural gas utilization.

Moreover, the cost structure of the purification process is optimized to ensure economic viability, striking a balance between operational efficiency and ecological responsibility. This balanced approach to natural gas purification is critical for ensuring that the process is adequate for contaminant removal, economically sustainable, and environmentally sound. Through a thorough evaluation of the purification methodologies and a reasonable selection of the most appropriate process, natural gas purification is achieved in a manner that adheres to environmental statutes, optimizes costs, and ensures the ecological responsibility of the operation.

In summary, the strategic integration of purification methodologies is necessary to achieve the requisite purity levels and operational efficiency in natural gas processing. Through a balanced approach that considers environmental compliance, cost optimization, and ecological responsibility, the purification process ensures the elimination of impurities from gas streams while meeting regulatory requirements and maintaining economic viability.

2.2 Membrane Gas Separation

Membrane separation technology is an efficient method in natural gas processing and is recognized as a mature and ubiquitously applied technique. At the core of this process lies the principle of selective permeation, where specific gas components are preferentially transported across the membrane interface. This separation is facilitated by a concentration gradient, meticulously maintained by imposing a high partial pressure of the targeted components against one face of the membrane barrier. The efficacy of this innovative procedure hinges critically on the characteristics of the membrane materials employed. An ideal membrane material is expected to possess a suite of desirable traits: high permeability to ensure robust gas flow, high selectivity to distinguish between different molecular species, robust mechanical strength to endure operational pressures, and chemical stability to resist degradation from gas constituents and the operational milieu [9].

The selection of membrane material for the separation process is vital to ensuring high separation performance and long-term stability. As shown in Fig. 2.2, membranes can be classified into distinct categories based on their material composition and functional properties. Membrane materials can be categorized as polymeric, inorganic, or hybrid. Each type of material offers its own set of advantages and disadvantages, which must be carefully considered when selecting the appropriate membrane for a specific application [10].

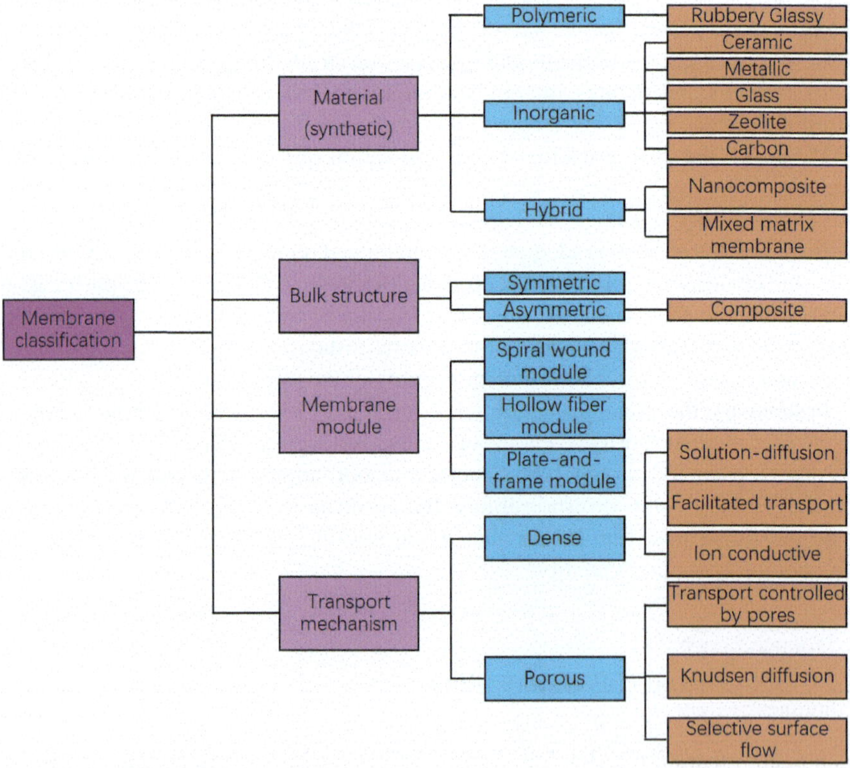

Fig. 2.2 Classifications of membranes

2.2.1 Membranes for Acid Gas Removal

Natural gas is classified as sour gas when it contains hydrogen sulfide (H_2S), carbon dioxide (CO_2), and other higher hydrocarbons. The purification of sour gas, crucial for enhancing its heating value and improving energy efficiency, is a complex process. This process requires careful consideration of the specific characteristics of the gas stream, including impurity concentration, pressure, temperature, the desired purity level, and the overall operational context [41].

Several CO_2-selective membrane systems have been implemented for CO_2 removal from natural gas, both onshore and topside on offshore platforms. These systems have proven particularly effective for small fields with low CO_2 content in the feed gas, such as the Qadirpur field, where the CO_2 content in the processed gas can meet pipeline specifications, often achieving levels below 2%. However, in larger fields with higher CO_2 content in the feed stream, such as those in Indonesia, Thailand, and Malaysia, membrane-treated streams often still contain elevated CO_2 levels (e.g., greater than 6%) and require additional treatment to meet quality standards. For smaller fields, membrane technologies offer a cost-effective alternative to absorption processes. In contrast, for larger fields, absorption remains the more economically favorable option [10].

Figure 2.3 illustrates the process of CO_2 extraction from natural gas using high-pressure membrane contactors, highlighting the effectiveness of membrane technology in selectively isolating CO_2 from natural gas mixtures. Using membrane contactors offers a robust solution for achieving high selectivity and permeability rates, making it an ideal choice for this application. This advanced technology ensures the efficient removal of harmful components such as CO_2 and plays a crucial role in enhancing the overall quality of natural gas. By effectively isolating CO_2, the membrane contactors help meet the stringent environmental regulations governing natural gas processing. The improved gas quality also contributes to greater operational efficiency, which is essential for the economic viability of natural gas production and processing. The deployment of membrane contactors offers a potent means of optimizing the environmental and operational aspects of natural gas handling. This approach ensures that the final gas product meets all necessary standards in terms of purity and regulatory compliance.

Key factors in selecting membrane material for CO_2 removal include high CO_2 permeability, CO_2–CH_4 selectivity, and robust thermal and mechanical stability [17]. Polymeric membranes are a focus of commercialization due to their ease of manufacture. Cellulose acetate, triacetate, and polyimide membranes are the most widely used for natural gas sweetening. These materials are favored for their effective separation capabilities and durability under operational conditions. Cellulose acetate membranes are composed of a blend of cellulose acetate, cellulose diacetate, and cellulose triacetate. These membranes dominate the natural gas processing market, representing up to 80% of the market share. Their widespread industrial use has established them as the industry standard against which other membrane technologies are typically compared [52]. Polyimides are a class of polymers that exhibit

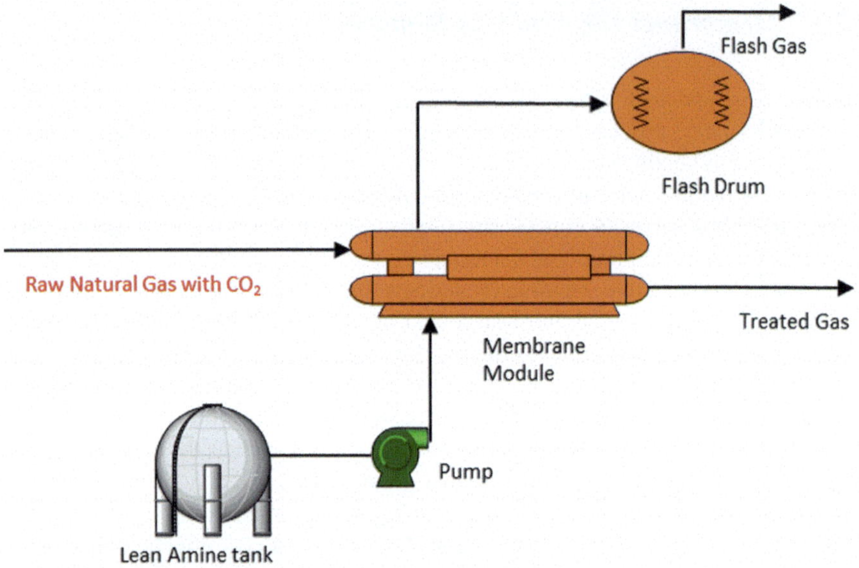

Fig. 2.3 High-pressure membrane processing for CO_2 removal from natural gas

high CO_2 permeability and excellent selectivity against CH_4. Their ease of fabrication into asymmetric membranes, coupled with their strong thermal and chemical stability, further enhances their appeal in gas separation applications [45]. Polyimides are formed through the polycondensation of diamine and dianhydride segments, resulting in a diverse range of polyimides with varying gas separation properties. The ability to modify both the diamine and dianhydride components allows for tailored gas separation performance, making polyimides a versatile option for CO_2 removal in natural gas processing [37]. Unlike polyimides and cellulose acetate, perfluoropolymers are not prone to the same degradation issues, thanks to the high energy of the carbon–fluorine (C–F) bonds present in the substituent groups attached to the polymer backbone. The most widely used perfluoropolymer is poly tetrafluoroethylene (PTFE). Therefore, they exhibit strong resistance to plasticization caused by CO_2 and hydrocarbons, making them particularly advantageous in natural gas processing applications [47]. However, conventional polymer membranes used in natural gas processing encounter several challenges, including contamination, physical aging, and plasticization [2]. Effective pretreatment of natural gas is essential to prevent performance degradation due to contamination. The primary pollutants include higher hydrocarbons, BTEX compounds (benzene, toluene, ethylbenzene, and xylene), and H_2S. Higher hydrocarbons in the membrane system reduce permeability, as these hydrocarbons gradually coat the membrane surface. Additionally, water vapor can cause the membrane to swell, leading to changes in performance.

2.2.2 Membranes for Water Separation

Generally, membranes are commonly employed for water removal in natural gas processing. Hydrophobic polymers, in particular, demonstrate higher water vapor diffusion rates compared to hydrophilic polymers, owing to their lower water repellency. Examples include hydrophilic rubbery polymers and hydrophilic glassy polymers. It has been estimated that a cost-effective membrane system can reduce the dew point of raw gas by 30–50 °C. Air Products has commercialized this method, utilizing their Prism polysulfone membrane materials.

In high-pressure natural gas (NG) processing, the gas stream is first directed into a membrane module, where water vapor is efficiently separated via the membrane [30]. The water vapor-rich permeate is subsequently recompressed, ensuring that the vapor is properly managed. After the separation of the condensate, it is reintegrated with the initial gas flow, thus completing the process. This method allows for effectively managing water vapor in high-pressure NG systems, maintaining the integrity and quality of the gas stream. In contrast, when dealing with low-pressure NG, the sequence of operations begins with the compression of the gas flow. This initial compression is essential to increase the gas pressure, making it more suitable for subsequent processing stages. Following compression, the condensed water is separated from the gas using a specialized separator. This step is critical to remove any moisture that may have formed during the compression stage, ensuring that the gas is sufficiently dry before it enters the next phase. Once the gas is free from condensation, it is introduced into the membrane assembly. The membrane refines the gas by selectively allowing certain components to pass through while retaining others. The permeate obtained after membrane separation, which now contains the water vapor and other light components, is then carefully amalgamated with the original gas flow. This final step ensures that the gas, now processed and purified, is ready for its intended application, completing the procedure with enhanced efficiency and effectiveness. By employing these tailored processes for both high-pressure and low-pressure NG, the overall efficiency of gas processing can be significantly improved, ensuring that the gas meets the required specifications for further use or transportation.

In these applications, the primary challenge is the loss of methane when membranes for water separation are used. Despite the membrane's potentially high selectivity, a significant amount of methane can permeate through due to the limitations imposed by the pressure ratio. For effective separation, the partial pressure of water in the feed gas must exceed that in the permeate stream. Consequently, regardless of the membrane's selectivity, the extent of separation achieved is inherently constrained by the pressure ratio. A more effective approach involves using a dry sweep gas on the permeate side, which enhances water permeation while mitigating methane losses. Figure 2.4 illustrates using a recycle stream as a dry sweep gas in the membrane separation process [41].

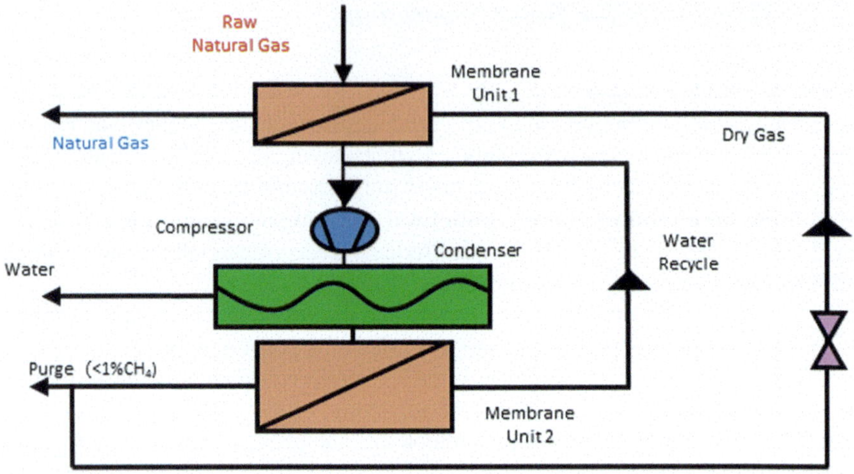

Fig. 2.4 Membrane process design for removal of water

2.3 Absorption Process

The absorption method has been widely used for natural gas dehydration since the 1930s. After decades of development and evolution, nearly 100 methods have been reported internationally, which can be primarily categorized into two types: chemical solvent absorption and physical solvent absorption [20].

The absorption method for water removal in natural gas processing can be performed using various agents. Each type of agent has its mechanism and characteristics:

Organic liquids, such as diethylene glycol (DEG) and triethylene glycol (TEG), absorb water vapor from the gas stream. These solvents have a high affinity for water and can effectively remove moisture from natural gas. They offer advantages such as high water absorption capacity, low volatility, and relatively low cost. However, their disadvantages include the need for energy-consuming regeneration and potential degradation over time.

Aqueous solutions of salts, such as lithium chloride, absorb water vapor through dissolution, with the salt dissolving in the vapor to create a concentrated solution. These solutions offer advantages such as high water absorption capacity and stability under varying conditions. However, their disadvantages include potential corrosiveness and the need for special handling and disposal procedures.

Solid absorbents such as fused anhydrous calcium sulfate, lime (calcium oxide), and magnesium perchlorate remove water vapor through chemical reactions. When water contacts these materials, they form hydrates or other stable compounds. These absorbents offer advantages such as high water absorption capacity, no need for regeneration, and reusability after drying. However, their disadvantages include a limited lifespan due to saturation, the need for replacement, and the potential release of dust during handling.

Deliquescent absorbents, such as calcium chloride and potassium hydroxide, achieve water removal through a combination of chemical reactions and dissolution. When exposed to water vapor, these absorbents dissolve and form a liquid solution. These absorbents offer advantages such as high water absorption capacity, rapid response to moisture, and suitability for portable or small-scale applications. However, their disadvantages include the need for frequent replacement due to saturation, heat generation during the absorption process, and the potential production of corrosive byproducts.

The absorption technique is grounded in the principle that gases have varying solubilities in liquids, which forms the basis for the selective removal of certain components from gas streams. This process may also involve chemical reactions that enhance gas purification efficiency. Central to the success of absorption methods is the absorbing material, which must exhibit high absorption capacity alongside excellent thermal stability to withstand the rigors of the process.

2.3.1 Chemical Solvent Processes

A widely adopted method for dehydration natural gas within the industry is the liquid desiccant contactor-regeneration process, as detailed in Fig. 2.5. This involves exposing a wet natural gas stream to a dry solvent with a shallow water content. The solvent, acting as a desiccant, absorbs the water from the gas, creating a concentrated aqueous solution while leaving behind a dehydrated gas stream. Subsequently, the solvent, now saturated with water, is directed to a second column where it undergoes regeneration. This process involves the removal of water from the solvent, restoring its ability to absorb moisture. Once regenerated, the solvent is then recycled back to the initial column to continue its role in removing water from the incoming gas stream. Using a liquid desiccant contactor-regeneration process is efficient in removing moisture and economical, as it minimizes the need for frequent replacement of the desiccant material. The process ensures that the solvent can be reused multiple times, significantly reducing operational costs and environmental impact.

In the gas industry, commercial amine-derivative compounds have been used, such as monoethanolamine (MEA), diethanolamine (DEA), triethanolamine (TEA), methyldiethanolamine (MDEA), di-isopropanolamine (DIPA), triethylene glycol (TEG), and diglycolamine (DGA) [25]. TEG is the predominant absorbent used for dehydration purposes, with other alternatives including calcium chloride, ethylene glycol, diethylene glycol, and tetraethylene glycol. The superior efficacy of glycols as liquid desiccants stems from their pronounced hygroscopic properties, low vapor pressures, high boiling points, and minimal solubility in natural gas. Among these, TEG stands out for its exceptional performance in lowering the dew point of gases. Its cost-effectiveness and operational reliability make it the preferred glycol for natural

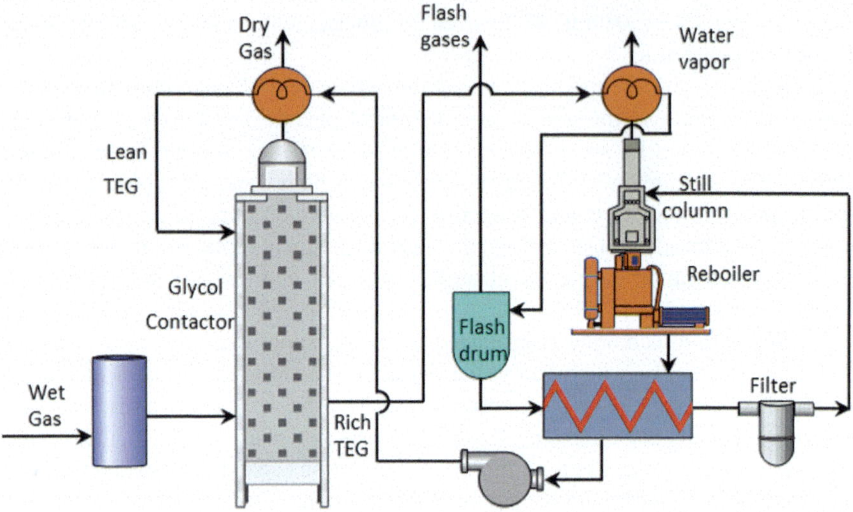

Fig. 2.5 Schematic of industrial absorption dehydration process using TEG

gas dehydration. However, glycol dehydration systems are not without their challenges. Issues arise from contaminants in glycol solutions, which can originate from suspended solids or foreign matter. Overheating of the solutions can lead to the formation of decomposition products, potentially compromising the integrity and effectiveness of the dehydration process. Foam formation in the glycol solution is another concern, as it can result in liquid carry-over, affecting the system's efficiency.

Moreover, environmental concerns associated with fugitive emissions are increasingly being addressed through measures aimed at mitigating their impact, reflecting a growing commitment to sustainable practices within the industry. Besides water vapor, natural gas may also contain natural gas liquids (NGLs), which are often removed to meet hydrocarbon dew-point specifications (HCDPA). This is critical for ensuring the quality and marketability of the gas, as NGLs have higher value when sold as separate commodities. Cryogenic processing, although more expensive, is favored for separating NGLs due to its effectiveness in meeting stringent dew-point requirements. The removal of NGLs reduces the heating value of the gas product, which can affect its commercial value. However, adherence to HCDPA specifications is imperative for natural gas's operational integrity and market acceptability, typically achieved through low-temperature separation techniques. These processes highlight the intricate balance between technical efficiency, economic viability, and environmental responsibility in the natural gas industry, with ongoing efforts to optimize each aspect for improved overall performance.

2.3.2 *Physical Solvent Processes*

Physical solvents possess several key characteristics that distinguish them from chemical solvents [11]:

(1) Chemical Inertia: Physical solvents do not undergo chemical reactions with CO_2, which prevents the formation of heat-stable salts. This simplifies the regeneration process and reduces the complexity of the solvent recovery system [15].
(2) High CO_2 Capacity: Physical solvents can absorb large quantities of CO_2 at high partial pressures and low temperatures. This makes them ideal for applications where CO_2 is present in high concentrations and needs to be efficiently removed.
(3) Low Energy Requirements: The regeneration of physical solvents typically requires less energy compared to chemical solvents. This is because the solvents can be regenerated by simply heating them to release the absorbed CO_2, without the need for complex chemical reactions.
(4) Broad Operating Range: Physical solvents can operate effectively over a wide range of temperatures and pressures, making them versatile for various processing conditions.

Two commonly used physical solvents in natural gas processing are Selexol (polyethylene glycol dimethyl ether) and Rectisol (methanol chilled to − 40 °C). Selexol (polyethylene glycol dimethyl ether) is based on the physical solubility of CO_2 in the solvent [46]. It works particularly well at low temperatures and high partial pressures of CO_2. It offers advantages such as high CO_2 absorption capacity, low energy consumption for regeneration, and good stability under process conditions. Selexol is especially suitable for natural gas streams with high CO_2 concentrations and for the purification of synthetic gases. Rectisol (chilled methanol) is methanol cooled to very low temperatures (− 40 °C or below). It absorbs CO_2 due to the increased solubility of CO_2 in cold methanol. Rectisol offers advantages such as high selectivity for CO_2, low energy requirements for regeneration, and the ability to operate at low temperatures. It is used in natural gas processing plants for the removal of CO_2 and other acid gases and is particularly effective in cryogenic processes and for the purification of gases containing high levels of CO_2 [46].

Meanwhile, studies have demonstrated that ionic liquids can effectively capture CO_2 from gas mixtures. Ionic liquids offer several advantages over traditional solvents [21], such as Low Energy Consumption: Ionic liquids require less energy for the regeneration process compared to conventional solvents. Easy Recycling: They can be easily recycled and reused multiple times without significant loss of performance. No Pollution: Ionic liquids are generally non-volatile and do not emit harmful vapors, making them environmentally friendly. Strong Functionality: They can be tailored with specific functionalities to enhance CO_2 absorption. High CO_2 Scrubbing Efficiency: Ionic liquids exhibit high selectivity and capacity for CO_2, making them highly effective for gas separation [43].

Several ionic liquids have been studied for their CO_2 capture abilities: 1-butyl-3-methylimidazolium tetrafluoroborate ([Bmim][BF_4]): This ionic liquid has a high affinity for CO_2 and can absorb significant quantities of the gas at low temperatures. It offers advantages such as good thermal stability and recyclability. 1-butyl-3-methylimidazolium hexafluorophosphate ([Bmim][PF_6]): Similar to [Bmim][BF_4], this ionic liquid effectively absorbs CO_2 and provides advantages like high CO_2 uptake and good thermal stability [54]. Tetrabutylphosphinoic acid ionic liquid ([P$(C_4)_4$][AA]) utilizes the acid–base interaction between its acidic group and CO_2 for absorption, offering advantages such as high selectivity for CO_2 and good thermal stability [55]. Bis-amino functionalized phosphine ionic liquid [56]: The amino groups in this ionic liquid facilitate CO_2 absorption through hydrogen bonding interactions, providing advantages such as high CO_2 uptake and enhanced selectivity.

Therefore, ionic liquids represent a promising alternative for CO_2 capture in natural gas processing. Their unique properties, including low energy consumption, easy recycling, no pollution, strong functionality, and high CO_2 scrubbing efficiency, make them attractive for industrial applications [8]. However, the high cost and scale-up challenges need to be addressed to fully realize their potential in commercial CO_2 capture technologies [42]. Ongoing research is focused on developing more cost-effective and scalable ionic liquid systems that can meet the demands of the natural gas industry [4].

2.4 Adsorption Process

Adsorption, as a fundamental process in the realm of mass transfer, is characterized by the adherence of gas molecules to solid surfaces due to intermolecular attractions or chemical bonding [16]. This phenomenon can manifest in two principal forms: chemisorption and physisorption. Chemisorption, or chemical adsorption, entails the formation of a chemical bond between the gas molecule and the solid surface, typically involving a more robust interaction and a higher activation energy than physisorption. Physisorption, conversely, is governed by relatively weaker van der Waals forces and does not entail the formation of a new chemical bond.

Adsorption is critical in achieving deficient concentrations of specific gases in various industrial applications. Within the gas industry, a range of solid adsorbents are employed, each tailored to the particular requirements of the process. Activated alumina, renowned for its high surface area and affinity for water and other polar molecules, is a staple in many dehydration processes. Silica gel, another prevalent adsorbent, is celebrated for its efficacy in moisture control owing to its porous structure, which facilitates the capture of water molecules. Molecular sieves, which are crystalline aluminosilicate materials, feature a precisely defined pore structure that enables them to selectively adsorb molecules based on their size and shape, making them exceptionally efficient for gas separation tasks.

The effectiveness of the adsorption process for water molecules, or any other target species, is heavily contingent upon the prevailing gas pressure and temperature. Elevated gas pressure augments the density of gas molecules near the adsorbent surface, thereby enhancing the adsorption rate. In contrast, increased temperatures provide additional kinetic energy upon the gas molecules, which can counteract the adsorption forces, potentially leading to desorption and consequently diminishing the adsorption capacity. Hence, the design and operation of adsorption systems necessitate meticulous balance of these variables to optimize the process performance.

In summary, adsorption, through its two modalities of chemisorption and physisorption, offers a versatile and robust method for the selective removal of components in natural gas. The choice of adsorbent and the operational conditions, including pressure and temperature, are crucial determinants of the process efficiency and must be judiciously selected to meet the specific demands of the application at hand.

In actual industrial applications, ensuring uninterrupted service is vital for maintaining production efficiency and quality standards. This is particularly true for adsorption systems designed for natural gas processing, where continuous operation is achieved by implementing dual or multi-bed configurations. Typically, at least two adsorption beds are utilized, allowing one bed to remain active in the adsorption process while the other undergoes regeneration. During operation, one bed actively captures water molecules from the incoming natural gas, effectively dehydrating it. Simultaneously, the second bed, having reached saturation after a previous cycle of adsorption, undergoes a regeneration phase. This process involves passing preheated gas or a portion of the already dehydrated natural gas through the spent bed. The elevated temperature of the regenerating gas provides the necessary energy to break the bonds holding the water molecules onto the adsorbent surface, resulting in the desorption of these molecules. Once the regeneration process is complete, the bed is ready to resume adsorption duties, while the previously active bed is switched to regeneration mode. This alternating cycle ensures that a bed is always available for adsorption, thus providing a continuous supply of dried and purified gas.

Figure 2.6 visually summarizes this cyclic operation, highlighting the seamless transition between adsorption and regeneration phases in dual-bed adsorption systems. This figure serves as a practical illustration of how adsorption technology is harnessed to deliver reliable and efficient gas drying and purification in various industrial contexts [19]. By leveraging the principles of adsorption and regeneration in a continuous cycle, industries can maintain high standards of gas quality, which is critical for processes that require precise control over moisture content and impurities. Based on the regeneration method, the adsorption process that can achieve CO_2 separation is divided into temperature swing adsorption (TSA) and pressure swing adsorption (PSA) processes [3, 40]. Continuous TSA and PSA processes operate with multiple beds containing a stationary adsorbent and use a manifold of valves to switch gas flow to the beds corresponding to adsorption and desorption steps [31]. The use of multiple beds allows for continuous operation, with one bed undergoing

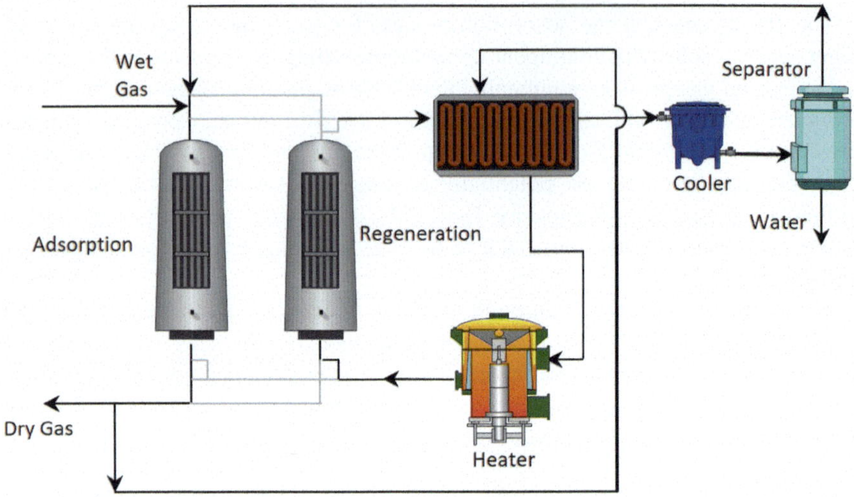

Fig. 2.6 Schematic of cyclic operation in adsorption dehydration

adsorption while another is being regenerated. This ensures that the gas processing does not need to be interrupted, providing a steady supply of purified gas [6].

2.4.1 Temperature Swing Adsorption (TSA)

Since the initial development of TSA technology in the 1950s, significant advancements have been made in both the theoretical understanding and practical application of these systems. Early patents and research laid the foundation for developing efficient TSA processes for gas purification [6].

Figure 2.7 depicts the working mechanism of a TSA plant, which plays a crucial role in natural gas processing by simultaneously achieving dehydration and hydrocarbon recovery. This plant employs three fixed beds to maintain a continuous production of natural gas that meets pipeline standards. During the adsorption phase in the first fixed bed, the adsorbent selectively captures components from the natural gas, with each component adhering to the adsorbent according to its specific affinity. While adsorption occurs in the first bed, the second bed undergoes regeneration. This involves passing a hot purge gas through the bed to remove the previously adsorbed components. The purge gas is heated to the necessary desorption temperature using a heater before entering the bed. As the temperature within the bed rises, the adsorbed molecules are released, forming a concentrated desorption stream that exits the bed. This stream, rich in heavy hydrocarbons and water, is subjected to multi-stage condensation at both high and low pressures. These stages effectively separate the heavy components, after which the purified desorption stream is reintroduced into the main feed stream. The third bed, which has completed its desorption cycle earlier, is cooled by the incoming feed gas to prepare it for the next adsorption cycle. Three beds allow for continuous operation, as one bed is always in the adsorption phase while the others are being regenerated or cooled. This carefully orchestrated sequence of operations ensures a steady output of purified natural gas. In practice, the cycle times for these processes can vary depending on several factors, including the type of adsorbent used, the concentration of components in the feed gas, and the specific requirements of the final product. For instance, silica gel-based quick cycle TSA systems can operate with cycle times as short as 0.5–2 h, while other applications may require cycle times extending beyond 16 h.

Due to their excellent renewability, polar oxidic adsorbents such as zeolites, silica gels, silica-alumina gels, and aluminum oxides are frequently employed in natural gas purification processes. These materials are particularly favored for their ability to undergo repeated cycles of adsorption and regeneration without significant performance loss, making them highly effective for long-term use in industrial settings. Their polar nature allows them to selectively adsorb polar impurities, such as water and certain hydrocarbons, from the natural gas stream, thereby ensuring

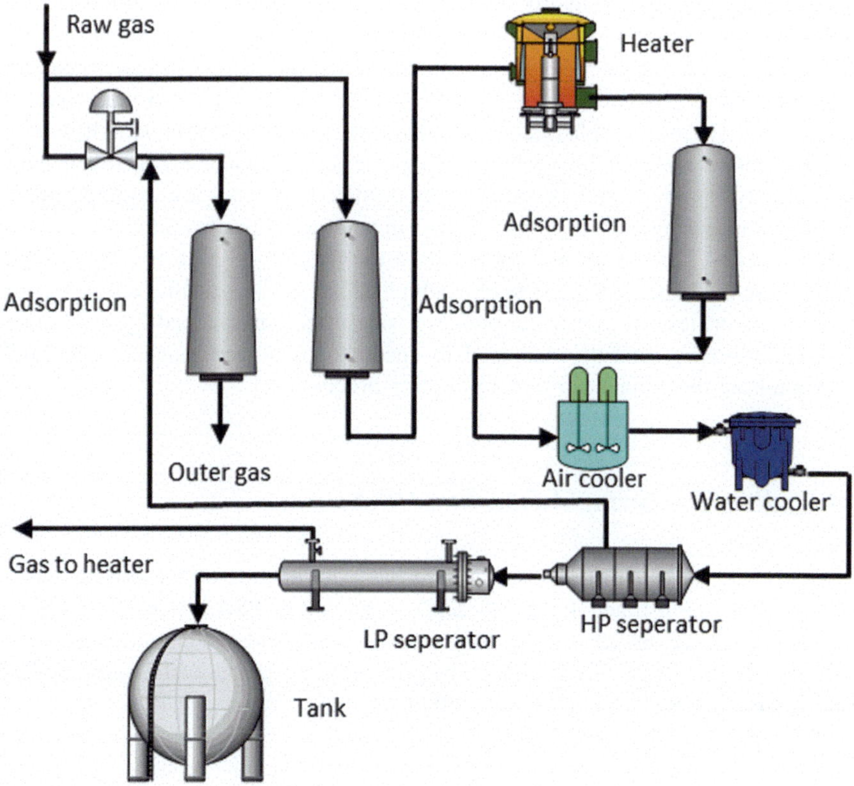

Fig. 2.7 Flow diagram of a TSA plant

the production of high-purity natural gas suitable for pipeline transmission and other applications.

For natural gas processing, zeolites are the most frequently used oxidic adsorbents, prized for their unique crystalline structure and versatile applications. Zeolites are crystalline, hydrated aluminosilicates of interconnected SiO_4 and AlO_4 tetrahedra, forming a rigid framework with well-defined pore sizes. These characteristics make them highly effective for selectively adsorbing specific molecules from natural gas streams. Among the various types of zeolites, Type A zeolites, specifically 3A, 4A, and occasionally 5A, are commonly employed for drying and purification processes. The 3A zeolite is predominantly chosen when removing carbonyl sulfide (COS) is a priority, as it effectively minimizes the presence of this impurity. The 4A zeolite is well-suited for applications where dehydration is required alongside the polishing of H_2S and CO_2. Meanwhile, the 5A zeolite is often utilized when there is a need for additional hydrocarbon removal, owing to its broader adsorption capabilities. Another widely used zeolite in the natural gas industry is the faujasite-type zeolite, known as type X. These variant features significantly larger pore sizes compared to Type A zeolites, making it ideal for the separation of heavier molecules, such as mercaptans and heavy hydrocarbons [50]. This property allows faujasite-type zeolites to play a crucial role in achieving the high-purity levels required for natural gas processing. To enhance zeolite adsorbents' mechanical strength and handling properties, the crystalline zeolite particles are often embedded in a macroporous binder material, forming spherical or cylindrical pellets with varying particle diameters. This binder material, however, can introduce certain challenges, as it may have varying porosity levels and may contain impurities that could lead to undesirable side reactions during the adsorption process in natural gas purification [18].

In summary, zeolites are indispensable in the natural gas industry due to their precisely defined microporous structure, which allows for the selective adsorption of targeted impurities. However, the potential side effects introduced by the binder materials must be carefully managed to ensure optimal performance and avoid compromising the efficiency of the adsorption process [16].

2.4.2 Pressure Swing Adsorption (PSA)

Pressure swing adsorption (PSA) technology has a rich history, with its origins dating back to 1960, as illustrated in Fig. 2.8. Since its inception, PSA systems have undergone significant advancements, leading to the development and commercialization of numerous sophisticated variations. The primary focus of these advancements has been optimizing the systems' energy efficiency, accelerating the PSA cycle, and engineering adsorbents with higher loading capacities. These improvements have garnered considerable attention, largely due to the substantial reduction in energy consumption, which, in turn, has lowered operational costs. Today, PSA technology is recognized as a mature and reliable process, extensively employed in various industrial applications, including natural gas processing, biogas upgrading, and beyond.

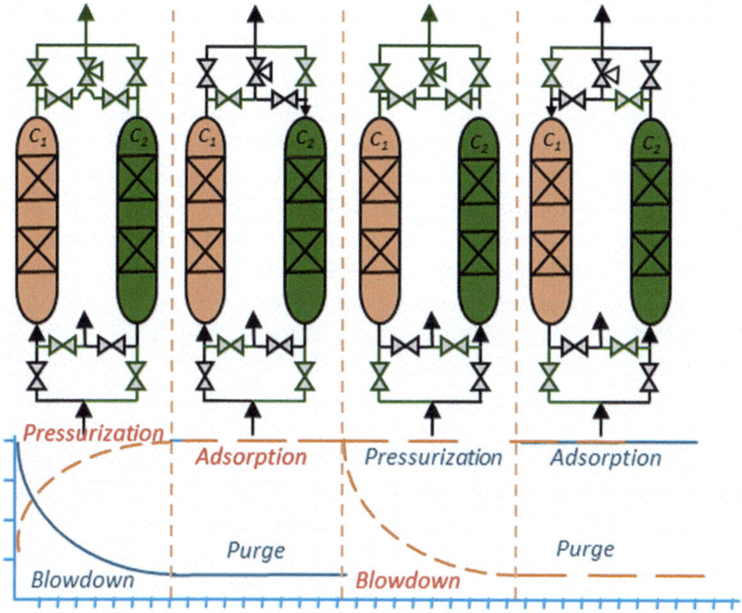

Fig. 2.8 Flow diagram of a PSA plant [1]

Its widespread adoption is a testament to its effectiveness and efficiency in these sectors.

The introduction of a vacuum PSA system represents a significant innovation aimed at enhancing the desorption process by lowering the bed pressure below atmospheric levels. While incorporating this vacuum step does result in an increase in the capital cost of the PSA system, it offers a crucial benefit: the vacuum action ensures the complete desorption of trapped gas molecules, which is particularly advantageous for adsorbents composed of small particles [51].

Selecting an appropriate adsorbent is the primary and most crucial step in designing and specifying a PSA unit. This task requires careful consideration of several criteria to ensure optimal performance. One key factor in selecting an adsorbent is its ability to selectively limit the adsorption of methane, especially at high partial pressures. The most effective way to achieve this is by restricting methane's access to the micropores of the adsorbent, where adsorption predominantly occurs. This selective exclusion is feasible because methane (CH_4) molecules are larger in size compared to carbon dioxide (CO_2) molecules.

Materials that restrict targeted molecules by size are known as "kinetic adsorbents." These adsorbents allow smaller molecules like CO_2 to enter the micropores while preventing the larger CH_4 molecules from doing so. Among the available options, a carbon molecular sieve is a readily accessible and practical material. Its unique pore structure makes it particularly suitable for applications where precise molecular separation is required, ensuring the efficient operation of the PSA system.

In addition to the vacuum step, implementing an equalization step plays a pivotal role in reducing energy consumption. This step involves recycling a portion of the captured gas from the desorption column back into the adsorption column for pressurization. This recycling process not only helps clean the adsorption bed of low-adsorptive gases but also aids in removing a portion of the light gases.

Regarding column design, adopting a tapered column configuration has shown remarkable effectiveness in managing thermal effects during adsorption. This design improvement has led to enhanced adsorption capacity and higher product recovery rates, making it a valuable advancement in PSA technology. The combined effects of these innovations contribute to more efficient and cost-effective PSA systems, further solidifying their role in various industrial applications [1].

2.5 Cryogenic Process of CO_2 Capture

Due to the high calorific value and relatively low cost of natural gas, Natural Gas Combined Cycle (NGCC) power plants have become a significant contributor to global electricity generation, accounting for over 20% of the total output [5, 27]. These plants are favored for their efficiency and cost-effectiveness,however, they are also major sources of CO_2 emissions, which pose a serious environmental challenge. Consequently, the implementation of CCS technologies is of paramount importance in reducing the carbon footprint of NGCC operations [22].

2.5.1 Conventional CCS Processes

The most widely used CCS method in NGCC plants is the post-combustion CCS process, which primarily involves MEA-based CO_2 absorption and subsequent CO_2 compression. As shown in Fig. 2.9, these factors contribute to a reduction in the overall efficiency of NGCC plants, creating what is often referred to as an "efficiency penalty" associated with traditional CCS processes [32, 53]. As shown in Fig. 2.10 [23, 24], this process is particularly advantageous because it can be retrofitted into existing power plants with relative ease, making it a practical solution for reducing emissions in established facilities [5, 38]. However, the MEA absorption process is not without its drawbacks. It requires considerable thermal energy to regenerate the CO_2 solvent, and the CO_2 compression process, which involves multi-stage compressors, is highly energy-intensive, consuming significant amounts of power [44].

In scenarios where NGCC power plants rely on natural gas sourced from distant locations, the gas must first be liquefied, reducing its volume to approximately 1/600th of its original size. This liquefaction is necessary to facilitate transportation via specialized LNG carriers to the power plant [26]. The liquefaction process is, however, extremely energy-intensive, largely due to the fact that natural gas has a

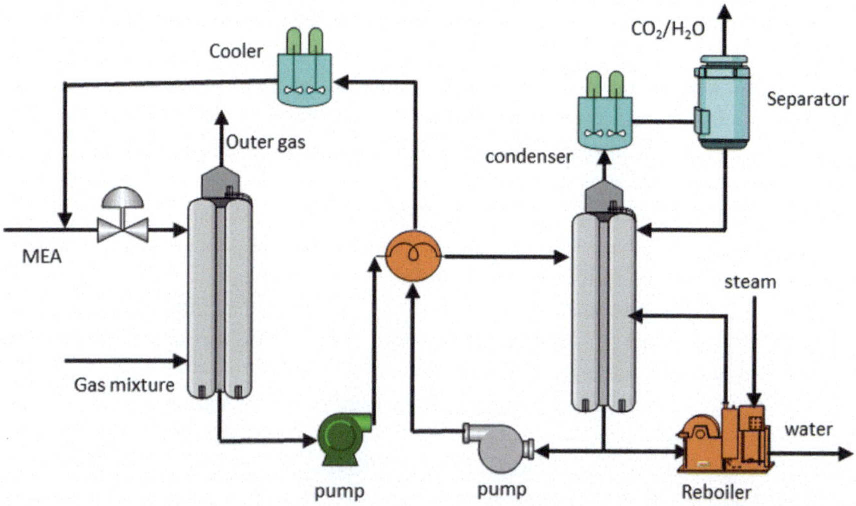

Fig. 2.9 Schematic of the conventional CCS process: MEA-based absorption process for CO_2 capture

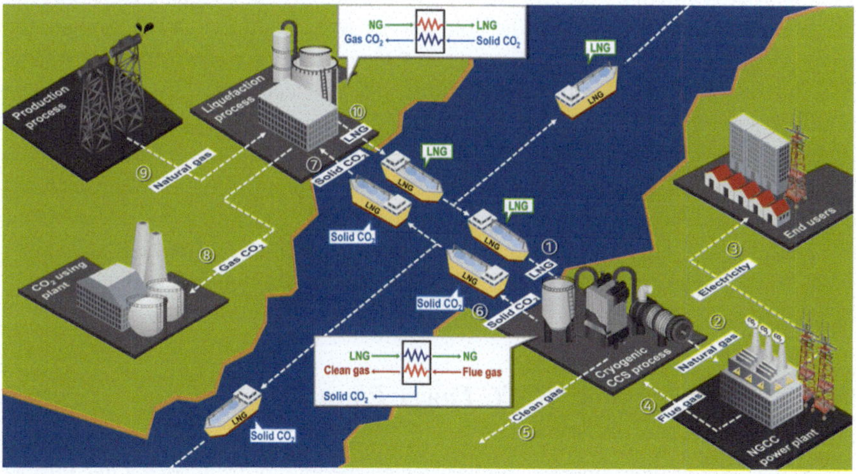

Fig. 2.10 Integration of natural gas liquefaction and regasification stages by the supply chain using solid CO_2 from the cryogenic CCS process [23, 24]

very low boiling point, typically below − 160 °C at atmospheric pressure, which requires significant energy input to achieve and maintain [28, 35]. Upon arrival at the destination, the LNG must be regasified before it can be used in the NGCC plant. This regasification process involves exchanging heat with seawater, which results in a substantial amount of cold energy being dissipated into the ocean, representing a significant loss of potential energy [29]. To address this inefficiency, there has been growing interest in the thermal integration of the liquefaction and regasification stages. By combining these processes, it may be possible to reduce the repeated energy losses that occur due to the dispersed nature of these operations. However, achieving such integration presents considerable challenges, particularly when the liquefaction and regasification facilities are located far apart, making direct energy transfer between them impractical [34].

As advancements in liquefaction technology continue, extensive research efforts have been directed toward optimizing the use of LNG cold energy across various applications. The application of LNG cold energy in conventional CCS processes has been a topic of exploration. Naveiro et al. [33] introduced an innovative approach wherein LNG cold energy is utilized to liquefy captured CO_2 through an absorption process, which simultaneously generates power. This method not only captures CO_2 but also effectively utilizes cold energy for energy recovery, enhancing the overall efficiency of the CCS process. Sultan et al. [44] took this a step further by investigating the liquefaction of gaseous CO_2, where LNG cold energy was employed to compress and cool CO_2 to its triple point, thereby eliminating the need for traditional multi-stage CO_2 compression. This approach offers a significant reduction in energy consumption, making the CCS process more economically viable. Furthermore, Kim et al. [23, 24] proposed a cutting-edge natural gas liquefaction and regasification process that integrates the LNG supply chain with a cryogenic CCS process. This advanced system leverages the cold energy from LNG to nearly eliminate the requirement for external thermal energy in CO_2 capture. By capturing and storing CO_2 in its solid form, this process circumvents the need for additional compression during storage, thereby addressing one of the major challenges of conventional CCS technologies—high energy consumption. Additionally, the cold energy from the captured solid CO_2 can be repurposed for LNG liquefaction, enabling seamless integration of the liquefaction and regasification stages. This not only prevents repetitive energy losses but also enhances the overall energy efficiency of the LNG supply chain, making it a more sustainable and cost-effective solution in the long term.

2.5.2 CCS Process in NGCC Power Plant

The NGCC power plant operates by harnessing the energy from both a gas turbine and a secondary steam turbine to generate electricity. The process begins with the combustion of natural gas and air, which produces high-temperature, high-pressure flue gas. This flue gas is directed through the gas turbine, driving it to generate electricity. Following this initial phase, the exhaust gases from the gas turbine, still containing substantial thermal energy, are utilized to produce steam. This steam powers a secondary steam turbine, enabling further electricity generation. This two-step approach maximizes the efficiency of the energy conversion process by utilizing the waste heat from the gas turbine to generate additional power. In addition to power generation, NGCC plants also address environmental concerns by incorporating a CCS process, designed to reduce the CO_2 emissions resulting from the combustion

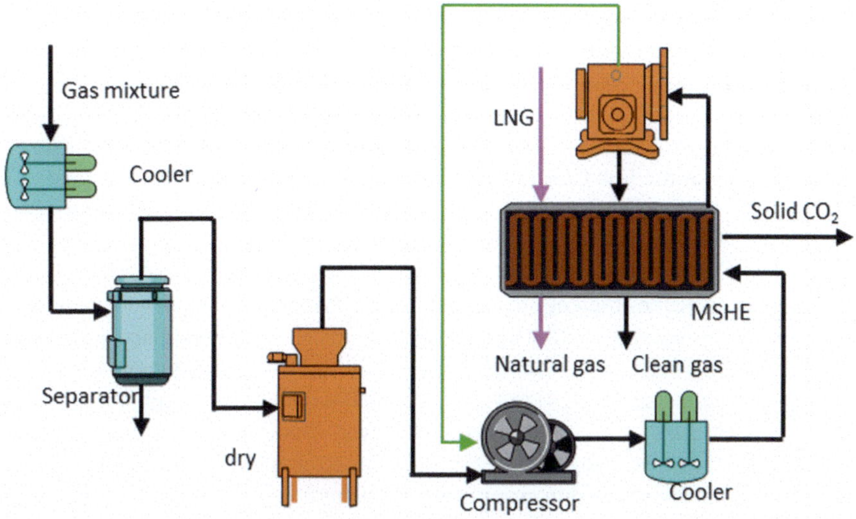

Fig. 2.11 Schematic of the cryogenic CCS process by NGCC power plant

of natural gas. Figure 2.11 presents a schematic of the traditional CCS process used in these plants.

Initially, the flue gas emitted by the NGCC power plant, which exits at a temperature of 127.5 °C, is cooled down to 40 °C through a heat exchange process with seawater. This cooling step is crucial as it prepares the flue gas for the subsequent CO_2 capture phase. The cooled flue gas is then channeled into an absorber, where CO_2 is absorbed by an MEA solvent under specific conditions—1.1 bar and 40 °C. This chemical absorption process is highly effective in capturing CO_2 from the flue gas, resulting in a CO_2-rich solvent. The next phase involves the regeneration of the MEA solvent. The CO_2-laden solvent is subjected to thermal treatment in a reboiler, operating at 2.0 bar and 120 °C. This thermal process releases the absorbed CO_2 from the solvent, allowing the MEA to be reused in the capture process. The separated CO_2, now in a gaseous state, undergoes a series of compressions using a four-stage compressor. This compression process increases the CO_2 pressure sequentially to 6, 18, 54, and 150 bar. Following compression, the CO_2 is cooled once again, this time to 30 °C, through another heat exchange with seawater, which liquefies the CO_2 for storage [23, 24]. Despite its effectiveness in capturing more than 90% of CO_2 emissions, the traditional CCS process is not without its drawbacks. One of the major challenges associated with this process is the significant energy penalty it imposes on the overall efficiency of the NGCC power plant. The energy required for the thermal regeneration of the MEA solvent and the multi-stage compression of CO_2 can result in a substantial reduction in the plant's power generation efficiency, often ranging from 8 to 14% [5, 44]. This efficiency loss represents a critical limitation, particularly as the industry strives to balance the need for effective CO_2 capture with the demand for high-efficiency power generation.

In comparison to conventional CCS processes, the utilization of the cold energy from the LNG regasification phase offers a unique opportunity to capture and store CO_2 in its solid state, taking advantage of the distinct sublimation characteristics of CO_2. This method not only maximizes CO_2 capture efficiency, achieving near-complete capture rates (close to 100%) but also operates with enhanced energy efficiency, even at the low CO_2 concentrations typically found in the flue gases of NGCC power plants. This stands in stark contrast to traditional compression-based gas–liquid separation techniques, which often suffer from significant energy penalties [7, 48, 49]. As a result, solid-phase CCS that harnesses LNG cold energy can effectively mitigate the efficiency losses commonly associated with power generation processes. This is because, unlike traditional methods, this innovative approach does not require additional steam or electricity to operate, all while maintaining high CO_2 capture rates. Kim et al. [23, 24] provide a detailed schematic of the cryogenic CCS process that leverages LNG cold energy. However, it is important to note that the implementation of this cryogenic CCS process is geographically limited to NGCC power plants that are situated in proximity to LNG terminals. In this advanced cryogenic CCS process, the flue gas emitted by the NGCC power plant first undergoes cooling to 30 °C via a seawater cooler. Following this, any condensed water within the flue gas is thoroughly removed using a two-phase separator and a dryer, ensuring that the gas is completely dried. The dried flue gas is then compressed to 1.1 bar using a compressor. This compression step is crucial as it increases the resublimation temperature of CO_2, thereby allowing for more effective CO_2 capture using the cold energy derived from LNG. The rationale behind compressing the dried flue gas is that a higher resublimation temperature directly correlates with a higher CO_2 capture rate. Subsequent to this compression, the dried flue gas is again cooled to 30 °C, this time through a seawater cooler. The compressed flue gas is then further cooled by exploiting the cold energy from both LNG and recirculated clean gas. During this cooling phase, CO_2 in the flue gas sublimates and is effectively separated from the gas stream. The resultant clean gas, now devoid of CO_2, is redirected into the turbine to help compensate for the energy consumption of the compressor. Finally, this clean gas, which has been further cooled due to the expansion process, is recirculated within the system to provide additional cooling energy to the dried flue gas before being safely released into the atmosphere. Moreover, the solid CO_2 captured and stored in this process is then transported via specialized ships to the site of the natural gas liquefaction facilities. Table 2.1 outlines the specific design parameters that govern the cryogenic CCS process, detailing the conditions and requirements necessary for its successful implementation.

2.6 Hybrid Process

The previous sections of this chapter introduced various methods used for natural gas processing. However, when any one of these methods is used alone, it often results in separation efficiency falling short of the target or being economically unfeasible.

Table 2.1 Cryogenic CCS process design parameters

Component	Parameter	Value	Unit
Flue gas	Feed temperature	127.5	°C
	Feed pressure	59.3	kPa
	Feed mass flow rate	649.8	kg/s
	Composition (mol%) N_2	74.45	%
	Composition (mol%) CO_2	3.98	%
	Composition (mol%) O_2	12.36	%
	Composition (mol%) H_2O	8.33	%
	Composition (mol%) Ar	0.88	%
LNG	Feed temperature	– 162	°C
	Feed pressure	101.3	kPa
	Feed mass flow rate	14.74	kg/s
	Composition (mol%) N_2	0.20	%
	Composition (mol%) CH_4	91.30	%
	Composition (mol%) C_2H_6	5.40	%
	Composition (mol%) C_3H_8	2.10	%
	Composition (mol%) i-C_4H_{10}	0.50	%
	Composition (mol%) n-C_4H_{10}	0.50	%
Solid CO_2	Product temperature	– 139.40	°C
	Product pressure	100.0	kPa
Discharge temperature of coolers 1 and 2		30.00	°C
Minimum temperature approach in the heat exchangers		3.00	°C
Discharge pressure of the compressor		110.0	kPa
Discharge pressure of the turbine		55.0	kPa
Isentropic efficiency of the compressor		80	%
Isentropic efficiency of the turbine		80	%

Therefore, in practical applications, multiple individual processes are often combined into a hybrid process for natural gas treatment. This section will introduce several commonly used hybrid processes.

2.6.1 *Hybrid Process Combining Membrane Separation and Chemical Absorption*

Chemical absorption is recognized as a highly effective method for treating natural gas, particularly in the removal of impurities such as carbon dioxide (CO_2). However, this process is not without its challenges. The primary drawback of chemical absorption lies in its high operational costs, which are driven by the expense of the solvents used and the difficulty in recovering these solvents for reuse. This difficulty in solvent recovery further escalates the overall cost of the process, making it a less economically viable option despite its effectiveness.

The performance of chemical absorption is highly dependent on several critical factors, including the concentration of the gas to be treated, the properties of the solvent used, and the prevailing environmental conditions. Given the substantial impact of these factors on the efficiency of the process, ongoing efforts to enhance the efficiency of chemical absorption are crucial. These efforts are aimed at optimizing the process to make it more cost-effective and sustainable in the long term.

In the context of natural gas power plants, chemical absorption is a widely employed technique for capturing carbon dioxide emissions. However, a significant challenge arises when comparing natural gas power plants to conventional coal-fired power plants. The carbon dioxide concentration in the exhaust gases of natural gas power plants is notably lower than that in coal-fired plants. This lower concentration results in a reduced capture efficiency, which in turn leads to a higher thermal load required for solvent regeneration. The increased thermal load negatively impacts the overall power generation efficiency of natural gas plants, making the process less efficient and more costly.

To address these challenges, Dong et al. [12] proposed an innovative hybrid approach that combines membrane separation with chemical absorption. In this approach, a CO_2 enrichment membrane is first used to increase the concentration of carbon dioxide in the gas stream. Following this enrichment step, chemical absorption is then applied to capture the CO_2. Figure 2.12 shows the diagram of the hybrid process of membrane separation and chemical absorption for CO_2 capture in a natural gas power plant. This hybrid method has been shown to improve the overall capture efficiency significantly. According to calculations, under conditions where 100% of the gas flow is treated and 20% of the CO_2 is recovered, the thermal load required for solvent regeneration is reduced by more than 20.7% compared to the traditional chemical absorption method.

2.6.2 Hybrid Process Combining Membrane Separation and Pressure Swing Adsorption

In addition to the hybrid process that combines membrane separation with chemical absorption, another widely adopted hybrid technique in the field of natural gas processing is the combination of membrane separation with pressure swing adsorption (PSA). This approach leverages the strengths of both technologies to achieve superior separation performance. As early as 2002, researchers Esteves and Mota [13] introduced a novel hybrid gas separation process that integrates membrane permeation with pressure swing adsorption. They developed a comprehensive integrated

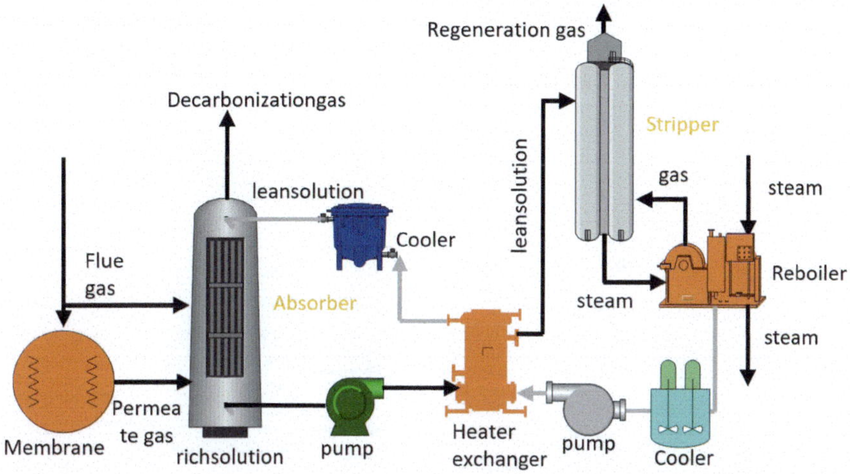

Fig. 2.12 Diagram of the hybrid process for CO_2 capture in a natural gas power plant

model that was able to accurately predict the characteristics and outcomes of the combined process. Their modeling efforts demonstrated that this hybrid approach notably improved the efficiency of both the pressurization and high-pressure adsorption phases when compared to using PSA alone. As a result, the overall separation performance of the system was significantly enhanced.

The innovative process introduced by Esteves and Mota [13] has been successfully applied to the bulk separation of gas mixtures, such as a 50/50 hydrogen and methane (H_2/CH_4) mixture. Additionally, they achieved promising preliminary results for the separation of carbon dioxide (CO_2) from methane (CH_4) and for mixtures involving hydrogen, carbon dioxide, and methane ($H_2/CO_2/CH_4$). The schematic of this hybrid process is depicted in Fig. 2.13, which outlines the different stages involved. In the diagram, the abbreviations PR, HPA, CD, BD, and PG correspond to pressurization, high-pressure adsorption, concurrent blowdown, countercurrent blowdown, and low-pressure purge, respectively. Each of these stages plays a critical role in optimizing the separation process and ensuring efficient gas purification.

Further building on their initial work, in 2007, Esteves and Mota [14] explored the potential of enhancing an existing PSA system by integrating it with an appropriate membrane separation module. The results of their study were encouraging, as they found that the integration of a suitable membrane separation module significantly improved the purity of the product gas and increased the recovery rates. This finding underscored the potential of hybrid systems to outperform conventional single-process methods, offering a more effective solution for complex gas separation challenges.

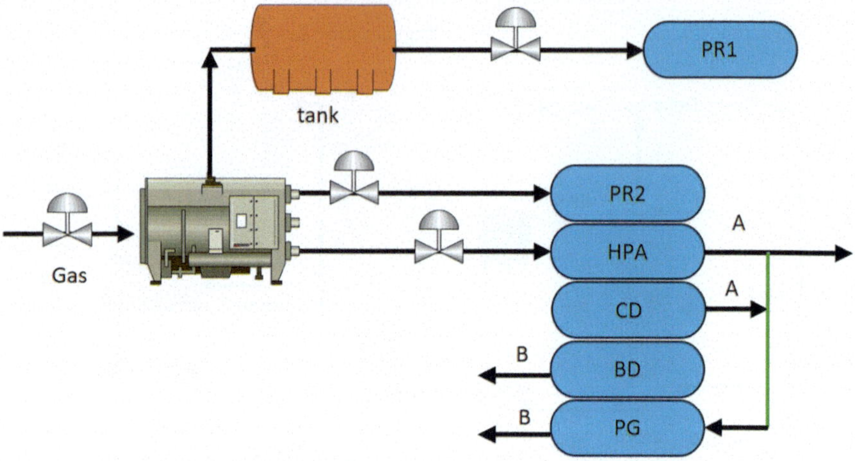

Fig. 2.13 Schematic of hybrid membrane-PSA process

2.6.3 Hybrid Process Combining Membrane Separation and Distillation

Distillation remains one of the most widely utilized methods in the chemical industry, particularly for the separation and purification of various chemical compounds. When combined with membrane separation technology, the efficiency and effectiveness of distillation can be significantly enhanced, yielding superior results in complex separation processes. In 2021, Quader et al. [39] conducted an extensive study using Aspen HYSYS simulation and techno-economic analysis to evaluate the potential of four different hybrid membrane-cryogenic distillation processes specifically designed for helium recovery from natural gas. The primary objective of their research was to improve energy efficiency and reduce the operational costs associated with helium recovery by optimizing the integration of the helium recovery unit (HeRU) with the nitrogen rejection unit (NRU) and methane production processes.

The study presented schematic diagrams of the distinct hybrid processes under investigation, as illustrated in Fig. 2.14. Among these designs, design (b) emerged as the most efficient and cost-effective. This particular hybrid process was able to produce grade A helium, with a purity level of 99%, by integrating a distillation column in the HeRU that generated 90% of the helium. This distillation process was then seamlessly combined with a single-stage membrane unit, which effectively purified the helium without the need for feed compression, further enhancing the overall efficiency of the system.

In another study, Petukhov et al. [36] explored a hybrid batch distillation and membrane separation process aimed at the removal of xenon from natural gas and the subsequent purification of high-purity xenon. This innovative process involved the initial extraction of xenon from natural gas using gas hydrate crystallization, followed

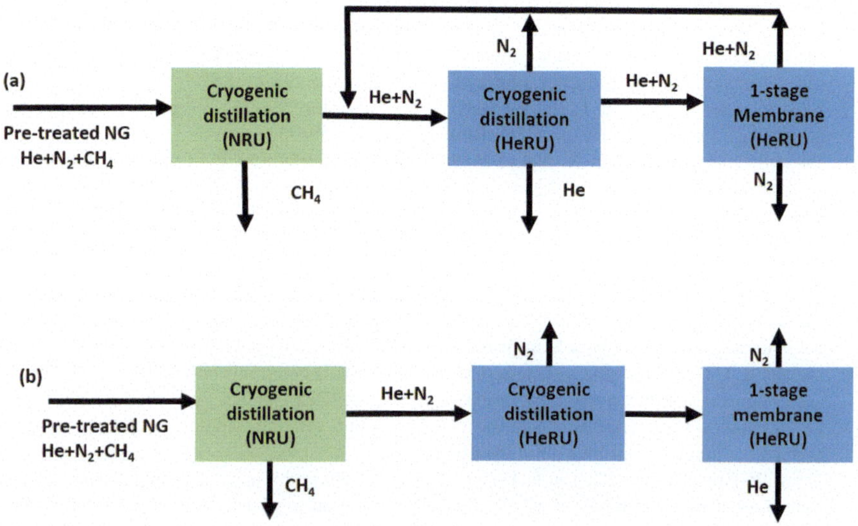

Fig. 2.14 Schematic diagrams of a kind of hybrid membrane-distillation processes

by the removal of impurities through membrane separation technology. Petukhov and colleagues conducted both experimental and simulation analyses, utilizing a range of membrane materials and operating conditions to assess the performance of the hybrid process. Their findings revealed that, in the scenarios studied, incorporating membrane separation into the batch distillation process led to a substantial increase in the batch product yield, with improvements of up to 30%. Furthermore, the hybrid process achieved a xenon recovery rate exceeding 99.8%, which was significantly higher than what could be accomplished with standalone batch distillation. The enhanced purity levels attained through this hybrid approach underscored the potential advantages of combining distillation with membrane separation for complex gas purification tasks.

References

1. Abd AA, Othman MR, Helwani Z, Kim J (2024) An overview of biogas upgrading via pressure swing adsorption: navigating through bibliometric insights towards a conceptual framework and future research pathways. Energy Convers Manage 306:118268
2. Adewole JK, Ahmad AL, Ismail S, Leo CP (2013) Current challenges in membrane separation of CO_2 from natural gas: a review. Int J Greenh Gas Control 17:46–65
3. An H, Feng B, Su S (2011) CO_2 capture by electrothermal swing adsorption with activated carbon fibre materials. Int J Greenh Gas Control 5(1):16–25
4. Anderson K, Atkins MP, Estager J, Kuah Y, Ng S, Oliferenko AA et al (2015) Carbon dioxide uptake from natural gas by binary ionic liquid–water mixtures. Green Chem 17(8):4340–4354

5. Bao J, Zhang L, Song C, Zhang N, Guo M, Zhang X (2019) Reduction of efficiency penalty for a natural gas combined cycle power plant with post-combustion CO_2 capture: integration of liquid natural gas cold energy. Energy Convers Manage 198:111852
6. Berg F, Pasel C, Eckardt T, Bathen D (2019) Temperature swing adsorption in natural gas processing: a concise overview. ChemBioEng Rev 6(3):59–71
7. Berstad D, Anantharaman R, Nekså P (2013) Low-temperature CO_2 capture technologies-applications and potential. Int J Refrig 36(5):1403–1416
8. Camper D, Bara JE, Gin DL, Noble RD (2008) Room-temperature ionic liquid–amine solutions: tunable solvents for efficient and reversible capture of CO_2. Ind Eng Chem Res 47(21):8496–8498
9. Chen GQ, Kanehashi S, Doherty CM, Hill AJ, Kentish SE (2015) Water vapor permeation through cellulose acetate membranes and its impact upon membrane separation performance for natural gas purification. J Membr Sci 487:249–255
10. Dalane K, Dai Z, Mogseth G, Hillestad M, Deng L (2017) Potential applications of membrane separation for subsea natural gas processing: a review. J Nat Gas Sci Eng 39:101–117
11. D'Alessandro DM, Smit B, Long JR (2010) Carbon dioxide capture: prospects for new materials. Angew Chem Int Ed 49(35):6058–6082
12. Dong W, Fang M, Wang T, Liu F, Yi N (2021) CO_2 capture by using a membrane-absorption hybrid process in the nature gas combined cycle power plants. Aerosol Air Qual Res 21(3):200374
13. Esteves IAAC, Mota JPB (2002) Simulation of a new hybrid membrane/pressure swing adsorption process for gas separation. Desalination 148(1–3):275–280
14. Esteves IA, Mota JP (2007) Gas separation by a novel hybrid membrane/pressure swing adsorption process. Ind Eng Chem Res 46(17):5723–5733
15. Faramawy S, Zaki T, Sakr AE (2016) Natural gas origin, composition, and processing: a review. J Nat Gas Sci Eng 34:34–54
16. Gandhidasan P, Al-Farayedhi AA, Al-Mubarak AA (2001) Dehydration of natural gas using solid desiccants. Energy 26(9):855–868
17. George G, Bhoria N, AlHallaq S, Abdala A, Mittal V (2016) Polymer membranes for acid gas removal from natural gas. Sep Purif Technol 158:333–356
18. Gleichmann K, Unger B, Brandt A (2017) Industrielle Herstellung von zeolithischen Molekularsieben Manufacturing of industrial zeolite molecular sieves. Chem Ing Tech (CIT) 89(7):851
19. Grande CA, Roussanaly S, Anantharaman R, Lindqvist K, Singh P, Kemper J (2017) CO_2 capture in natural gas production by adsorption processes. Energy Procedia 114:2259–2264
20. Hicks RL, Senules EA (1991) New gas-water-TEG equilibria. Hydrocarb Process 70(4):55–58
21. Karadas F, Atilhan M, Aparicio S (2010) Review on the use of ionic liquids (ILs) as alternative fluids for CO_2 capture and natural gas sweetening. Energy Fuels 24(11):5817–5828
22. Kazemi A, Moreno J, Iribarren D (2022) Techno-economic comparison of optimized natural gas combined cycle power plants with CO_2 capture. Energy 255:124617
23. Kim Y, Lee J, Cho H, Kim J (2023a) Novel cryogenic carbon dioxide capture and storage process using LNG cold energy in a natural gas combined cycle power plant. Chem Eng J 456:140980
24. Kim Y, Lee J, An N, Kim J (2023b) Advanced natural gas liquefaction and regasification processes: liquefied natural gas supply chain with cryogenic carbon capture and storage. Energy Convers Manage 292:117349
25. Kohl AL, Nielsen R (1997) Gas purification. Elsevier
26. Kumar S, Kwon HT, Choi KH, Lim W, Cho JH, Tak K, Moon I (2011) LNG: an eco-friendly cryogenic fuel for sustainable development. Appl Energy 88(12):4264–4273
27. Lee S, Kim JK (2020) Process-integrated design of a sub-ambient membrane process for CO_2 removal from natural gas power plants. Appl Energy 260:114255
28. Lee I, Tak K, Lee S, Ko D, Moon I (2015) Decision making on liquefaction ratio for minimizing specific energy in a LNG pilot plant. Ind Eng Chem Res 54(51):12920–12927

29. Luo X, Wang J, Dooner M, Clarke J (2015) Overview of current development in electrical energy storage technologies and the application potential in power system operation. Appl Energy 137:511–536
30. Miroshnichenko D, Teplyakov V, Shalygin M (2022) Recovery of methanol during natural gas dehydration using polymeric membranes: modeling of the process. Membranes 12(12):1176
31. Moreira MA, Ribeiro AM, Ferreira AF, Rodrigues AE (2017) Cryogenic pressure temperature swing adsorption process for natural gas upgrade. Sep Purif Technol 173:339–356
32. Muhammad HA, Sultan H, Lee B, Imran M, Baek IH, Baik YJ, Nam SC (2020) Energy minimization of carbon capture and storage by means of a novel process configuration. Energy Convers Manage 215:112871
33. Naveiro M, Gómez MR, Arias-Fernández I, Insua ÁB (2022) Thermodynamic and environmental analyses of a novel closed loop regasification system integrating ORC and CO_2 capture in floating storage regasification units. Energy Convers Manage 257:115410
34. Park J, You F, Mun H, Lee I (2021) Liquefied natural gas supply chain using liquid air as a cold carrier: novel method for energy recovery. Energy Convers Manage 227:113611
35. Park J, Mun H, Kim J, Lee I (2022) Advanced natural gas liquefaction process on LNG supply chain with liquid air: from design to thermodynamic and techno-economic analyses. Energy Convers Manage 252:115107
36. Petukhov AN, Shablykin DN, Trubyanov MM, Atlaskin AA, Zarubin DM, Vorotyntsev AV et al (2022) A hybrid batch distillation/membrane process for high purification part 2: removing of heavy impurities from xenon extracted from natural gas. Sep Purif Technol 294:121230
37. Powell CE, Qiao GG (2006) Polymeric CO_2/N_2 gas separation membranes for the capture of carbon dioxide from power plant flue gases. J Membr Sci 279(1–2):1–49
38. Qi M, Lee J, Hong S, Kim J, Liu Y, Park J, Moon I (2022) Flexible and efficient renewable-power-to-methane concept enabled by liquid CO_2 energy storage: optimization with power allocation and storage sizing. Energy 256:124583
39. Quader MA, Rufford TE, Smart S (2021) Integration of hybrid membrane-distillation processes to recover helium from pre-treated natural gas in liquefied natural gas plants. Sep Purif Technol 263:118355
40. Rufford TE, Smart S, Watson GC, Graham BF, Boxall J, Da Costa JD, May EF (2012) The removal of CO_2 and N_2 from natural gas: a review of conventional and emerging process technologies. J Petrol Sci Eng 94:123–154
41. Scholes CA, Qader A, Stevens GW, Kentish SE (2014) Membrane gas-solvent contactor pilot plant trials of CO_2 absorption from flue gas. Sep Sci Technol 49(16):2449–2458
42. Shannon MS, Bara JE (2012) Reactive and reversible ionic liquids for CO_2 capture and acid gas removal. Sep Sci Technol 47(2):178–188
43. Shiflett MB, Yokozeki A (2005) Solubilities and diffusivities of carbon dioxide in ionic liquids: [bmim][PF_6] and [bmim][BF_4]. Ind Eng Chem Res 44(12):4453–4464
44. Sultan H, Muhammad HA, Bhatti UH, Min GH, Baek IH, Baik YJ, Nam SC (2021) Reducing the efficiency penalty of carbon dioxide capture and compression process in a natural gas combined cycle power plant by process modification and liquefied natural gas cold energy integration. Energy Convers Manage 244:114495
45. Tanaka K, Okamoto KI (2006) Structure and transport properties of polyimides as materials for gas and vapor membrane separation. Materials science of membranes for gas and vapor separation, 271–291
46. Tay WH, Lau KK, Lai LS, Shariff AM, Wang T (2019) Current development and challenges in the intensified absorption technology for natural gas purification at offshore condition. J Nat Gas Sci Eng 71:102977
47. Tokarev A, Friess K, Machkova J, Šipek M, Yampolskii Y (2006) Sorption and diffusion of organic vapors in amorphous Teflon AF2400. J Polym Sci Part B Polym Phys 44(5):832–844
48. Tuinier MJ, Hamers HP, van Sint Annaland M (2011a) Techno-economic evaluation of cryogenic CO_2 capture—a comparison with absorption and membrane technology. Int J Greenh Gas Control 5(6):1559–1565

49. Tuinier MJ, van Sint Annaland M, Kuipers JAM (2011) A novel process for cryogenic CO_2 capture using dynamically operated packed beds—an experimental and numerical study. Int J Greenh Gas Control 5(4):694–701
50. Unger B, Schumann K, Brandt A (2010) Zeolithe als Sorptionsmittel. Chem Ing Tech 82(6):929–940
51. Vakylabad AB (2024) Process modeling and simulation of nitrogen separation from natural gas. In: Advances natural gas: formation, processing, and applications. Volume 8: natural gas process modelling and simulation. Elsevier, pp 335–366
52. White LS (2010) Evolution of natural gas treatment with membrane systems. Membrane gas separation, 313–332
53. Zaman M, Lee JH (2013) Carbon capture from stationary power generation sources: a review of the current status of the technologies. Korean J Chem Eng 30:1497–1526
54. Zhang J, Zhang S, Dong K, Zhang Y, Shen Y, Lv X (2006) Supported absorption of CO_2 by tetrabutylphosphonium amino acid ionic liquids. Chem Eur J 12(15):4021–4026
55. Zhang Z, Xie Y, Li W, Hu S, Song J, Jiang T, Han B (2008) Hydrogenation of carbon dioxide is promoted by a task-specific ionic liquid. Angew Chem 120(6):1143–1145
56. Zhang Y, Zhang S, Lu X, Zhou Q, Fan W, Zhang X (2009) Dual amino-functionalised phosphonium ionic liquids for CO_2 capture. Chem Eur J 15(12):3003–3011

Chapter 3
Natural Gas Processing with Supersonic Separator

3.1 Introduction

Around 80% of the world's energy demands are currently met by power plants that rely on fossil fuels. Human actions have played a major role in boosting levels of greenhouse gases, leading to a 1 °C uptick in Earth's average temperature from preindustrial times [9]. The environmental and health issues such as global warming, extreme climate changes, rising sea levels, declining air quality, and worsening poverty are becoming increasingly serious [16]. It is essential to control CO_2 emissions as these gases are the main cause of worldwide climate change. Carbon capture and storage (CCS) is becoming a popular method for addressing global warming [45]. Nations worldwide are actively exploring models for sustainable energy development, aiming to facilitate the swift transition from traditional fossil fuels to new energy sources [30]. To reduce greenhouse gas emissions by 50–85% by the middle of this century, various measures and policies to accelerate the energy revolution have been implemented, including the Paris Agreement, the European Green Deal, and China's dual carbon goals [26, 28]. The incorporation of CCS is seen as crucial in strategies to reduce climate change because it serves as a connection between conventional fossil fuels and sustainable energy options. More important, the significance of low-carbon fossil energy has heightened as a crucial element in addressing climate change and ensuring energy sustainability [50]. Natural gas, recognized as the cleanest fossil energy resource, has evolved into a vital bridge facilitating the transformation of global energy structures. Compared with coal and oil, it demonstrates a reduction in CO_2 emissions ranging from 29 to 44% [14]. Consequently, many nations are actively exploring and developing natural gas resources. Projections indicate a potential 30% increase in demand for natural gas by 2040 [38].

At present, unprocessed natural gas is obtained from three different types of wells: oil wells, gas wells, and condensate wells. Associated gas is the name commonly used for natural gas extracted from oil wells. This gas can exist on its own in the reservoir as free gas or be mixed into the crude oil as dissolved gas. On the other hand,

H. Ding et al., *Supersonic Separators for Sustainable Gas Purification*, Green Energy and Technology, https://doi.org/10.1007/978-3-031-85213-8_3

non-associated gas comes from wells that primarily produce gas and condensate without much or any crude oil. Gas wells normally produce just raw natural gas, whereas condensate wells yield both free natural gas and semi-liquid hydrocarbon condensates [5].

Regardless of where it comes from, natural gas is frequently discovered in combination with other hydrocarbons like ethane, propane, butane, and pentanes after being separated from crude oil. Moreover, unprocessed natural gas consists of a variety of other substances such as H_2O, H_2S, CO_2, He, N_2, among others. In this blend, natural gas liquids (NGLs) have substantial worth as byproducts of processing natural gas. NGLs, which include ethane, propane, butane, iso-butane, and natural gasoline, are sold individually and have various uses. These objectives consist of improving oil recovery in oil wells, serving as materials for oil refineries or petrochemical plants, and acting as sources of energy [59] .

Natural gas obtained directly from the wellhead, while rich in methane, is not pure enough for various applications. To produce pure methane and high-molecular-weight hydrocarbons suitable for diverse uses, a series of purification steps, known as natural gas processing, are employed. These steps involve the separation of different hydrocarbons and fluids from the raw natural gas. Additionally, natural gas contains water vapor, with its concentration increasing with temperature or pressure. The presence of water and various gas components can lead to hydrate formation at different temperatures and pressures. It is crucial to develop techniques to minimize hydrate formation to ensure smooth operations and prevent potential issues. Before transportation through pipelines, natural gas with impurities undergoes drying and sweetening processes. Dehydration is a vital step during the gas pipeline gas-up phase to prevent hydrate formation and protect pipelines from fouling and corrosion. The removal of impurities from natural gas is essential for efficient combustion, safeguarding equipment, enhancing heat transfer efficiency, controlling emissions, and maintaining the effectiveness of catalysts. These purification processes contribute to improved energy conversion, reduced environmental impact, and enhanced overall system performance.

Traditional process methods for natural gas include molecular sieve adsorption [39], triethylene glycol (TEG) absorption [33], and Joule–Thomson expansion [3]. Compared with the aforementioned methods, the supersonic separator (SS) emerges as a cutting-edge technology for the dehydration of natural gas, distinguished by its outstanding features such as high efficiency, environmental friendliness, low energy consumption, simple structure, and easy maintenance [7, 53].

The supersonic separator has undergone comprehensive assessments by researchers, highlighting its superior performance in gas processing concerning energy consumption, environmental sustainability, and economic efficiency. Additionally, the high flow velocity of the device minimizes fouling or solid deposition, eliminating the need for cryogenic cooling units. The natural cooling effect in the device allows it to reach temperatures of up to 60 °C. Moreover, this apparatus can be used in various environments, including land, sea, and deep water applications. Arinelli et al. demonstrated that, in the processing of CO_2-rich natural gas,

SS could achieve a 44% reduction in power consumption compared with the established membrane permeation (MP) route [20] along with a 7.5% increase in net value compared with non-intensified methods). Teixeira et al. found that SS could significantly reduce methanol losses by 91.9% in high-pressure natural gas processing [42]. Additionally, Teixeira et al. substantiated that the SS process excels on both environmental and economic fronts [44], enhancing gas-plant profitability through reduced power consumption and increased revenues from liquefied petroleum gas [43]. Furthermore, de Melo et al. demonstrated that exclusive reliance on SS can yield a 33% higher net value, a 40% boost in oil production, and a 10% reduction in investment costs [19]. Arinelli et al. provided evidence that the process using a supersonic separator results in the lowest carbon emissions at − 28.3%, the lowest power consumption at − 21.3%, and the best economic performance, boasting the highest 20-year net value of + 860 MMUSD [22]. This suggests that using supersonic separators in natural gas processing provides significant advantages in environmental impact, economic benefits, and power consumption efficiency.

In summary, the comprehensive advantages offered by supersonic separators, coupled with their favorable comparison to other purification technologies in terms of cost-effectiveness, environmental impact, and operational flexibility, underscore their prominence as a promising and preferred technique for natural gas purification. The ongoing scientific exploration of their design and applications further solidifies the potential of supersonic separation in addressing the evolving needs of the natural gas industry.

3.2 Supersonic Separation Processing

Supersonic separators represent a significant advancement in natural gas processing technology, offering a more energy-efficient alternative to conventional separation methods. These separators exploit the unique dynamics of gas flow at supersonic velocities to achieve enhanced separation efficiency. Unlike traditional technologies, supersonic separation boasts a simple mechanism, simplified equipment design, the absence of moving parts, reduced maintenance needs, and zero emissions, making them an attractive option for modern, environmentally conscious operations. Figure 3.1 illustrates the configuration of a supersonic dehydration line and outlines the configuration of a supersonic dehydration line, illustrating the integration of these advanced separators within the natural gas conditioning process. The novel design of the supersonic separator enables it to perform multiple functions, including dehumidification and the extraction of heavy hydrocarbons from natural gas streams. This capability is particularly advantageous in gas conditioning applications where the removal of moisture and heavier components is critical for improving the quality of the gas and protecting downstream equipment from damage caused by condensation or fouling.

Correcting dew points for water and heavy hydrocarbons is a critical process in the preparation of natural gas for transportation and optimal economic utilization.

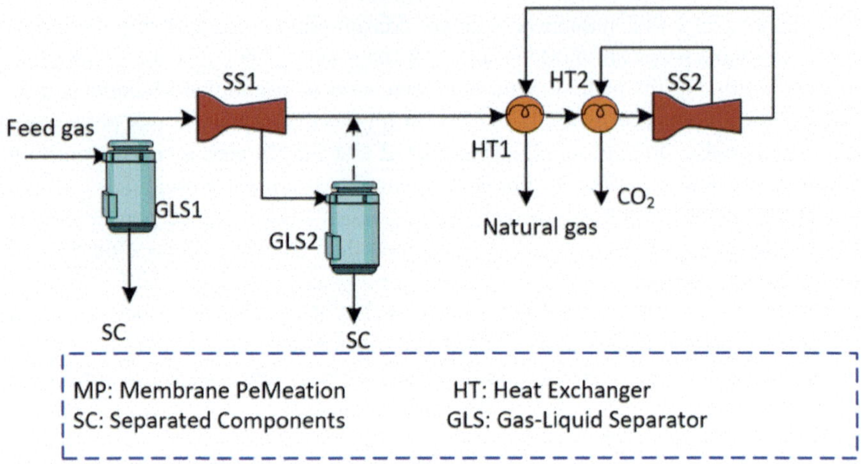

Fig. 3.1 Schematic of a supersonic processing

Supersonic separators (SS units) have emerged as a versatile and efficient solution in the natural gas industry, not only capable of removing carbon dioxide and hydrogen sulfide but also applicable in diverse scenarios such as the production of liquefied natural gas (LNG), hydrogen generation, biogas upgrading, industrial emission control, and petrochemical refining enhancement [2]. Traditional methods for dew-point correction, such as refrigeration, membranes, adsorption, and absorption, can be resource-intensive and environmentally burdensome due to their complexity, substantial facility requirements, and reliance on chemicals that may have adverse environmental impacts. In contrast, supersonic separators offer a more sustainable and efficient approach to purifying natural gas streams by removing a broad spectrum of impurities.

One of the key advantages of supersonic separators is their robust reliability, achieving separation efficiencies of up to 99% [40]. Unlike conventional separators, which often require the input of a multiphase mixture, SS units can operate effectively with a single-phase gas stream. Through the process of adiabatic expansion, the temperature drop can cause water or heavy hydrocarbons to condense into a liquid phase at low temperatures. Upon pressure recovery, the gas and liquid phases separate, and a clean, single-phase gas stream exits the SS unit. Unlike conventional separators, which often require the input of a multiphase mixture, SS units can operate effectively with a single-phase gas stream. Through the process of adiabatic expansion, the temperature drop can cause water or heavy hydrocarbons to condense into a liquid phase at low temperatures. Upon pressure recovery, the gas and liquid phases separate, and a clean, single-phase gas stream exits the SS unit.

The compact tubular structure of supersonic separators contributes to their stability and efficiency, making them lightweight and space-saving. Their design lacks any moving or rotating parts, which significantly reduces maintenance needs and increases operational reliability. Moreover, the absence of chemical requirements for pollutant discharge makes supersonic separators an environmentally responsible choice, aligning with contemporary efforts to reduce the ecological footprint of industrial processes. In essence, supersonic separators represent a significant leap in natural gas processing technology, offering a streamlined, reliable, and eco-friendly solution for dew-point correction and gas purification. Their ability to handle a wide range of impurities and operate efficiently in various applications underscores their importance in advancing the sustainability and economic viability of the natural gas industry.

3.3 Typical Structures of the Supersonic Separator

Supersonic separators (SS) are compact devices equipped with convergent–divergent Laval nozzles and stationary vanes at the inlet to generate swirls. Their refrigeration effect surpasses that of expanders, vortex tubes, and Joule–Thomson (J-T) devices. The typical structures of supersonic separator are shown in Chap. 1, comprising essential components such as inlet, swirl generator, Laval nozzle, cyclone separation section, drainage section, wet gas outlet, diffuser, and dry gas outlet. Meanwhile, two of them are characterized by a guide cone or an inner body that stabilizes the flow field.

Swirl generator is located either before or after the Laval nozzle, creating a strongly rotating flow field when natural gas flows through them. The Laval nozzle uses pressure-to-cold energy conversion to condense condensable components into liquid droplets. In the cyclone separation section, a thin liquid film is formed through the physical deposition of droplets, aided by the swirling effect in the flow field. This film is then separated in the drainage section. The resulting dry natural gas undergoes pressure energy recovery in the diffuser before being discharged through the dry gas outlet. The key determinants of supersonic separation performance are the condensation effect of vapor and the swirl strength generated by the swirl generator [10].

A significant advantage of supersonic separators is their short residence time, which effectively prevents hydrate formation, eliminating the need for regeneration devices and inhibitors. This process offers environmental benefits. Simulation studies on supersonic separators require an integrated approach that considers problem formulation and all performance-affecting factors, such as the number of phases, the equation of state, the type of shockwave, the nozzle geometry, and the working fluid selection. Designing or analyzing the performance and operating conditions of a supersonic separator involves making several key decisions. Effective design of supersonic separators relies on a comprehensive understanding of physical phenomena and precise control of various design parameters. For example, different equations of state significantly impact predicted flow behavior, while nozzle geometry directly affects separation efficiency. Combining numerical simulations and experimental data allows for more accurate prediction and optimization of supersonic separator performance, leading to more efficient and greener operations.

The performance and design of supersonic separators have been extensively studied. The supersonic separator enhances the centrifugal separation of the condensed phase by converting axial velocity into angular velocity. Compared to traditional technology, supersonic separators offer the following significant advantages:

(1) Compact size: Supersonic separators are smaller and require less installation space. This design simplifies installation and improves transportation and handling.
(2) Cost-effective: Their simplicity and small size make supersonic separators cheaper to install than traditional systems. This cost advantage makes them a top choice, especially for projects with tight budgets.
(3) Lower operating costs: Supersonic separators have a simple design and low energy consumption, resulting in generally low operating costs. They effectively separate heavy components like C^{3+} from natural gas streams, leading to significant long-term cost savings.
(4) Environmentally friendly: Supersonic separators enhance environmental sustainability by removing impurities and heavy components from the gas stream, producing cleaner gas and reducing environmental impact.

In general, supersonic separators offer clear advantages in space efficiency, transportation convenience, cost-effectiveness, reduced operating expenses, and environmental protection. These characteristics make them an attractive choice across multiple industries and applications, strongly supporting efficient, economical, and sustainable separation processes. They not only solve many issues in traditional separation technology but also bring more possibilities to industrial production through innovative design and efficient operation. In particular, as international environmental protection requirements increase, equipment that provides cleaner production processes will become increasingly popular. Therefore, with future technological advancements and expanding applications, supersonic separators are expected to play a significant role in a wider range of fields.

3.3.1 Laval Nozzle

In supersonic separators, Laval nozzles (converging–diverging nozzles) condense condensable components into droplets by converting pressure into cold energy. This process is critical to the design and optimization of supersonic separators because it plays a key role in accelerating the fluid to supersonic speeds. Designing and optimizing a Laval nozzle for a specific separation task requires careful consideration of multiple factors, including inlet and outlet conditions, fluid properties, nozzle shape, shock wave structure, and reliability requirements. Selecting and optimizing nozzle type and geometry is critical to achieving efficient particle separation.

A Laval nozzle consists of a converging section, a throat, and a diverging section that can accelerate flows to sonic or supersonic speeds. For applications that require acceleration from subsonic to sonic speeds, a simpler converging nozzle may be sufficient. The shape of the converging section directly affects how quickly and efficiently the fluid is accelerated, while the diverging section is critical to maintaining supersonic flow and preventing shock waves from forming. Optimizing these nozzles requires a variety of strategies, such as tailoring the nozzle to achieve a specific Mach number at the exit, controlling shock waves, adjusting the divergence angle, optimizing the throat size, ensuring uniform flow distribution, and considering droplet behaviors.

Matching the nozzle design to the fluid properties, density, and particle size is critical to achieving effective separation. Balancing the nozzle size with varying inlet conditions and droplet properties is essential for operational flexibility in separator design. Achieving the desired Mach number and pressure ratio at the outlet is critical to effective separation while avoiding flow boundary separation, shock wave boundary layer interactions (SWBLI), and flow instabilities are key to maintaining performance and reliability.

In the design of a Laval nozzle, the effects of fluid properties such as compressibility, viscosity, density, and latent heat on flow behavior and separation efficiency should be considered. By comprehensively considering these factors, the nozzle design can be optimized, thereby improving the overall performance and reliability of the supersonic separator. Moreover, historical contributions to the field by researchers like Oswatitsch and studies by Wyslouzil and colleagues [37, 55] have further advanced the understanding of Laval nozzle principles.

Figure 3.2 shows a schematic of a Laval nozzle. As the gas expands and cools in the nozzle, the supersaturation ratio rapidly increases. When the supersaturation exceeds a critical value on a microsecond timescale, a phase transition occurs. It is observed that when the gas mixture undergoes a phase transition, its temperature and pressure are higher than when the same gas mixture is expanded without condensation due to isentropic expansion [56].

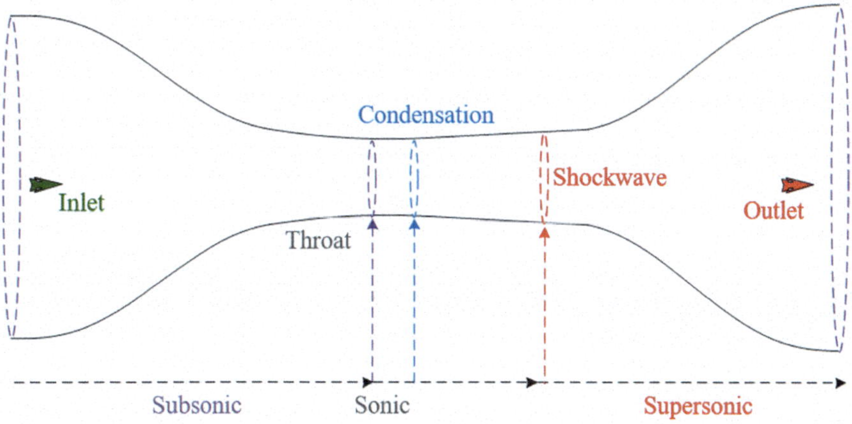

Fig. 3.2 Schematic of a Laval nozzle

As shown in Fig. 3.2, a Laval nozzle can be segmented into three regions based on flow speed: subsonic, sonic, and supersonic. Because the thermodynamic properties within the nozzle change along its length, precise positioning of the nozzle during simulation is essential. The geometry of each nozzle design varies and is typically defined by a series of convergent–divergent equations. To model the nozzle effectively, it is beneficial to establish a general positional framework that accommodates different nozzle geometries. This approach ensures that the nozzle's thermodynamic properties and flow behavior are accurately represented in simulations, which is crucial for optimizing performance and achieving efficient separation.

In a Laval nozzle, the temperature of the fluid drops sharply as it expands to supersonic speeds, often leading to the condensation of vapor. The flow characteristics are described by the Mach number (Ma), which is the ratio of the fluid's axial velocity (v) to the speed of sound (c). The behavior of the flow in a Laval nozzle is as follows:

Initially, the flow is subsonic ($\mathrm{Ma} < 1$) in the converging section, where the nozzle's cross-sectional area decreases gradually. As the flow approaches the nozzle's throat, where the cross-sectional area is at its minimum, the velocity reaches the speed of sound ($\mathrm{Ma} = 1$). Beyond the throat, in the diverging section of the nozzle, the cross-sectional area increases rapidly, causing the flow to expand and accelerate. As a result, the flow becomes supersonic ($\mathrm{Ma} > 1$) and reaches extremely high velocities.

To ensure a smooth increase in flow rate without causing fluid segregation, the profile curve of the nozzle's convergent section should be smoothly contoured and avoid abrupt changes. Once the inlet and outlet dimensions of the nozzle are established, it is crucial to maintain a uniform mass flow distribution throughout the flow field. This uniformity is essential for maintaining both the stability and efficiency of the fluid flow. The structural profile models for the convergent section are as follows:

(1) Witoszynski curve

$$D = \frac{D_t}{\sqrt{1 - \frac{(1-x^2)}{\left(1+\frac{x^2}{3L_{con}^2}\right)^3}\left[1 - \left(\frac{D_t}{D_{in}}\right)^2\right]}} \tag{3.1}$$

(2) Bicubic curve

$$\frac{D - D_t}{D_{in} - D_t} = \begin{cases} 1 - \frac{(x/L_{con})^3}{x_m^2}, & (x/L_{con}) \le x_m \\ \frac{[1-(x/L_{con})]^3}{(1-x_m)^2}, & (x/L_{con}) > x_m \end{cases} \tag{3.2}$$

where x_m is the connection point at both ends of bicubic curve equation, which meets the condition of $x_m/L_{con} = 0.52$.

(3) Batchelor–Shaw curve

$$\frac{D}{D_{th}} = \sqrt[4]{\left[1 - \left(\frac{D_{th}}{D_{in}}\right)^4\right]\left(\frac{x}{L_{con}} - \frac{1}{2\pi}\sin\frac{2\pi x}{L_{con}}\right) + \left(\frac{D_{th}}{D_{in}}\right)^4} \tag{3.3}$$

(4) Quintic profile curve

$$\frac{D - D_{th}}{D_{in} - D_{th}} = 1 - 10\left(\frac{x}{L_{con}}\right)^3 + 15\left(\frac{x}{L_{con}}\right)^4 - 6\left(\frac{x}{L_{con}}\right)^5 \tag{3.4}$$

(5) Improved quintic profile curve

$$\frac{D}{D_{in}} = \left\{\eta\left[1 - \left(\frac{D}{D_{in}}\right)^{\frac{1}{f(\varepsilon)}}\right] + 1\right\}^{f(\varepsilon)} \tag{3.5}$$

where $f(\varepsilon)$ is 0.1, 0.3, 0.5, 0.7, 1.0, $\varepsilon^{1/4}$, $\varepsilon^{1/3}$, $\varepsilon^{1/2}$, ε, ε^2, and ε^3.

$$\eta = -10\varepsilon^3 + 15\varepsilon^4 - 6\varepsilon^5 \tag{3.5a}$$

$$\varepsilon = \frac{x}{L_{con}} \tag{3.5b}$$

(6) Area ratio method

$$\frac{D}{D_{th}} = \sqrt{\frac{1}{\mathrm{Ma}(x)}\left[\left(1 + \frac{\gamma - 1}{2}\mathrm{Ma}(x)^2\right)\left(\frac{2}{\gamma + 1}\right)\right]^{\frac{\gamma+1}{2(\gamma-1)}}} \tag{3.6}$$

where

$$\mathrm{Ma}(x) = (1 - \mathrm{Ma}_{in})\left(x - \frac{1}{2\pi}\sin 2\pi x\right) + \mathrm{Ma}_{in} \tag{3.6a}$$

(7) Linear profile curve
A simple straight line is used to replace the profile curve.

Figure 3.3 shows the profile curve corresponding to the equations for the convergent section. It is important to consider the effects of the internal body and guide cone when designing the nozzle's convergence profile.

In the diverging section of a Laval nozzle, an irreversible normal shock can occur as supersonic flow transitions to subsonic flow. This adiabatic shock leads to an increase in entropy, pressure, and temperature, while energy, momentum, and mass flow rate remain constant. To minimize separation losses caused by re-evaporation, it is crucial to collect condensate before the shock wave forms. After the shock, the subsonic flow recovers its temperature and pressure as it continues through the nozzle's exit.

The common structural profiles for the convergent section are as follows:

(1) Equal slope method

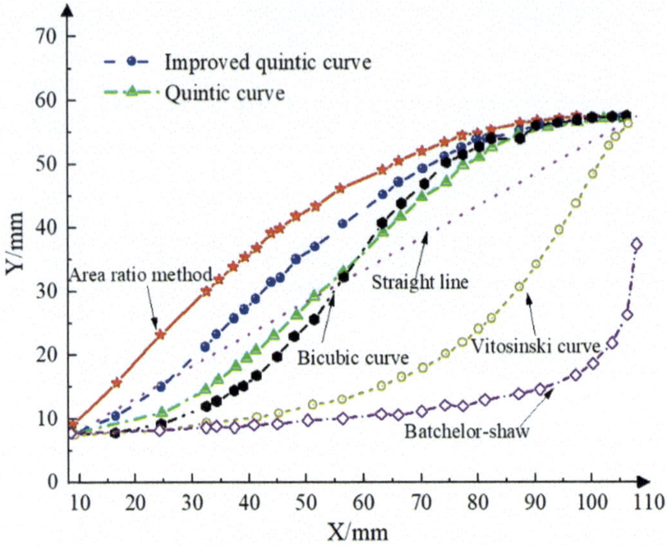

Fig. 3.3 Profiles curve of convergent section

$$L_{\text{div}} = \frac{D_{\text{div}} - D_{\text{th}}}{2\tan\frac{\phi}{2}},\ \phi = 8^\circ \sim 12^\circ \tag{3.7}$$

$$\frac{D_{\text{div}}}{D_{\text{th}}} = \frac{1}{\sqrt{\text{Ma}_{\text{div}}}}\left[\frac{1 + \frac{1}{2}(\gamma - 1)\text{Ma}_{\text{div}}^2}{1 + \frac{1}{2}(\gamma - 1)}\right]^{\frac{\gamma+1}{4(\gamma-1)}} \tag{3.8}$$

(2) Area ratio method

$$\frac{D}{D_{\text{th}}} = \sqrt{\frac{1}{\text{Ma}(x)}\left[\left(1 + \frac{\gamma - 1}{2}\text{Ma}(x)^2\right)\left(\frac{2}{\gamma + 1}\right)\right]^{\frac{\gamma+1}{2(\gamma-1)}}} \tag{3.9}$$

where

$$\text{Ma}(x) = (1 - \text{Ma}_{\text{div}})\left(x - \frac{1}{2\pi}\sin 2\pi x\right) \tag{3.9a}$$

(3) Foelsch method

$$y = \frac{D_{\text{th}}}{2} + \frac{(n-1)x_{\text{B}}\tan\beta_{\text{B}}}{2(n-2)}\left(\frac{x}{x_{\text{B}}}\right)^2 - \frac{x_{\text{B}}\tan\beta_{\text{B}}}{2(n-2)}\left(\frac{x}{x_{\text{B}}}\right)^n \tag{3.10}$$

where

$$x_B = \frac{2n}{n+1}\frac{D_{div}}{2}\frac{(A_B/A_t)[180\sin\beta_B(\pi\beta_B)] - 1}{(A_T/A_t)\tan\beta_B} \tag{3.10a}$$

The n is the degree of the curve, which is the cubic curve when $n = 3$.

(4) Linear profile curve

A simple straight line is used to replace the profile curve.

The throat transition between the convergent and divergent sections is critical for the performance of a Laval nozzle. The design of this transition directly impacts the nozzle's overall efficiency. An improperly designed or rough throat transition can cause fluid separation, leading to unstable flow and reduced efficiency. A well-designed transition minimizes energy loss and ensures smooth fluid flow through both the throat and the divergent section. The circular arc transition curve is the commonly used arc transition curve at the throat as shown in Fig. 3.4. Its arc transition curve equation is as follows:

The ratio of source flow r_0 is expressed as

$$r_0 = \frac{y^*}{\beta_B} \tag{3.11}$$

where y^* is the half height of throat, β_B is the expansion angle at the turning point B. According to design experience, when the Mach number of the test section is given, the value of the β_B angle can be determined by referring to Table 3.1.

The Prandt–Meyer angle at turning point α_B is expressed as

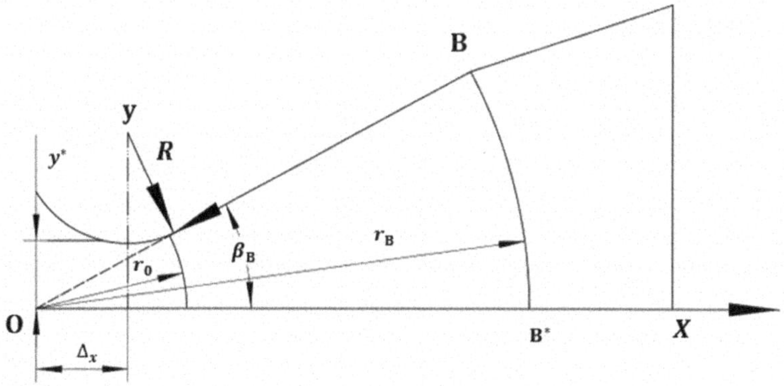

Fig. 3.4 Circular arc transition curve

Table 3.1 Expansion angle β_B more dependent on Mach number

Ma	1.5 ~ 2	3 ~ 4	5 ~ 7	10
β_B	3° ~ 5°	8°	12°	13°

$$\alpha_{\mathrm{B}} = \frac{\nu_1 - \beta_{\mathrm{B}}}{1+\sigma} \tag{3.12}$$

For axisymmetric nozzle, $\sigma = 1$. For two-dimensional nozzle $\sigma = 0$.

$$\left(\frac{r_{\mathrm{B}}}{r_0}\right)^{1+\sigma} = \frac{1}{\mathrm{Ma_B}}\left[\left(\frac{2}{\gamma+1}\right)\left(1+\frac{\gamma-1}{2}\mathrm{Ma_B^2}\right)\right]^{\frac{\gamma+1}{2(\gamma-1)}} \tag{3.13}$$

where γ is specific heat ratio. Thus, the vertical position of turning point B is

$$y_{\mathrm{B}} = r_{\mathrm{B}} \sin \beta_{\mathrm{B}} \tag{3.14}$$

The lateral position of turning point B is

$$x_{\mathrm{B}} = r_{\mathrm{B}} \cos \beta_{\mathrm{B}} - \Delta x \tag{3.15}$$

where

$$\Delta x = y^* \cot \beta_{\mathrm{B}} - R \tan \frac{\beta_{\mathrm{B}}}{2} \tag{3.15a}$$

According to experience, in this case of $R \geq 8y^*$, the uniform sonic flow at the throat can be transformed into the source flow.

3.3.2 *Swirl Generator*

The swirl generator is a crucial part of the supersonic separator, essential for removing natural gas components. These vanes impart centrifugal force to the fluid, enabling the separation of condensed droplets. The droplets then collide with the separator wall, forming a liquid film. Currently, there are two primary installation positions for the swirling vanes: at the rear or the front of the Laval nozzle. The position of the swirl generator greatly impacts the temperature distribution within the nozzle. When installed at the rear of the Laval nozzle, the swirl generator can easily produce shock waves. These shock waves increase temperature, causing secondary evaporation of condensed droplets, reducing the overall efficiency of the separation process. Consequently, axis flow swirl generators, installed at the front of the Laval nozzle, are now more commonly used to avoid these issues and provide more stable temperature control [47].

This section discusses the forming method and structural cross-sectional model of the swirl blades installed on the axial flow swirl generator. Designing these vanes involves intricate considerations to ensure they provide the desired centrifugal force while maintaining structural integrity under high-pressure and high-velocity conditions typical of supersonic separators. The vanes' profile must be optimized to achieve

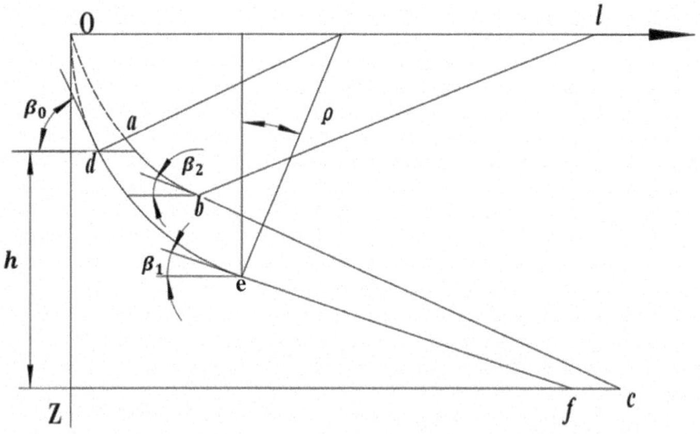

Fig. 3.5 Profiles of swirling vanes

efficient separation without causing excessive turbulence or pressure drops. The vanes' structural profile model includes detailed geometrical features designed to maximize effectiveness in creating necessary centrifugal forces while minimizing adverse effects on flow dynamics within the nozzle.

Therefore, carefully considering the position and design of swirling vanes allows supersonic separators to achieve improved efficiency in removing natural gas components, ensuring better performance and reliability in industrial applications. Figure 3.5 shows the unfolded drawings of swirling vanes on the three-dimensional axial flow swirl generator. The structure and the forming principle of four kinds of swirling vanes are described as follows, and Fig. 3.6 displays the corresponding profile curve of the swirling vanes.

(1) Circular arc swirling vanes

The parameter equation of the curve arc of inside directrix is shown in:

$$z = \sqrt{r_{\mathrm{a}}^2 - (r_{\mathrm{a}} - l)^2} \tag{3.16}$$

where r_{a} radius of circular arc.

$$l = \frac{\pi r_0 \alpha_1 t}{180}, \quad (0 \le t \le 1) \tag{3.16a}$$

$$r_{\mathrm{a}} = \frac{\pi r_0 \alpha_1}{180}(1 - \sin \beta_1) \tag{3.16b}$$

The parameter equation of the tangential path of inside directrix is shown:

$$z = \left(l - \frac{\pi r_0 \alpha_1}{180}\right) \tan \beta_1 + r_{\mathrm{a}} \cos \beta_1 \tag{3.17}$$

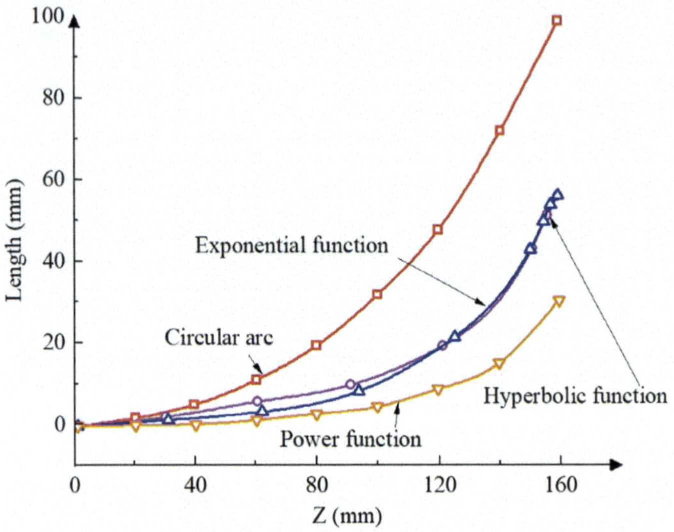

Fig. 3.6 Profiles of four different functions

where

$$l = \frac{\pi r_0(\alpha_1 + \alpha_2 t)}{180}, \quad (0 \leq t \leq 1) \tag{3.17a}$$

The parameter equation of the curve arc for the outside directrix is shown in:

$$z = \sqrt{r_a^2 - \frac{(kr_a - l)^2}{k^2}} \tag{3.18}$$

where $k = r_1/r_0$

$$l = \frac{\pi r_1 \alpha_{1s} t}{180}, \quad (0 \leq t \leq 1) \tag{3.18a}$$

The parameter equation of the tangential path for the outside directrix is shown in:

$$z = \left(1 - \frac{\pi r_1 \alpha_{1s}}{180}\right) \tan \beta_2 + \sqrt{r_a^2 - \frac{\left(kr_a - \frac{\pi r_1 \alpha_{1s}}{180}\right)^2}{k^2}} \tag{3.19}$$

where $k = r_1/r_0$

$$l = \frac{\pi r_0(\alpha_{1s} + \alpha_{2s} t)}{180}, \quad (0 \leq t \leq 1) \tag{3.19a}$$

where α_{1s} is the enveloping angle of the arc segment for outside directrix, α_{2s} is the enveloping angle of the tangential path for outside directrix.

(2) Power function swirling vanes

The parameter equation of the curve arc of inside directrix is shown in:

$$x = AZ^B \tag{3.20}$$

where

$$B = \frac{h_0 \cot \beta_1}{l_0} \tag{3.20a}$$

$$A = \frac{l_0}{h_0^{h_0 \cot \beta_1 / l_0}} \tag{3.20b}$$

$$l = \frac{\pi r_0 \alpha_1 t}{180}, \quad (0 \le t \le 1) \tag{3.20c}$$

$$Z = \sqrt[B]{\frac{l}{A}} \tag{3.20d}$$

The parameter equation of the tangential path of inside directrix is shown in:

$$Z = l \tan \beta_1 + h_0 - l_0 \tan \beta_1 \tag{3.21}$$

where

$$l = \frac{\pi r_0(\alpha_{1s} + \alpha_{2s} t)}{180}, \quad (0 \le t \le 1) \tag{3.21a}$$

$$\frac{l - l_0}{z - z_0} = \frac{1}{\tan \beta_1} \tag{3.21b}$$

The parameter equation of the curve arc for the outside directrix is shown in:

$$x = kAB^Z \tag{3.22}$$

where

$$l = \frac{\pi r_1 \alpha_{1s} t}{180}, \quad (0 \le t \le 1) \tag{3.22a}$$

$$Z = \sqrt[B]{\frac{l}{kA}} \tag{3.22b}$$

$$B = \frac{h_0 \cot \beta_1}{l_0} \tag{3.22c}$$

$$A = \frac{l_0}{h_0^{h_0 \cot \beta_1 / l_0}} \tag{3.22d}$$

The parameter equation of the tangential path for the outside directrix is shown in:

$$Z = \frac{\pi r_1 \alpha_{2s} t}{180} \tan \beta_2 + h_0 \tag{3.23}$$

where

$$l = \frac{\pi r_1 (\alpha_{1s} + \alpha_{2s} t)}{180}, \quad (0 \leq t \leq 1) \tag{3.23a}$$

$$\frac{l - l_0}{z - z_0} = \frac{1}{\tan \beta_2} \tag{3.23b}$$

(3) Exponential function swirling vanes

The parameter equation of the curve arc of inside directrix is shown in:

$$x = AB^Z, \tag{3.24}$$

where

$$B = e^{\frac{\cot \beta_1}{l_0}} \tag{3.24a}$$

$$A = e^{\left(\ln l_0 - \frac{h_0 \cot \beta_1}{l_0}\right)} \tag{3.24b}$$

$$l_0 = \frac{\pi r_0 \alpha_1}{180} \tag{3.24c}$$

$$l = \frac{\pi r_0 \alpha_1 t}{180}, \quad (0 \leq t \leq 1) \tag{3.24d}$$

$$h_0 = \frac{\ln l_0 - \ln A}{\ln B} \tag{3.24e}$$

$$Z = \frac{\ln l - \ln A}{\ln B} \tag{3.24f}$$

The parameter equation of tangential path of inside directrix is shown in:

$$Z = \frac{\pi r_0 \alpha_2 t}{180} \tan \beta_1 + h_0 \tag{3.25}$$

where

$$l = \frac{\pi r_0 (\alpha_1 + \alpha_2 t)}{180}, \quad (0 \le t \le 1) \tag{3.25a}$$

$$\frac{l - l_0}{z - z_0} = \frac{1}{\tan \beta_1} \tag{3.25b}$$

The parameter equation of the curve arc for the outside directrix is shown in:

$$x = kAB^z \tag{3.26}$$

where

$$z = \frac{\ln \frac{l}{kA}}{\ln B} \tag{3.26a}$$

$$l = \frac{\pi r_1 \alpha_{1s} t}{180}, \quad (0 \le t \le 1) \tag{3.26b}$$

$$B = \frac{h_0 \cot \beta_1}{l_0} \tag{3.26c}$$

$$A = \frac{l_0}{h_0^{h_0 \cot \beta_1 / l_0}} \tag{3.26d}$$

The parameter equation of the tangential path for the outside directrix is shown in:

$$z = \frac{\pi r_1 \alpha_{2s} t}{180} \tan \beta_2 + h_0 \tag{3.27}$$

where

$$l = \frac{\pi r_1 (\alpha_{1s} + \alpha_{2s} t)}{180}, \quad (0 \le t \le 1) \tag{3.27a}$$

$$\frac{l - l_0}{z - z_0} = \frac{1}{\tan \beta_2} \tag{3.27b}$$

(4) Hyperbolic function swirling vanes

The parameter equation of the curve arc of inside directrix is shown in:

$$x = \frac{a}{b}\sqrt{z^2 + b^2} \tag{3.28}$$

where

$$b = \sqrt{\frac{l_0 h_0 - h_0^2 \cot \beta_1}{\cot \beta_1}} \tag{3.28a}$$

$$a = \sqrt{l_0 (l_0 - h_0 \cot \beta_1)} \tag{3.28b}$$

$$l = \frac{\pi r_0 \alpha_1 t}{180}, \quad (0 \le t \le 1) \tag{3.28c}$$

$$l_0 = \frac{\pi r_0 \alpha_1}{180} \tag{3.28d}$$

$$z = \frac{a}{b}\sqrt{l^2 - a^2} \tag{3.28e}$$

The parameter equation of the tangential path of inside directrix is shown in:

$$Z = \frac{\pi r_0 \alpha_2 t}{180} \tan \beta_1 + h_0 \tag{3.29}$$

where

$$l = \frac{\pi r_0 (\alpha_1 + \alpha_2 t)}{180}, \quad (0 \le t \le 1) \tag{3.29a}$$

$$\frac{l - l_0}{z - z_0} = \frac{1}{\tan \beta_1} \tag{3.29b}$$

The parameter equation of the curve arc for the outside directrix is shown in:

$$x = k\frac{a}{b}\sqrt{z^2 + b^2} \tag{3.30}$$

where

$$l = \frac{\pi r_1 \alpha_{1s} t}{180}, \quad (0 \le t \le 1) \tag{3.30a}$$

$$z = \frac{b}{a} \frac{\sqrt{r_0^2 l^2 - r_1^2 a^2}}{r_1} \tag{3.30b}$$

The parameter equation of the tangential path for the outside directrix is shown in:

$$z = \frac{\pi r_1 \alpha_{2s} t}{180} \tan \beta_2 + h_0 \tag{3.31}$$

where

$$l = \frac{\pi r_1(\alpha_{1s} + \alpha_{2s}t)}{180}, \quad (0 \le t \le 1) \tag{3.31a}$$

$$\frac{l - l_0}{z - z_0} = \frac{1}{\tan \beta_2} \tag{3.31b}$$

3.3.3 Drainage Section

In a supersonic separator, condensation occurs when the fluid's temperature drops below its condensation point following expansion. This process results in the formation of condensate. The centrifugal force generated by the swirl generator directs these condensate droplets toward the separator's wall, where they coalesce into a liquid film. This film then gradually progresses toward the drain port. Most of the condensate is collected in the drain cavity and is ultimately discharged from the separator.

The design of the drainage section plays a crucial role in the overall performance of the separator. An ineffective drainage structure can disrupt the fluid dynamics within the nozzle's divergent section. This disruption may lead to an advancement of the shock wave position, an increase in temperature and pressure, and potentially cause secondary vaporization of the condensate. Therefore, the drainage design must facilitate efficient removal of the condensate while minimizing any adverse effects on the flow field.

Traditionally, conical drainage structures are commonly used in supersonic separators. Figure 3.7 illustrates two variations of this design: the flush type and the internal extension type [47]. The flush type features a drainage profile that is level with the surrounding wall, while the internal extension type includes an extended drainage component designed to improve condensate collection. Both designs have their advantages and are selected based on specific operational requirements and performance objectives.

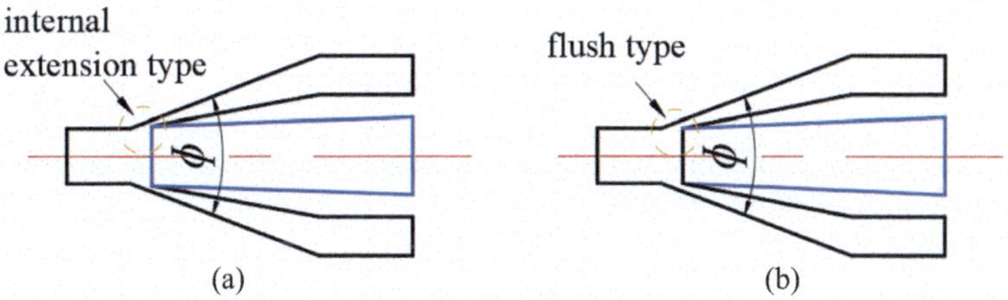

Fig. 3.7 Two different structures of liquid discharge port: **a** internal extension type; **b** flush type

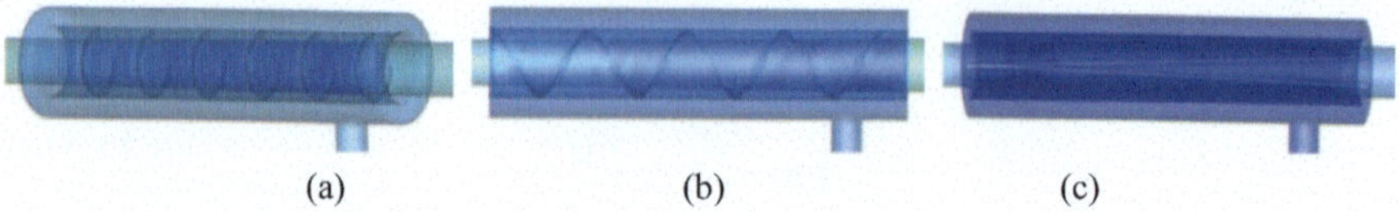

Fig. 3.8 Schematic diagram of cylindrical drainage structure: **a** circular opening; **b** spiral opening; **c** oblique opening [49]

The key parameters of the drainage structure include the outer angle, insertion depth, and the width of the drainage gap. Variations in these structural parameters can significantly affect the flow field characteristics and separation performance of the supersonic separator. Research has shown that while a flush drainage structure may allow some condensed droplets to escape, it offers several advantages. This design reduces the impact of the drainage opening on the internal flow field of the separator. Additionally, the internal flow field of a flush drainage structure generally performs better compared to an inserted drainage structure.

Optimal drainage structural parameters exist for different operational conditions. These parameters can maximize separation efficiency while minimizing any adverse effects on the flow field [12]. Identifying these optimal parameters is crucial for achieving the best performance in supersonic separators.

In addition to conical designs, cylindrical drainage structures are also a highly effective method for liquid discharge in supersonic separators. Figure 3.8 illustrates three different cylindrical drainage structures [48]. Each design offers unique benefits and operates differently based on its specific geometry and configuration. Research has demonstrated that the efficiency of these cylindrical drainage structures varies depending on the size of the droplets being handled. For example, some cylindrical designs may be more effective at separating larger droplets, while others may be optimized for smaller droplets. The choice of cylindrical drainage structure thus needs to be carefully aligned with the expected droplet sizes and the overall design requirements of the separator.

Therefore, selecting the appropriate drainage structure is essential for optimizing performance. The choice of structure should be carefully matched to the specific operational conditions and performance requirements of the application. Different cylindrical drainage designs can influence the efficiency of liquid discharge and the overall separation process. For instance, the geometry and dimensions of the cylindrical structure can impact how effectively it handles droplets of varying sizes and how well it integrates with the separator's flow dynamics.

A well-chosen drainage structure will enhance the separator's ability to efficiently remove condensate while maintaining stable flow conditions. This optimization ensures that the separator operates at peak performance, reducing potential issues such as excess condensation or inadequate discharge, which can compromise the efficiency and effectiveness of the separation process. Therefore, a thorough analysis of operational parameters and design considerations is critical for selecting the most suitable drainage structure for any given application.

3.3.4 *Diffuser*

In a supersonic separator, the diffuser plays a crucial role in converting kinetic energy into increased static temperature and recovery pressure energy. As the supersonic gas flows through the diffuser, it encounters a shock wave, which causes a transition from supersonic to subsonic flow. This shock wave results in a significant increase in static pressure by converting the remaining kinetic energy in the separated flows.

In an idealized model of an inviscid gas, the thickness of the shock wave is comparable to the molecular mean free path, allowing it to be approximated as a discontinuous surface. Due to the minimal thickness of the shock wave, the cross-sectional areas upstream and downstream of the shock are nearly equal, and energy or heat loss across the shock is typically negligible. It is important to maintain the divergence angle of the diffuser within an appropriate range. An excessive divergence angle can lead to interactions between the shock wave and the boundary layer, which may further delay the transition from supersonic to subsonic flow. This interaction can adversely affect the overall efficiency and performance of the separator. Three types of diffusers with distinct profiles are described as follows [52].

(1) Conical diffuser

For the conical diffuser, both the structural and flow parameters can be determined using the flow function. This function is essential for calculating various characteristics of the diffuser, such as pressure recovery and flow stability. The flow function is expressed as:

$$q(\mathrm{Ma}) = \frac{A_{\mathrm{cr}}}{A} = \frac{\mathrm{Ma}}{\left[\left(\frac{2}{k+1}\right)\left(1 + \frac{k-1}{2}\mathrm{Ma}^2\right)\right]^{\frac{k+1}{2(k-1)}}} \tag{3.32}$$

where q is the gas flow rate, k is adiabatic exponent, Ma is the Mach number, A is the sectional area.

The axial length L of the conical diffuser can be written as:

$$L = \frac{r_2 - r_1}{\tan \beta} \tag{3.33}$$

where r_1 and r_2 are the inlet and outlet radius of the diffuser, respectively, β is the divergence angle.

(2) Curved wall diffuser

In a conical diffuser, if the divergence angle β is set to a low value, the axial length of the diffuser can become excessively long. This extended length can pose practical challenges in design and space constraints. To address this issue, a curved wall diffuser is often employed as an alternative. The curved wall design effectively mitigates the problem associated with long axial lengths by providing a more compact and efficient configuration. To calculate the sectional radius of the curved wall diffuser, the following equation is used:

$$r = \frac{r_1}{\sqrt{1 + \left(\frac{r_1^2}{r_2^2} - 1\right)\frac{x}{L}}} \tag{3.34}$$

where x is the distance between the arbitrary cross-section and the inlet, r is the radius at an arbitrary cross-section of x.

(3) Second throat diffuser

Shock waves are crucial in transitioning the flow from supersonic to subsonic speeds in diffuser. These shock waves significantly contribute to the recovery of pressure, which is essential for maintaining efficient performance. However, the presence of shock waves can introduce disturbances in the fluid flow, potentially affecting the stability and efficiency of the separator. To mitigate the negative impacts of these disturbances, the concept of a second throat is employed. This design feature helps to reduce the adverse effects on the flow by providing an additional stage that smooths out the transition and minimizes flow disruptions. The area of the second throat is determined using the following equation:

$$A_{\text{cr2}}^* = A_{\text{cr1}}^* \left(\frac{2k}{k+1}\text{Ma}_1^2 - \frac{k-1}{k+1}\right)^{\frac{1}{k-1}} \left[\frac{(k+1)\text{Ma}_1^2}{(k-1)\text{Ma}_1^2 + 2}\right]^{-\frac{k}{k-1}} \tag{3.35}$$

where A_{cr1}^* and A_{cr2}^* are the areas of the nozzle throat and second throat in the supersonic separator, respectively, Ma_1 is gas inlet Mach number.

Effective pressure recovery is critical in the operation of supersonic separators, particularly for applications in natural gas processing. For a supersonic separator used in natural gas processing, the primary objective of the diffuser is to efficiently convert kinetic energy into pressure energy. Given the superior pressure recovery performance of the conical diffuser, as demonstrated by the above analysis, it is the preferred choice for such applications. Selecting the conical diffuser ensures optimal recovery of pressure energy, thereby enhancing the overall efficiency and effectiveness of the supersonic separator.

3.4 Condensation and Liquefaction Theory

In the flow of natural gas through a supersonic separator, the presence of a Laval nozzle causes the natural gas to experience accelerated expansion and cooling. This leads the condensable components to reach a state of supersaturation, initiating spontaneous condensation and the formation of condensation nuclei. These newly formed condensation nuclei subsequently grow into droplets. During this process, the droplets exchange heat and mass with the surrounding supersonic gas in the flow field. Non-equilibrium phase changes occur in tandem with the supersonic flow and mass transfer processes of the condensing gas. Consequently, the spontaneous condensation process of vapor in the flow field of a supersonic separator involves two

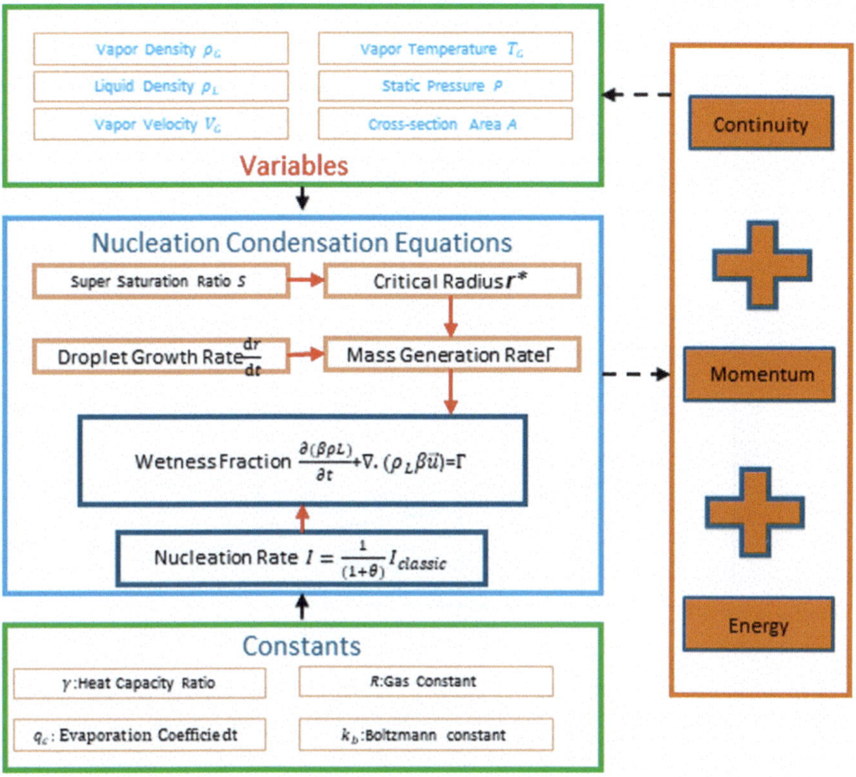

Fig. 3.9 Strategy for implementing the condensation model

primary stages: nucleation and droplet growth. Figure 3.9 illustrates the operational approach employed in the condensation model [51], providing an overview of the simulation methodology used in studying this phenomenon.

3.4.1 Supersaturation and Metastable State

In the process of supersonic processing of natural gas, supersaturation is one of the most important indicators related to condensation. It is the driving force for droplet nucleation and growth, determines the final droplet size distribution, and is also an indicator to measure the degree to which fluid properties deviate from thermodynamic equilibrium. It is usually represented by a saturation line. When steam flows through a nozzle, a phase change occurs, transitioning supercritical steam into subcooled water. This phase transition involves spinodal decomposition, producing dry steam and liquid water from wet steam. It also involves phase separation, taking place in the unstable region depicted in Fig. 3.10. Spinodal decomposition occurs when a fluid

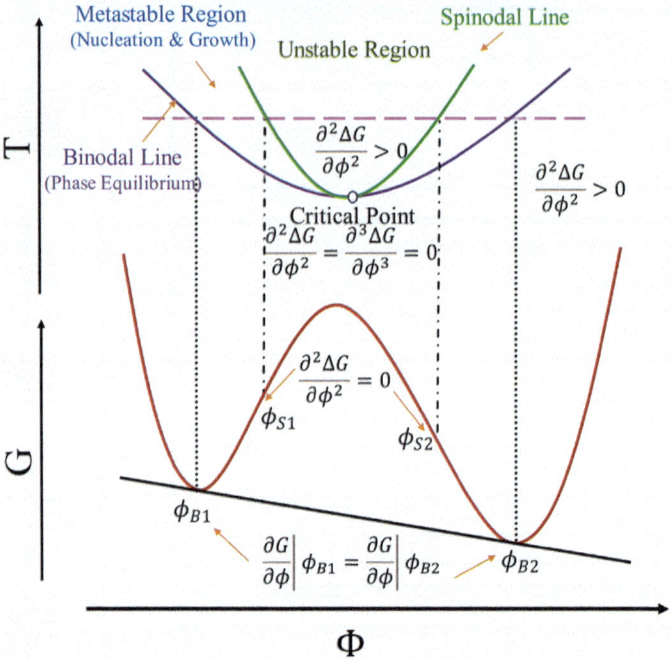

Fig. 3.10 Fluid transition path along the spinodal and binodal lines, metastable, and unstable regions. Reprinted from Ochi et al. [35], an open access article from IOP Science

enters the spinodal region of a phase diagram. In the case of wet steam within the metastable zone, between the binodal and spinodal lines, homogeneous nucleation occurs [35].

The phenomenon involves the precipitation of nano-sized micro-droplets from supersaturated steam. This triggers a supersonic condensation and growth process, which then changes from supercritical steam to supercooled water. However, so far, although researchers have used optical and electrical methods, it has been quite difficult to measure this non-equilibrium condensation process, and interpreting the metastable thermodynamic properties remains a major challenge. It is defined by Brazhkin that metastable phases are non-equilibrium states of matter with reversible properties throughout an experiment [8]. Metastable regions on phase diagrams, located between the spinodal line and phase boundary, can have a finite fluctuation that can make a solution unstable. It is this fluctuation, known as a nucleus, and how much energy is expended in creating such a nucleus that determines the phase's metastability. The theoretical calculations of this energy value show a decrement toward zero closer to the spinodal line. In the analysis of the macroscopic system's behavior, Kaplun and Meshalkin find no fundamental difference between stable and metastable states, except for a limited lifetime in metastable states [29].

For the condensation of some typical gases such as water vapor, nitrogen, and CO_2, some good research results have been obtained in Laval Nozzle. The general experimental method is to measure the static pressure of the wall, which can determine the starting position and condensation strength of condensation [58]. The condensation of steam in the Laval nozzle is carried out in a supersaturated state. This process will release a certain amount of energy and cause the pressure in the contraction section to drop and the pressure in the divergence section to rise [23].

3.4.2 *Classical Nucleation Theory*

In the field of supersonic separators, the behavior of fluids under rapid expansion has been the subject of extensive investigation. As demonstrated through experimentation, the formation of liquid droplets does not take place under conditions of saturation; rather, the onset of nucleation occurs under subcooled conditions within high-speed flows [6]. Herein, the energy barrier to phase transformation is overcome, initiating a cascade of processes that lead to the spontaneous condensation of the fluid. Consequently, the system evolves toward a state of near equilibrium. It is noteworthy that the initial phase of this phenomenon is characterized by nucleation-driven condensation, a process that precedes the creation of the first droplets of the liquid phase. Following this stage, the evolution of the phase transition is governed by the growth of supercritical droplets—those exceeding a critical radius. These droplets, by their very nature, suppress further nucleation events, thereby restoring a semblance of equilibrium within the system. This dynamic equilibrium, defined by the interplay between nucleation and droplet growth, underscores the intricate nature of phase transitions in supersonic environments.

The Wilson line, a critical characteristic delineating the boundary of condensing vapor, exhibits significant dependence on the rate of expansion. Higher expansion rates induce a deeper penetration into the metastable region, a condition characterized by a shift of the Wilson line toward zones of higher subcooling [32]. This observation highlights the pivotal role of expansion rate in determining the extent of deviation from equilibrium conditions, influencing the likelihood of droplet formation and growth. The utilization of Laval nozzles in experimental setups is found to be particularly advantageous for the study of these phenomena. Through the manipulation of the length of the nozzles, while maintaining a constant pressure ratio, researchers can modulate both the rate of expansion and the propensity for nucleation.

During the condensation process, the induction of thermodynamic non-equilibrium states within gas mixtures is often precipitated by dramatic fluctuations in temperature and pressure. Such states necessitate a return to thermodynamic equilibrium, a process that involves nucleation within the gas mixture, culminating in the growth of droplets to a size detectable by current analytical techniques. Exploring the onset of nucleation, crystallization, droplet growth measurements, and the delineation of key phase boundaries such as the Wilson line or spinodal is crucial. These phenomena are typically studied in detail using sophisticated laser transmission and scattering techniques such as shearing interferometry.

In the context of supersonic condensation within the Laval nozzle, the occurrence of spontaneous nucleation is significantly contingent upon the attainment of supersaturation. Supersaturation dictates the capacity of the flow to engender new nuclei. As the flow nears saturation, vapor molecules do not undergo instantaneous condensation. This is attributed to the existence of a free energy barrier, a composite of volume and surface free energies (as illustrated in Fig. 3.11), which must be overcome before condensation can ensue. Consequently, the vapor continues its expansion as superheated steam. Yet, upon reaching a certain level of supersaturation, the critical radius is achieved. This milestone triggers the formation of a predetermined quantity of condensation nuclei, each of a specified size, within the steam. It is noteworthy that the degree of supersaturation is directly proportional to the velocity of nuclei formation. The kinetic energies intrinsic to vapor molecules, both individually and on average, are intimately tied to the probability of nucleation events. Molecules with sufficient kinetic energy can overcome the free energy barrier, facilitating the aggregation of molecules around a nucleus. This process is the foundation of nucleation, a critical step in the transition from a vapor to a liquid state within the confines of supersonic flows.

The rate of nucleation formation can be characterized by Gibbs free energy, which is shown as [25]

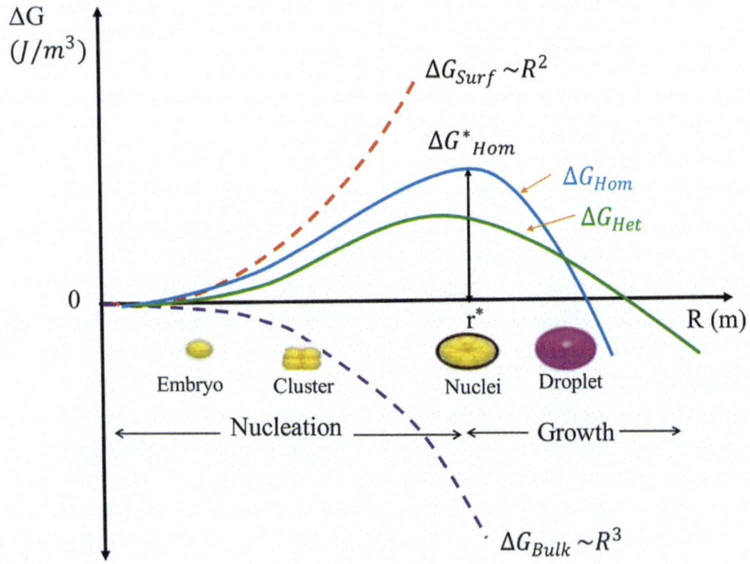

Fig. 3.11 Radius of particles and the changes of Gibbs free energy. Reprinted from Taqieddin et al. [41], an open access article from IOP Science

$$\Delta G = 3\pi r^2 \sigma - \frac{4}{3}\pi r^3 \rho_l R T_g \ln S \tag{3.36}$$

where r is the droplet radius, σ is surface tension, ρ is density, and S is the supersaturation.

In the two two-phase flows, the formation of droplets is dictated by nucleation rates under conditions of supersaturation. These rates, crucial for understanding the dynamics of phase transition, are quantifiable through the application of the classical nucleation theory (CNT):

$$J = \frac{\rho_g^2}{\rho_l}\sqrt{\left(\frac{2\sigma_r}{\pi m_m^3}\right)}\mathrm{EXP}\left[-\frac{\Delta G^*}{k_B T_G}\right] \tag{3.37}$$

where m_m is the molecule mass, k_B is the Boltzmann's constant.

Within the natural gas supersonic separation, the elucidation of non-equilibrium condensation phenomena, particularly within the nanoscale dimension, assumes paramount importance for enhancing the collective understanding of critical processes such as nucleation and droplet growth. Nucleation, which is the initial step in condensation, plays a significant role in quantifying the condensation process. However, despite experimental efforts, accurately measuring the nucleation rate remains a challenge due to limitations and discrepancies between experimental results and classical theoretical predictions [61]. To address this challenge, researchers have been exploring effective methods to describe the nucleation rate in supersonic flows. One approach involves using optical methods and advanced imaging techniques to visualize the nucleation process in real time. By analyzing the data obtained from these experiments, the dynamics of nucleation and droplet growth can be gained.

Another method being explored is the use of computational models to simulate the nucleation process. These models can help researchers understand the factors that influence nucleation and predict the flow behavior under different conditions. Over recent years, molecular dynamics (MD) simulations have ascended to prominence as a potent instrument in narrowing the knowledge gap concerning nucleation processes [24]. These computational models furnish investigators with a microscopic perspective, enabling the observation of non-equilibrium dynamics at a molecular scale and the tracking of their temporal evolution. Extensive studies have been conducted on models of water [4] and Lennard–Jones fluids [57], illustrating the versatility and power of MD simulations in elucidating complex fluid behaviors. However, the exploration of nucleation in carbon dioxide remains relatively uncharted territory [60].

The MD simulations have proven instrumental in quantifying nucleation rates through sophisticated methodologies, such as the calculation of average first-pass times or the application of predetermined thresholds. It is crucial to underscore, however, that nucleation is inherently a stochastic process, characterized by its unpredictability and variability. Consequently, the accurate prediction of nucleation rates necessitates the execution of numerous independent MD simulations, each contributing to a more comprehensive and nuanced understanding of the underlying mechanisms. By adopting this approach, researchers can gain valuable insights into the behavior of CO_2 nucleation, thereby advancing our collective knowledge in this specialized domain. The integration of MD simulations into the study of supersonic condensation, therefore, represents a significant step forward in elucidating the complex dynamics governing this phenomenon.

More importantly, a supersonic separator has received significant scholarly attention, promising solutions for the extraction of CO_2 from natural gas. Despite this, the applicability of classical nucleation theory (CNT) to elucidate the condensation mechanisms of CO_2 within natural gas environments remains ambiguous. For the binary mixture of CH_4 and CO_2, the investigations have ventured into the exploration of CO_2 condensation characteristics using CFD and MD simulations [11]. These studies have pivotal insights into the molecular-scale dynamics of CO_2 nucleation and droplet growth. The creation of conducive conditions for CO_2 liquefaction is achieved through the Laval nozzle, which facilitates the reduction of temperature to a point where condensation becomes favorable. Optimization of condensation parameters is attainable by adjusting the inlet temperature and pressure. A detailed examination of the nucleation stage via MD simulation has illuminated the intricate interplay between CO_2 gas molecules, the release of latent heat, and the stabilization of clusters, all influenced by energetic interactions with adjacent molecules.

It is worthy that the comparison between CNT predictions and MD simulation outcomes has revealed a substantial disparity, differing by several orders of magnitude. This discrepancy underscores the necessity for revisions to classical nucleation theory, prompting a reconsideration of established principles in light of the nuanced realities uncovered by modern computational methods. As a consequence, molecular simulations, with a particular emphasis on MD simulations, emerge as a pivotal tool in deciphering the complexities of non-equilibrium condensation phenomena in supersonic separators at the nanoscopic level. Beyond furnishing a microscopic comprehension of nucleation mechanisms, these simulations serve as a theoretical benchmark for the refinement of CO_2 separation efficiency. The fusion of diverse research findings, the modification of classical theories, and the enhancement of simulation methodologies promise to make substantial contributions to the evolution of supersonic gas separation techniques and their practical implementations in the processing of natural gas resources.

3.4.3 Droplet Growth Rate

During the actual process of droplet growth, numerous vapor molecules cluster around the condensation nuclei, contributing to the continuous enlargement of the droplets. This phenomenon occurs as vapor molecules adhere to the surfaces of the nuclei, resulting in a steady accumulation and expansion of the droplets. A critical aspect of this process involves the release of latent heat, which is an inherent part of the phase transition from vapor to liquid. This latent heat transfer occurs as vapor molecules condense onto the nuclei, and the energy released is absorbed by the surrounding gas. This exchange is a fundamental mechanism for heat and mass transfer in multiphase systems. The concept of slip velocity, which refers to the relative motion between gas and liquid phases, is typically of lesser concern during droplet growth. This is primarily due to the exceedingly small diameters of the droplets at the initial stages of formation, often on the order of nanometers [62]. At such minuscule scales, the influence of slip velocity tends to be negligible compared to the effects of surface tension, viscosity, and other forces acting at the interface between gas and liquid.

In the context of droplet growth within a gaseous environment, these minute droplets may initially form as a result of heterogeneous nucleation on preexisting particles or through homogeneous nucleation when the vapor reaches a critical level of supersaturation. Once formed, the droplets grow by attracting more vapor molecules, which condense and release latent heat, leading to an increase in droplet size until a steady state is reached or until further growth is inhibited by external conditions such as temperature, pressure, or the availability of vapor. A droplet growth model was developed by Gyarmathy [54] by considering capillary influence and heat-mass transfer. The droplet's energy balance is written as [31]:

$$h_{\mathrm{v}}\frac{\mathrm{d}m}{\mathrm{d}t} = mc_{\mathrm{l}}\frac{\mathrm{d}T_{\mathrm{l}}}{\mathrm{d}t} + 4\pi r^2\frac{p}{\sqrt{2\pi RT_{\mathrm{g}}}}\left(\frac{\lambda_{\mathrm{v}}+1}{2\lambda_{\mathrm{v}}}\right)c_{\mathrm{v}}(T_{\mathrm{l}} - T_{\mathrm{g}}) \tag{3.38}$$

where h_{v} is latent heat, c_{v} and c_{l} represent the specific heat capacities in the vapor and liquid phases, respectively, λ_{v} is the thermal conductivity of the vapor.

When molecules condense onto a droplet, they release their latent heat of condensation. This energy can affect the droplet and its surroundings in two primary ways: Heat transfer to vapor: Some of the energy released during condensation can be transferred back into the vapor phase, potentially increasing the temperature of the vapor or affecting the thermodynamic equilibrium of the system. Increase in droplet temperature: The energy can also raise the temperature of the droplet itself. However, in many cases, especially with very small droplets, the thermal inertia of the droplet is considered negligible. Thus, Eq. (3.38) is simplified as follows:

$$h_{\mathrm{v}}\frac{\mathrm{d}m}{\mathrm{d}t} = 4\pi r^2\frac{p}{\sqrt{2\pi RT_{\mathrm{g}}}}\left(\frac{\lambda_{\mathrm{v}}+1}{2\lambda_{\mathrm{v}}}\right)c_{\mathrm{v}}(T_{\mathrm{l}} - T_{\mathrm{g}}) \tag{3.39}$$

Gyarmathy has presented a relationship in the following form to calculate the temperature of the droplet despite the change in surface tension coefficient and enthalpy of gas between T_{l} and T_{g}.

$$T_{\mathrm{l}} = T_{\mathrm{g}} + \left[1 - \frac{r^*}{r}\right](T_{\mathrm{s}}(p) - T_{\mathrm{g}}) \tag{3.40}$$

where T_{s} is the saturation temperature corresponding to the current vapor pressure.

Once nucleation has occurred and droplets are formed, further growth is achieved by the continuous condensation of vapor onto these droplets. This process is influenced by various factors including the supersaturation of the vapor, the concentration of condensation nuclei, and the thermodynamic conditions of the environment. The exchange of mass and energy between droplets and their surroundings is crucial for droplet growth. When vapor molecules condense onto a droplet, they transfer their latent heat of condensation to the droplet. The growth rate of droplets can be described mathematically using different models, depending on the assumptions made about the system. One common growth rate of the droplet is obtained based on the above two relations.

3.5 Economic and Environmental Impact

Supersonic separator technology indeed represents a significant leap beyond the constraints of traditional separation methods, offering numerous advantages that cater to contemporary industrial needs and environmental concerns. Unlike conventional technologies, supersonic separators do not require external power sources for operation, making them a more sustainable and eco-friendly option. Moreover, the absence of moving parts in supersonic separators translates to increased reliability and durability. The static nature of these separators minimizes wear and tear, substantially reducing the likelihood of mechanical failures and the need for frequent maintenance. This not only leads to lower maintenance costs but also ensures consistent performance over extended periods. The eco-friendly profile of supersonic separators is further enhanced by their non-leaking and non-polluting characteristics. The technology's minimal ecological footprint resonates with industries committed to sustainability, as it helps mitigate environmental impact without compromising operational efficiency.

Safety is another paramount consideration, especially in sectors with stringent safety protocols. The chemical-free operation of supersonic separators eliminates potential hazards associated with hazardous substances, contributing to a safer working environment. This is particularly valuable in industries where the handling of chemicals poses significant risks to personnel and operations. The straightforward design of supersonic separators simplifies the dehydration process. This uncomplicated structure translates to lower operational expenses and autonomous functionality, making it particularly appealing for industries requiring continuous and unsupervised operation [13].

In addition, recently research shows that the sustainability of supersonic separation technology is quantifiably superior to conventional methods across multiple dimensions. Environmental impact, efficiency, health and safety, and economic viability are all areas where supersonic separation excels. The Sustainable Plant-Wide Index for supersonic separation, rated at 0.99, clearly outshines the 0.86 index of conventional routes, demonstrating its superior sustainability credentials [17].

3.5.1 Cost Analysis

For natural gas processing and achieving HCDPA, the Joule–Thomson Expansion (JTE) method emerges as a straightforward and commonly applied technique. This process utilizes a heat exchanger to cool the gas, an isenthalpic valve to reduce pressure, and a vessel to collect the extracted natural gas liquids (NGLs). The JTE process is thermodynamically driven by the Joule–Thomson effect, where the temperature of a gas changes as it passes through a throttling valve or porous plug under conditions of no work, heat exchange, or change in enthalpy [1]. The total cost analysis must encompass not just the initial investment but also the operational complexities. Other techniques, such as absorption or adsorption, might have their own unique complexities and costs associated with installation, operation, and maintenance.

Teixeira et al. [43] showed that SS-methanol-recovery entails an economic leverage that affords a post-capture plant abating 43% of emitted CO_2,i.e., such SS processing is a cleaner gas production compared to the conventional counterpart. For safe transportation and to preserve the heating value of natural gas streams, dew-point corrections of water and heavy hydrocarbons are essential. A supersonic separator can achieve both tasks with a high degree of reliability, minimal operating costs, and minimal requirements for operational facilities. The ability of supersonic separation to handle a wide range of impurities and operating conditions can be advantageous in certain applications. Flexibility in adapting to varying feed gas compositions and impurity levels may offer cost benefits compared to other purification techniques that are more specialized. Comparing the SS process with a conventional sequence of operations—WDPA through water absorption by triethylene glycol (TEG) followed by HCDPA with Joule–Thomson (JT) and low-temperature separator (LTS)—revealed nuanced findings [18]. The SS process exhibited higher capital expenditure (CAPEX) when compared to the TEG + JT/LTS process.

The economic evaluation of supersonic separation (SS) technology versus conventional processes, specifically TEG dehydration combined with Joule–Thomson (JT) or low-temperature separation (LTS), reveals important insights into their respective operational and financial performances. Despite the conventional process having a slight edge in operational expenses (OPEX) due to its superior availability, leading to higher annual natural gas (NG) supply costs, the SS process demonstrates a clear advantage in terms of revenue generation. This revenue superiority is attributed to the higher availability of the SS process, enabling it to operate more consistently and efficiently. Additionally, the SS process excels in natural gas liquid (NGL) recovery, which significantly boosts the overall economic performance of the plant. The enhanced NGL recovery is particularly advantageous as NGLs are valuable commodities in the petrochemical industry, contributing significantly to the revenue stream of a gas processing plant. The economic performance of a process is often summarized by its net present value (NPV), a key metric for evaluating the profitability of an investment over time. It is found that the NPV of the SS process was found to exceed that of the TEG + JT/LTS process, indicating that the SS process is more financially viable in the long term.

To visualize these financial aspects, Fig. 3.12 presents a comparative analysis of the fixed capital investment and discounted cumulative cash flow between conventional and supersonic separation gas plants. The data shows that although the SS process requires a higher initial capital investment, it compensates for this with a superior discounted cumulative cash flow, showcasing its economic benefits over the conventional route. This highlights the trade-off between the higher upfront costs of supersonic separation technology and its potential for generating greater revenues and higher NPVs. For investors and operators in the gas processing industry, these findings suggest that the increased initial expenditure on supersonic separation could be justified by its superior long-term financial returns and operational efficiency.

The economic comparison between the SS process and the conventional TEG dehydration combined with JT or LTS process is analyzed. The incremental cumulative discounted cash flow (in millions of US dollars) illustrates the initial disadvantage of the SS process due to its higher capital expenditure (CAPEX), mainly attributed to the requirement for additional compression. However, this extra compression also results in an advantageous side effect: the export gas pressure of the SS process reaches significantly higher than that achieved by the TEG + JT/LTS process. A key financial benefit of the SS process is its augmentation of NGL production, which translates into higher revenues. This increased income, combined with the SS process's ability to achieve early payback of the initial investment and generate a higher NPV, makes it a compelling choice despite the higher upfront costs. Depreciation is a significant factor in any investment decision, and the supersonic separation technology shines in this regard. The demonstrated 99% uptime over six years, alongside virtually negligible maintenance costs and inspection intervals of only once every six years, indicates lower opportunity costs compared to traditional TEG + JT/LTS technology. This reliability and low maintenance requirement translate to reduced downtime and lower operational costs over the lifetime of the equipment.

Another important distinction between the SS process and conventional methods lies in the environmental impact. Traditional glycol regeneration processes involve the release of stripping gas to the atmosphere after flaring, which contains not only carbon dioxide (CO_2) emissions due to incomplete combustion but also potentially hazardous components such as C^{3+} alkanes and benzene–toluene–xylene (BTX) aromatics. The SS process, on the other hand, significantly reduces major equipment requirements for natural gas dew-pointing and completely eliminates the need for chemical additives, thus minimizing environmental pollution and operational hazards.

In summary, the cost–benefit analysis of supersonic separation technology compared to conventional methods presents a detailed picture of its economic and environmental superiority. Although the SS process initially demands higher investments, its long-term economic advantages, operational efficiency, and environmental benefits make it a competitive and sustainable option for various industrial applications. The technology's resilience, reduced maintenance, and lower environmental impact contribute to its attractiveness in today's market, where sustainability and operational excellence are increasingly valued.

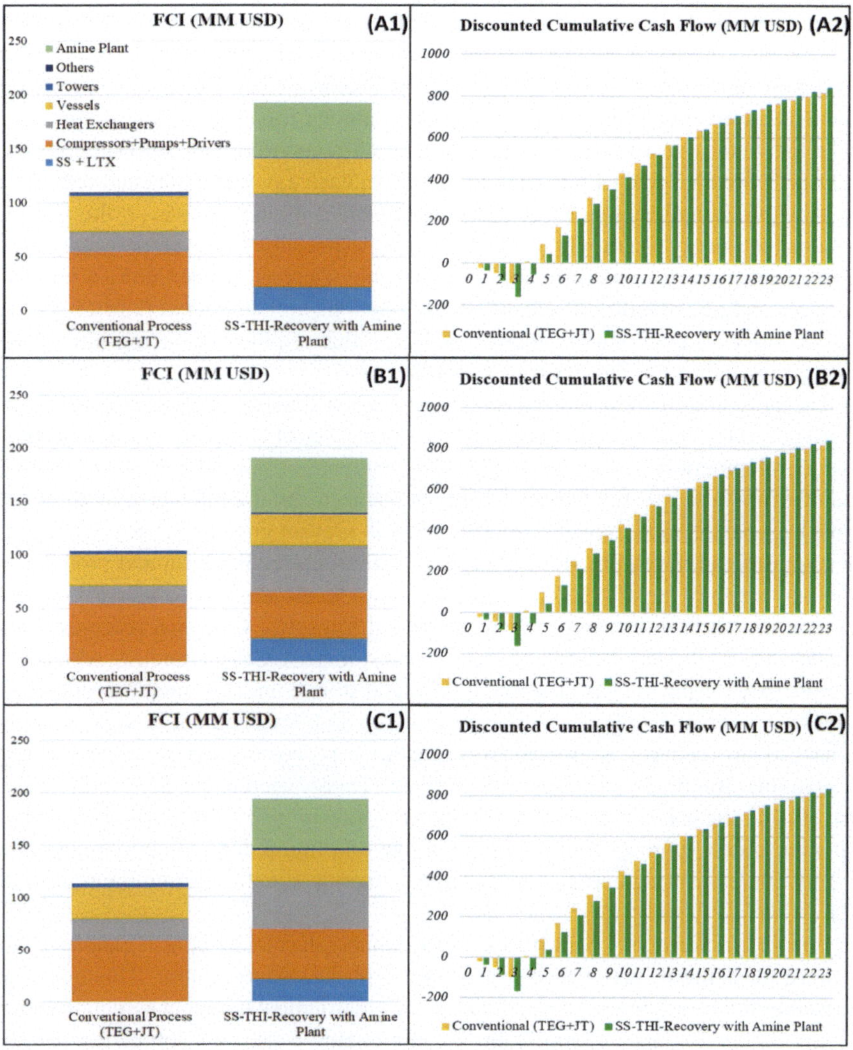

Fig. 3.12 Fixed capital investment and discounted cumulative cash flow versus year for conventional and SS gas plants [44]

3.5.2 *Sustainable Development*

Sustainable development is indeed a widely discussed concept, aiming to ensure that the needs of the present are met without compromising the ability of future generations to meet their own needs [36]. This philosophy is particularly relevant in the context of energy production and consumption, where the transition from high-carbon intensity sources to lower ones is a critical step toward sustainability. Natural gas, with its relatively higher hydrogen-to-carbon (H/C) ratio compared to other fossil fuels like coal and oil, has emerged as a preferred transitional fuel in the journey toward decarbonization. Its lower carbon dioxide emissions per unit of energy produced make it a more environmentally benign option among fossil fuels, contributing to a reduction in the overall carbon footprint of energy generation. This shift supports the industry's broader goal of phasing out carbon-intensive power plants and replacing them with alternatives that minimize environmental impact. However, the path to sustainable energy is fraught with challenges, especially in developing countries where energy access, intermittency of renewable sources, infrastructure for transmission, and affordability of cleaner energy options are significant hurdles. These nations often face a complex energy landscape that requires balancing immediate energy demands with long-term sustainability goals.

The presence of high concentrations of carbon dioxide in certain natural gas reserves (up to 80% in some cases) adds another layer of complexity to the sustainable development equation. Such high CO_2 content not only affects the economic viability of these gas fields but also poses environmental challenges, as the CO_2 must be separated and managed appropriately to prevent greenhouse gas emissions. In response, the development of innovative natural gas exploration and production methods becomes imperative. Advanced technologies are needed to efficiently extract natural gas while mitigating the environmental impact of CO_2-rich reserves. This includes the deployment of carbon capture and storage (CCS) techniques, which can capture CO_2 during the gas processing stage and store it underground, preventing its release into the atmosphere. Moreover, research and development in areas such as enhanced gas recovery, improved CO_2 separation processes, and integrated energy systems that combine natural gas with renewable energy sources can help optimize the use of natural gas in a manner that supports the transition to a low-carbon economy. This integrated approach can provide a bridge to a future dominated by renewable energy while ensuring energy security and economic stability during the transition [21].

Energy efficiency is a cornerstone of sustainable development as it enables us to achieve the same or better levels of service with less energy input. This not only conserves finite resources but also reduces the environmental impact associated with energy production and consumption. By minimizing waste and optimizing performance, energy-efficient practices and technologies can significantly cut down on greenhouse gas emissions and other pollutants, leading to cleaner air, water, and ecosystems. Efforts to enhance energy efficiency should not be viewed solely as a temporary measure until alternative energy sources become viable. Instead, they should be considered an ongoing process integral to sustainable development. This is because even with abundant and clean energy sources, the principles of efficiency will remain vital to ensure the optimal use of resources and to minimize ecological footprints.

Projected until 2050, a steady growth rate of 1.1% per annum is anticipated for natural gas, contrasting with the decline forecasted for petroleum-based liquid fuels, traditionally the dominant energy source. Natural gas's ascendancy is bolstered by vast reserves and low-carbon emissions, notably the nascent field of natural gas hydrates [15]. Among fossil fuels, natural gas stands out for its lower CO_2 emissions per unit of energy generated, attributing to a cleaner combustion profile. Recognized as a pivotal energy resource globally, natural gas is poised to maintain a substantial role throughout the current century. Nonetheless, the challenges inherent in natural gas extraction and production, particularly within the context of remote offshore fields, merit attention. Characteristically, these locales feature power generation via gas-fired turbines that exhibit suboptimal efficiency, leading to the emission of hot flue gas. This phenomenon exacerbates resource depletion, elevates carbon emissions, and undermines sustainability [27]. Essential to the enhancement of exploration and production efficiency for natural gas, the identification of innovative solutions to these obstacles becomes paramount. Critical awareness is drawn to the environmental repercussions associated with offshore oil and gas rigs, which are notable emitters of CO_2 and CH_4 through on-site power generation, flare systems, and processing facilities. Under end-of-life scenarios for platforms, these environmental impacts are magnified [34]. The on-site presence of generators, flares, and handling facilities contributes substantially to the production of CH_4 and CO_2.

Meeting the escalating demand for natural gas while upholding global sustainability mandates calls for the innovation of advanced offshore processing designs. Such designs aim to optimize resource utilization, ensuring that the benefits of natural gas are harnessed in a manner consistent with environmental stewardship and long-term energy security. Through the implementation of these forward-thinking approaches, the industry can strive toward a more sustainable and efficient future in the realm of natural gas production.

3.5.3 Environmental Impact

Environmental factors must be considerate, due to the truth that sustainable development is deeply rooted in environmental stewardship. Continuous environmental degradation, driven by human activities, is unsustainable in the long term, given the cumulative effects on ecosystems that can culminate in multifaceted health, ecological, and socioeconomic challenges. The quantum of energy consumed by a civilization bears a profound imprint on the environment. Ideally, a society pursuing sustainable development would leverage energy sources devoid of environmental repercussions. Efforts geared toward enhancing energy efficiency hold promise in alleviating certain concerns regarding sustainable growth constraints linked to environmental emissions.

In the preceding decades, the mitigation of CO_2 emissions has emerged as a paramount concern, prompting the development of both chemical and physical methodologies for their capture and sequestration. However, these measures often entail substantial economic costs for emission-producing processes. A potential resolution to this predicament lies in the valorization of CO_2, converting it into valuable products such as polymers, methanol, and other chemical commodities. Beyond merely curtailing emissions, this endeavor promises economic dividends, thereby harmonizing environmental protection with economic viability.

Natural gas occupies a unique position within the spectrum of energy sources, serving dual roles in the transition toward a sustainable energy economy. On one hand, it acts as a stepping stone to reduce carbon emissions, particularly when displacing coal and petroleum in power generation and industrial applications. This is underscored by its comparative advantage as a cleaner-burning fossil fuel, with lower emissions of greenhouse gases (GHGs) and pollutants during combustion. On the other hand, natural gas itself is a source of GHG emissions, chiefly methane, which has a potent warming effect on the atmosphere. The trend toward increased reliance on natural gas as a primary fuel is prevalent among developed nations, influenced by the desire to mitigate environmental, economic, and social repercussions associated with traditional energy sources. The surge in natural gas production has coincided with a decline in national GHG emissions, positioning natural gas as a potentially cleaner and more efficient substitute for coal and petroleum. This transformation has been facilitated by its status as a mature and cost-effective energy source, adaptable to diverse sectors such as electricity generation, transportation, and manufacturing.

Nevertheless, the composition of natural gas extracted from reservoirs often includes significant quantities of contaminants and heavy hydrocarbons, suggesting that a substantial portion of the raw gas comprises these elements. Critics argue that the promotion of natural gas as a bridge fuel could hinder the advancement and adoption of more sustainable, "terminal" technologies. Moreover, sourcing natural gas from unconventional reservoirs might introduce unacceptable environmental risks, including potential leakage of methane and other pollutants, which could negate the purported benefits of its lower emissions profile compared to coal. The debate surrounding natural gas as a bridge fuel thus highlights the complex trade-offs involved in the energy transition. While it offers a practical short-term solution for reducing carbon emissions, its long-term viability as part of a sustainable energy mix remains contentious. Policymakers and industry leaders must navigate these challenges carefully, balancing the immediate environmental gains against the potential impediments to achieving a fully sustainable energy future.

To address these issues, sophisticated technologies have been devised to refine natural gas, a process known as gas processing, which encompasses various stages including separation, dehydration, sweetening, and fractionation. These processes are essential for removing unwanted substances from the natural gas stream, ensuring that the final product is suitable for use in a wide array of applications, from residential heating to industrial processes. The shift toward natural gas as a principal energy source is widely regarded as a positive development, as it helps mitigate the detrimental environmental and health impacts associated with the combustion of dirtier fossil fuels. However, the efficacy of this transition is contingent upon the implementation of robust gas processing methodologies that effectively minimize the presence of contaminants and hydrocarbons in the gas supply. Among these methodologies, supersonic separation stands out as an innovative technique for gas purification. Supersonic separation can significantly enhance the environmental profile of natural gas utilization. Key environmental benefits include:

(1) Reduced emissions: Supersonic separation contributes to the removal of particulate matter, liquid droplets, and other potentially harmful substances from the natural gas stream. This results in cleaner combustion and lower emissions of pollutants, leading to improved air quality and reduced environmental impact.
(2) Greenhouse gas emissions: Given that natural gas is primarily composed of methane, a potent greenhouse gas, its efficient utilization after undergoing purification processes like supersonic separation can lead to reduced emissions of carbon dioxide and methane. This can contribute to efforts aimed at mitigating climate change.
(3) Air quality improvement: Supersonic separation plays a critical role in refining the quality of natural gas used for energy production. By removing impurities such as sulfur dioxide (SO_2), nitrogen oxides (NO_x), and volatile organic compounds (VOCs), it helps prevent the emission of harmful pollutants that can contribute to smog formation and adverse health effects.
(4) Reduced environmental contamination: Natural gas extracted from underground reservoirs may contain impurities that can contaminate the environment if not properly managed. Supersonic separation aids in the removal of these impurities, reducing the potential for soil, water, and ecosystem contamination during gas processing and utilization.
(5) Conservation of natural resources: The efficient utilization of natural gas facilitated by supersonic separation can lead to better conservation of natural resources. By maximizing the energy value obtained from each unit of natural gas, less gas is wasted or lost during the utilization process, reducing the overall demand for additional resource extraction.
(6) Water resource conservation: Supersonic separation can also assist in conserving water resources by reducing the water content in natural gas. This not only minimizes the water requirement for gas processing but also helps prevent water contamination during gas utilization.

In conclusion, the adoption of advanced gas processing techniques, such as supersonic separation, is crucial for harnessing the full potential of natural gas as a cleaner and more sustainable energy source. These technologies not only refine the quality of natural gas but also play a pivotal role in mitigating environmental impacts, improving air quality, and promoting the conservation of natural resources.

References

1. AlNouss A, Ibrahim M, Al-Sobhi SA (2018) Potential energy savings and greenhouse gases (GHGs) emissions reduction strategy for natural gas liquid (NGL) recovery: process simulation and economic evaluation. J Clean Prod 194:525–539
2. Álvarez-Gutiérrez N, Gil MV, Rubiera F, Pevida C (2016) Adsorption performance indicators for the CO_2/CH_4 separation: application to biomass-based activated carbons. Fuel Process Technol 142:361–369
3. Arinelli LDO, Araújo OQ, de Medeiros JL (2015) Dynamic simulation and analysis of slug flow impact on offshore natural gas processing: TEG dehydration, Joule–Thomson expansion and membrane separation. Computer aided chemical engineering 37:1775–1780
4. Ayuba S, Suh D, Nomura K, Ebisuzaki T, Yasuoka K (2018) Kinetic analysis of homogeneous droplet nucleation using large-scale molecular dynamics simulations. J Chem Phys 149(4):044504
5. Bains P, Psarras P, Wilcox J (2017) CO_2 capture from the industry sector. Prog Energy Combust Sci 63:146–172
6. Baltadjiev ND, Lettieri C, Spakovszky ZS (2015) An investigation of real gas effects in supercritical CO_2 centrifugal compressors. J Turbomach 137(9):091003
7. Bian J, Liu Y, Zhang X, Li Y, Gong L, Cao X (2023) Co-condensation and interaction mechanism of acidic gases in supersonic separator: a method for simultaneous removal of carbon dioxide and hydrogen sulfide from natural gas. Sep Purif Technol 322:124296
8. Brazhkin VV (2006) Metastable phases and 'metastable' phase diagrams. J Phys Condens Matter 18(42):9643
9. Brethomé FM, Williams NJ, Seipp CA, Kidder MK, Custelcean R (2018) Direct air capture of CO_2 via aqueous-phase absorption and crystalline-phase release using concentrated solar power. Nat Energy 3(7):553–559
10. Brigagão GV, de Oliveira Arinelli L, de Medeiros JL, Ofélia de Queiroz FA (2020) Low-pressure supersonic separator with finishing adsorption: higher exergy efficiency in air pre-purification for cryogenic fractionation. Sep Purif Technol 248:116969
11. Cao H, Cao X, Li H, Zhao X, Cai W, Guo D et al (2023) Nucleation and condensation characteristics of carbon dioxide in natural gas: a molecular simulation perspective. Fuel 342:127761
12. Chen GQ, Kanehashi S, Doherty CM, Hill AJ, Kentish SE (2015) Water vapor permeation through cellulose acetate membranes and its impact upon membrane separation performance for natural gas purification. J Membr Sci 487:249–255
13. Chen J, Jiang W, Lai X, Cao X, Bian J, Bi Z (2020) Study on the influence of wall-mounted cyclone on the purification and separation performance of supersonic separator. Chem Eng Process Process Intensif 150:107898
14. Chen G, Wei F, Zhang K, Xiao R, Wang Z, Yang S (2023) Investigation on combustion characteristics and gas emissions of a high-pressure direct-injection natural gas engine at different combustion modes. Energy Convers Manage 277:116617
15. Chong ZR, Yang SHB, Babu P, Linga P, Li XS (2016) Review of natural gas hydrates as an energy resource: prospects and challenges. Appl Energy 162:1633–1652
16. Chu H, Huang Z, Zhang Z, Yan X, Qiu B, Xu N (2024) Integration of carbon emission reduction policies and technologies: research progress on carbon capture, utilization and storage technologies. Sep Purif Technol 127153
17. de Faria DR, de Oliveira Arinelli L, de Medeiros JL, Araújo ODQF (2020) Novel ethylene oxide production with improved sustainability: loss prevention via supersonic separator and carbon capture. J Environ Manag 269:110782
18. de Medeiros JL, de Oliveira Arinelli L, Teixeira AM, Araújo ODQF, Araújo ODQF, de Medeiros JL (2019) Overview of natural gas processing with supersonic separator. Offshore processing of CO_2-rich natural gas with supersonic separator: multiphase sound speed, CO_2 freeze-out and HYSYS implementation 41–53

19. de Melo DC, Arinelli LDO, de Medeiros JL, Teixeira AM, Brigagão GV, Passarelli FM et al (2019) Supersonic separator for cleaner offshore processing of supercritical fluid with ultra-high carbon dioxide content: economic and environmental evaluation. J Clean Prod 234:1385–1398
20. de Oliveira Arinelli L, Trotta TAF, Teixeira AM, de Medeiros JL, Ofélia de Queiroz FA (2017) Offshore processing of CO_2 rich natural gas with supersonic separator versus conventional routes. J Nat Gas Sci Eng 46:199–221
21. de Oliveira Arinelli L, de Medeiros JL, de Melo DC, Teixeira AM, Brigagão GV, Passarelli FM et al (2019) Carbon capture and high-capacity supercritical fluid processing with supersonic separator: natural gas with ultra-high CO_2 content. J Nat Gas Sci Eng 66:265–283
22. de Oliveira Arinelli L, Teixeira AM, de Medeiros JL, Ofélia de Queiroz FA (2019) Supersonic separator for cleaner offshore processing of natural gas with high carbon dioxide content: environmental and economic assessments. J Clean Prod 233:510–521
23. Duff KM (1964) Condensation of carbon dioxide in laval nozzles. Massachusetts Institute of Technology
24. Dumitrescu LR, Smeulders DM, Dam JA, Gaastra-Nedea SV (2017) Homogeneous nucleation of water in argon. Nucleation rate computation from molecular simulations of TIP4P and TIP4P/2005 water model. J Chem Phys 146(8)
25. Ebrahimi-Fizik A, Lakzian E, Hashemian A (2019) Entropy generation analysis of wet-steam flow with variation of expansion rate using NURBS-based meshing technique. Int J Heat Mass Transf 139:399–411
26. Ellis PW, Page AM, Wood S, Fargione J, Masuda YJ, Carrasco Denney V et al (2024) The principles of natural climate solutions. Nat Commun 15(1):547
27. Interlenghi SF, de Medeiros JL, Ofélia de Queiroz FA (2020) Protected supersonic separator performance against variable CO_2 content on natural gas processing: energy and sustainability analyses. J Nat Gas Sci Eng 78:103282
28. Jia W, Gong C, Pan K, Yu S (2023) Potential changes of regional natural gas market in China amidst liberalization: a mixed complementarity equilibrium simulation in 2030. Energy 284:129254
29. Kaplun AB, Meshalkin AB (2007) Calculation of thermodynamic properties of substances in metastable and labile regions of real gas states. J Eng Thermophys 16(4):259–269
30. Keiner D, Gulagi A, Breyer C (2023) Energy demand estimation using a pre-processing macro-economic modelling tool for 21st century transition analyses. Energy 272:127199
31. Lakzian E, Yazdani S, Mobini R, Abadi MHME, Ramezani A, Yahyazadeh M, Rashedi Tabar M (2022) Investigation of the effect of water droplet injection on condensation flow of different nozzles geometry. Eur Phys J Plus 137(5):613
32. Lettieri C, Paxson D, Spakovszky Z, Bryanston-Cross P (2018) Characterization of nonequilibrium condensation of supercritical carbon dioxide in a de Laval nozzle. J Eng Gas Turbines Power 140(4):041701
33. Netusil M, Ditl P (2011) Comparison of three methods for natural gas dehydration. J Nat Gas Chem 20(5):471–476
34. Nguyen TV, Tock L, Breuhaus P, Maréchal F, Elmegaard B (2014) Oil and gas platforms with steam bottoming cycles: system integration and thermoenvironomic evaluation. Appl Energy 131:222–237
35. Ochi Y, Kawakubo R, Van-Pham DT, Kitamura Y, Nakanishi H, Norisuye T, Tran-Cong-Miyata Q (2015) Phase separation of polymer mixtures induced by light and heat: a comparative study by light scattering. Adv Nat Sci Nanosci Nanotechnol 6(4):045002
36. Omer AM (2009) Energy use and environmental impacts: a general review. J Renew Sustain Energy 1(5)
37. Oswatitsch K (1942) Kondensationserscheinungen in Überschalldüsen. ZAMM J Appl Math Mech/Z Angew Math Mech 22(1):1–14
38. Pereira MA, Santos LF, Caballero JA, Ravagnani MA, Costa CB (2022) Energy and economic comparison of five mixed-refrigerant natural gas liquefaction processes. Energy Convers Manage 272:116364

39. Santos MG, Correia LM, de Medeiros JL, Ofélia de Queiroz FA (2017) Natural gas dehydration by molecular sieve in offshore plants: impact of increasing carbon dioxide content. Energy Convers Manage 149:760–773
40. Shooshtari SHR, Shahsavand A (2014) Predictions of wet natural gases condensation rates via multi-component and multi-phase simulation of supersonic separators. Korean J Chem Eng 31:1845–1858
41. Taqieddin A, Allshouse MR, Alshawabkeh AN (2018) Critical review—mathematical formulations of electrochemically gas-evolving systems. J Electrochem Soc 165(13):E694
42. Teixeira AM, de Oliveira Arinelli L, de Medeiros JL, Ofélia de Queiroz FA (2018) Recovery of thermodynamic hydrate inhibitors methanol, ethanol and MEG with supersonic separators in offshore natural gas processing. J Nat Gas Sci Eng 52:166–186
43. Teixeira AM, de Oliveira Arinelli L, de Medeiros JL, Ofélia de Queiroz FA (2019) Economic leverage affords post-combustion capture of 43% of carbon emissions: supersonic separators for methanol hydrate inhibitor recovery from raw natural gas and CO_2 drying. J Environ Manage 236:534–550
44. Teixeira AM, de Oliveira Arinelli L, de Medeiros JL, Ofélia de Queiroz FA (2022) Sustainable offshore natural gas processing with thermodynamic gas-hydrate inhibitor reclamation: supersonic separation affords carbon capture. Chem Eng Res Des 181:55–73
45. Theo WL, Lim JS, Hashim H, Mustaffa AA, Ho WS (2016) Review of pre-combustion capture and ionic liquid in carbon capture and storage. Appl Energy 183:1633–1663
46. Wang Y (2020) Analysis for spiral vortex and effect of profile of nozzle and swirler on performance of supersonic separator. Chem Eng Process Process Intensif 147:107676
47. Wang Y, Hu D (2018) Experimental and numerical investigation on the blade angle of axial-flow swirling generator and drainage structure for supersonic separators with diversion cone. Chem Eng Res Des 133:155–167
48. Wang Y, Yu Y, Hu D, Xu D, Yi L, Zhang Y, Zhang S (2020) Improvement of drainage structure and numerical investigation of droplets trajectories and separation efficiency for supersonic separators. Chem Eng Process Process Intensif 151:107844
49. Wang Y, Yu Y, Hu D (2020) Experimental investigation and numerical analysis of separation performance for supersonic separator with novel drainage structure and reflux channel. Appl Therm Eng 176:115111
50. Wang Q, Duan L, Zheng N, Lu Z (2023) 4E analysis of a novel combined cooling, heating and power system coupled with solar thermochemical process and energy storage. Energy 275:127498
51. Wegener PP, Wu BJ (1976) Homogeneous and binary nucleation: new experimental results and comparison with theory. Faraday Discuss Chem Soc 61:77–82
52. Wen C, Cao X, Yang Y, Li W (2012) Numerical simulation of natural gas flows in diffusers for supersonic separators. Energy 37(1):195–200
53. Wiesberg IL, de Oliveira Arinelli L, Ofélia de Queiroz FA, de Medeiros JL (2021) Upgrading exergy utilization and sustainability via supersonic separators: offshore processing of carbonated natural gas. J Clean Prod 310:127524
54. Wiśniewski P, Dykas S, Miyazawa H, Furusawa T, Yamamoto S (2023) Modified heat transfer correction function for modeling multiphase condensing flows in transonic regime. Int J Heat Mass Transf 201:123597
55. Wu BJ, Wegener PP, Stein GD (1978) Homogeneous nucleation of argon carried in helium in laval nozzle flow. J Chem Phys 69(4):1776–1777
56. Wyslouzil BE, Wölk J (2016) Overview: homogeneous nucleation from the vapor phase—the experimental science. J Chem Phys 145(21)
57. Yasuoka K, Matsumoto M (1998) Molecular dynamics of homogeneous nucleation in the vapor phase. II. Water. J Chem Phys 109(19):8463–8470
58. Gyarmathy G (2005) Nucleation of steam in high-pressure nozzle experiments. Proc Inst Mech Eng Part A J Power Energy 219(6):511–521
59. Liu Q, Liu Y, Xu J, Teng Y, Ling Z, Zhang Y et al (2023) A review of the gas hydrate phase transition with a microfluidic approach. Energy Rev 2(1):100011

60. Niknam PH, Fiaschi D, Mortaheb HR, Mokhtarani B (2019) Numerical investigation of multiphase flow in supersonic separator considering inner body effect. Asia-Pac J Chem Eng 14(6):e2380
61. Bian J, Cao X, Yang W, Guo D, Xiang C (2020) Prediction of supersonic condensation process of methane gas considering real gas effects. Appl Therm Eng 164:114508
62. Cao X, Bian J (2019) Supersonic separation technology for natural gas processing: a review. Chem Eng Process Process Intensif 136:138–151

Chapter 4
Homogeneous Condensation of Water Vapor in Supersonic Flows

4.1 Introduction

The supersonic separator is designed for natural gas processing, emphasizing the removal of water vapor [5, 34]. A significant characteristic of this separator is the simultaneous occurrence of non-equilibrium condensation and a shock wave in supersonic flows. While the primary operational principle of the supersonic separator involves the non-equilibrium condensation of water vapor in supersonic flows, it also benefits from the shock wave as supersonic flows transition to subsonic flows [38], thereby restoring pressure to maximize energy efficiency [4].

For condensation in pure water vapor, several researchers have delved into non-equilibrium condensation in supersonic nozzles over the years. Dykas and Wróblewski [14] introduced a wet steam gas–liquid model, revealing that the single-fluid model underestimated droplet sizes by around 40% compared to their two-fluid model. Abadi et al. [1] utilized a two-fluid model in high-pressure nozzles, noting significant slip velocities and temperature disparities between liquid and vapor phases. Pillai and Prasad [32] explored how the wall surface roughness affects condensation, finding that the boundary layer thickness increased by 33% with varying roughness heights from 1 to 1000 μm. Ding et al. [11] assessed the impact of surface roughness on condensing flow in turbine blades using a single-fluid wet steam model. It is indicated a slight decrease in the liquid fraction at the outlet with increasing roughness height. Liang et al. [24] addresses non-equilibrium condensation in high-speed steam flow through turbine final stages. It is indicated that the adjustments of inlet steam superheat can achieve the effective suppression of condensation and improved turbine efficiency. Hamidi and Kermani [19] investigated the influence of the normal aerodynamic shocks on the condensation jump and entropy generations of non-equilibrium condensation.

H. Ding et al., *Supersonic Separators for Sustainable Gas Purification*, Green Energy and Technology, https://doi.org/10.1007/978-3-031-85213-8_4

For condensation in moist-air, water content is typically denoted by relative humidity, impacting gas turbine and electrical machinery efficiency [33]. Latent heat release from condensation affects the flow field in supersonic nozzle, and the wide ranges of temperature, pressure of carrier gas and water vapor make the complicate condensation phenomena [25]. Aliabadi et al. [3] explored a 20 nm heterogeneous droplet's impact on saturated vapor condensation which finds with an increase in the number of droplets, condensation shock and the nucleation phenomenon will not occur, yet few researchers investigate temperature and pressure effects on condensation processes and performance. In additional, many regions experience average relative humidity above 70%, with Houston exceeding 90% [16] , thus, it is significant important to study the high-relative humidity moist air flows. Zhang et al. [42] A blend condensation model is introduced to examine the impact of the full range of Knudsen numbers on droplet growth, aiming to enhance the understanding of condensation flow dynamics. Additionally, the model analyzes how relative humidity (RH) influences nucleation, droplet growth, and the formation of condensation shock waves.

Numerical solutions offer the advantage of calculating comprehensive flow property details during the condensation process. However, a significant drawback is the substantial computational demand in terms of grid resolution, computing resources, and processing time required to track droplet growth and characterize nucleation in complex multi-dimensional flow fields. Moreover, numerical simulations cannot capture the dominant physical effects during nucleation conveniently [13]. Furthermore, due to limited understanding of nucleation theory, numerical calculations often fail to valid the detailed changes of fluid properties during nucleation and droplet growth in supersonic nozzle.

Given these challenges, there is a strong desire to derive an algebraic expression that specifically addresses the nucleation process, particularly the flow properties at the Wilson point. Previous attempts by Dobbins [13] and Huang and Young [21] introduced algebraic methods for determining the Wilson point under low pressure conditions. However, Dobbins' approach did not account for the variations in expansion rate proposed by Gyarmathy, making direct solution overly complex. Huang's method ignored the influence of Rayleigh flow relationships on the expansion rate. Additionally, Delale et al. [8] proposed asymptotic methods, but their expressions are too intricate for generalization and calculation purposes. Ding et al. [9] presented analytical solution effectively determines the position and flow properties of the Wilson point in nozzle flow under low pressure conditions, incorporating classical nucleation and droplet growth theories, Lagrangian equations for momentum and energy, and Clausius–Clapeyron relationships. Bolaños-Acosta et al. [6] proposed two semi-analytical approaches models for gas condensation in supersonic flow where one based on Rankine–Hugoniot and Rayleigh equations for weak condensation shocks, and another derived from homogeneous nucleation theory. Both models agreed closely with experimental data from literature, affirming their robustness in predicting condensation behaviors.

4.2 Mathematical Model for Non-equilibrium Condensation

The three-dimensional compressible Navier–Stokes equations are applied to characterize the non-equilibrium condensation phenomenon of water vapor in steam flow and moist air flows. A non-equilibrium condensation phase change model, encompassing nucleation and droplet growth is developed to forecast the condensation phenomenon in gas liquid two phase flow. It assumes a no-slip velocity condition between the vapor and liquid phases, justified by the small size of droplets, typically less than 1 μm [15].

4.2.1 Governing Equations

The mass conservation equation of gas phase in tensor form can be written as follows:

$$\frac{\partial \rho_{\mathrm{g}}}{\partial t} + \frac{\partial}{\partial x_j}(\rho_{\mathrm{g}} u_j) = S_{\mathrm{m}} \tag{4.1}$$

where ρ_{g} (kg m^{-3}) is the density of gas phase, x_j (m) and u_j (m s^{-1}) represent the j-wise Cartesian coordinate and j-wise velocity. S_{m} is the mass source due to phase change.

The momentum conservation equation of gas phase in tensor form is computed by

$$\begin{aligned}\frac{\partial}{\partial t}(\rho_{\mathrm{g}} u_i) + \frac{\partial}{\partial x_j}(\rho_{\mathrm{g}} u_i u_j) &= -\frac{\partial p}{\partial x_i} + \Delta\rho_{\mathrm{g}} g_j + S_{ui} \\ &\quad + \frac{\partial}{\partial x_j}\left(\mu_{\mathrm{eff}}\left(\frac{\partial u_j}{\partial x_i} + \frac{\partial u_i}{\partial u_j} - \frac{2}{3}\frac{\partial u_j}{\partial x_j}\right)\right)\end{aligned} \tag{4.2}$$

where u_i (m s^{-1}) and x_i (m) represent the i-wise velocity and Cartesian coordinate, μ_{eff} (Pa s) is effective dynamic viscosity, g_j (m s^{-2}) is the j-wise acceleration of gravity. S_{ui} is the momentum source of gas phase due to phase change.

The energy conservation equation of gas phase is given by

$$\frac{\partial}{\partial t}(\rho_{\mathrm{g}} E) + \frac{\partial}{\partial x_j}(\rho_{\mathrm{g}} u_j E) = \frac{\partial}{\partial x_j}\left(\lambda_{\mathrm{eff}}\frac{\partial T}{\partial x_j}\right) + S_{\mathrm{E}} \tag{4.3}$$

where T (K) represent static temperature, λ_{eff} (W m^{-1} K^{-1}) is effective thermal conductivity. S_{E} is the source of energy. E (J kg^{-1}) is total energy and computed by

$$E = h - \frac{p}{\rho} + \frac{u^2}{2} \tag{4.4}$$

μ_{eff} and λ_{eff} can be calculated by

$$\mu_{\text{eff}} = \mu + \frac{\mu_t}{\sigma_k} \tag{4.5}$$

$$\lambda_{\text{eff}} = \lambda + \frac{c_p \mu_t}{\text{Pr}_t} \tag{4.6}$$

where σ_k turbulent Prandtl numbers for k. Pr_t is the turbulent Prandtl number for energy. c_p is specific heat of fluid. The turbulent viscosity μ_t for $k - \varepsilon$ turbulence model is calculated by

$$\mu_t = \rho C_\mu \frac{k^2}{\varepsilon} \tag{4.7}$$

where $C_\mu = 0.09$. If using the $k - \omega$ turbulence model, turbulent viscosity μ_t is rewritten as

$$\mu_t = \frac{\rho k}{\omega} \frac{1}{\max\left[\frac{1}{\alpha^*}, \frac{\Omega F_2}{a_1 \omega}\right]} \tag{4.7a}$$

where Ω is the strain rate magnitude [29], $\Omega = \sqrt{2S_{i,j}S_{i,j}}$. α^* damps the turbulent viscosity causing a low-Reynolds number correction.

The turbulence kinetic energy k and turbulence dissipation ε equations are described by

$$\frac{\partial(\rho k)}{\partial t} + \frac{\partial(\rho k u_i)}{\partial x_i} = \frac{\partial}{\partial x_j}\left(\left(\mu + \frac{\mu_t}{\sigma_k}\right)\frac{\partial k}{\partial x_j}\right) + G_k - \rho\varepsilon \tag{4.8}$$

$$\frac{\partial(\rho \varepsilon)}{\partial t} + \frac{\partial(\rho \varepsilon u_i)}{\partial x_i} = \frac{\partial}{\partial x_j}\left(\left(\mu + \frac{\mu_t}{\sigma_\varepsilon}\right)\frac{\partial \varepsilon}{\partial x_j}\right) + \frac{\varepsilon}{k}(C_{1\varepsilon}G_k - C_{2\varepsilon}\rho\varepsilon) \tag{4.9}$$

where $C_{1\varepsilon}$, $C_{2\varepsilon}$, σ_k and σ_ε are equal to 1.44, 1.92, 1.0 and 1.3 respectively. G_k is the generation of turbulence kinetic energy due to the mean velocity gradients.

The turbulence kinetic energy k and the specific dissipation rate ω are as follows:

$$\frac{\partial}{\partial t}(\rho k) + \frac{\partial}{\partial x_i}(\rho k u_i) = \frac{\partial}{\partial x_j}\left[\left(\mu + \frac{\mu_t}{\sigma_k}\right)\frac{\partial k}{\partial x_j}\right] + G_k - Y_k \tag{4.10}$$

$$\frac{\partial \rho \omega}{\partial t} + \frac{\partial}{\partial x_i}(\rho \omega u_i) = \frac{\partial}{\partial x_j}\left[\left(\mu + \frac{\mu_t}{\sigma_\omega}\right)\frac{\partial \omega}{\partial x_j}\right] + G_\omega - Y_\omega \tag{4.11}$$

where G_k represents the generation of k due to the mean velocity gradient. Y_k and Y_ω are the values of the dissipations of k and ω produced by turbulence.

If the transition of the boundary layer should be taken into account, the Transition SST model (also known as the $\gamma - \tilde{Re}_{\theta t}$ model) is a better choice, based on the coupling of the SST $k - \omega$ transport equations with two other transport equations, one for the intermittency γ and another for transition onset the momentum-thickness Reynolds number $\tilde{Re}_{\theta t}$.

The transport equation for the intermittency γ and transition momentum thickness Reynolds number $\tilde{Re}_{\theta t}$ can be calculated by Eqs. (4.12)–(4.13).

$$\frac{\partial(\rho\gamma)}{\partial t}+\frac{\partial(\rho u_j\gamma)}{\partial x_j}=P_{\gamma 1}-E_{\gamma 1}+P_{\gamma 2}-E_{\gamma 2}+\frac{\partial}{\partial x_j}\left[\left(\mu+\frac{\mu_t}{\sigma_\gamma}\right)\frac{\partial\gamma}{\partial x_j}\right] \tag{4.12}$$

$$\frac{\partial(\rho\tilde{\mathrm{Re}}_{\theta t})}{\partial t}+\frac{\partial(\rho u_j\tilde{\mathrm{Re}}_{\theta t})}{\partial x_j}=P_{\theta t}+\frac{\partial}{\partial x_j}\left[\sigma_{\theta t}(\mu+\mu_t)\frac{\partial\tilde{\mathrm{Re}}_{\theta t}}{\partial x_j}\right] \tag{4.13}$$

where $P_{\gamma 1}, E_{\gamma 1}, P_{\gamma 2}, E_{\gamma 2}$ represent the transition sources. The constant for the intermittency equation $\sigma_\gamma = 1.0$. $P_{\theta t}$ is the source term in transport equation for $\tilde{\mathrm{Re}}_{\theta t}$. The boundary condition for $\tilde{\mathrm{Re}}_{\theta t}$ at a wall is zero flux. The boundary condition for $\tilde{\mathrm{Re}}_{\theta t}$ at the inlet is calculated from the empirical correlation based on the inlet turbulence intensity, $\mathrm{Tu} = \frac{100}{U}\sqrt{\frac{2k}{3}}$.

The wall roughness will affect the transition process, thus the empirical correlation for the transition momentum thickness Reynolds number is calculated by [35]

$$\tilde{\mathrm{Re}}_{\theta t,\mathrm{rough}}=\begin{cases}\tilde{\mathrm{Re}}_{\theta t,s} & \frac{k}{\delta_1}\le 0.01\\ \left(\frac{1}{\tilde{\mathrm{Re}}_{\theta t,s}}+0.0061f_\Lambda\left(\frac{K_s}{\delta_1-0.01}\right)^{f_{\mathrm{Tu}}}\right)^{-1} & \frac{k}{\delta_1}>0.01\end{cases} \tag{4.14}$$

where K_s represent average height of sand-grain roughness element. δ_1 is the displacement thickness. f_{Tu} is as function of the effective turbulence intensity. The correction function f_Λ is to account for roughness topographies density outside the range $4 < \Lambda_R < 7$.

By using the automatic wall treatment of $k-\omega$ equations, the wall shear stress τ_w is computed by $\tau_w = \rho u^* u_\tau$. Two friction velocities u^* and u_τ are blended between the viscous sublayer and the logarithmic region. The former friction velocity u^* is expressed by [30]

$$u^*=\left(\frac{\nu u_p}{y_p}+C_\mu^{\frac{1}{4}}k\right)^{\frac{1}{2}} \tag{4.15}$$

where y_p is the distance between wall-adjacent cell centroid P and the wall node. The default constant value C_μ is equal to 0.09 and ν is kinematic viscosity. The friction velocity u_τ is defined by

$$u_\tau = u_p \left[\left(y^+\right)^{-4} + \left(\frac{1}{\kappa} \ln\left(Ey^+\right) + \Delta B \right)^{-4} \right]^{\frac{1}{4}} \tag{4.16}$$

where κ is the von Kármán constant, $\kappa = 0.4187$. The empirical constant E is equal to 9.793. ΔB is the roughness correction. y^+ is normalized distance. To avoid the singularity when the roughness height is large, a virtually shift the wall to 50% of the height of the roughness element is better. The corrected y^+ at wall-adjacent cell centroid is $y^+ = y^+ + \frac{K_s^+}{2}$, where the nondimensional distance $y^+ = \sqrt{\tau_w/\rho}\frac{y_p}{\nu}$ and roughness $K_s^+ = \sqrt{\tau_w/\rho} K_s/\nu$. In addition, the empirical values of ΔB in the hydro-dynamically smooth regime, transitional regime and fully rough regime is different.

In the hydro-dynamically smooth regime when $K_s^+ < 2.25$,

$$\Delta B = 0 \tag{4.17}$$

In the transitional regime when $2.25 \le K_s^+ \le 90$,

$$\Delta B = \frac{1}{\kappa} \ln\left[\frac{K_s^+ - 2.25}{87.75} + C_s K_s^+ \right] \sin\left[0.4258\left(\ln K_s^+ - 0.811\right)\right] \tag{4.17a}$$

In the fully rough regime when $K_s^+ \ge 90$,

$$\Delta B = \frac{1}{\kappa} \ln\left(1 + C_s K_s^+\right) \tag{4.17b}$$

where roughness constant $C_s = 0.5$. The boundary conditions at the wall boundary are $\frac{\partial k}{\partial n} = 0$ and $\frac{\partial \omega}{\partial n} = 0$, where n represents the normal to the wall.

The value of ω at wall is specified as

$$\omega_{\text{wall}} = \left(\omega_{\text{vis}}^2\left(y^+\right) + \omega_{\log}^2\left(y^+\right) \right)^{\frac{1}{2}} \tag{4.18}$$

where ω_{vis} and $\omega_{\log}$ are friction velocity in linear near-wall region and the logarithmic near-wall region can be computed by $\omega_{\text{vis}} = 6/\left[\beta_\infty^*\left(y_p\right)^2\right]$, and $\omega_{\log} = u_\tau/\left(C_\mu^{0.5} \kappa y_p\right)$, where the constant value $\beta_\infty^* = 0.075$.

Species transport equation for mixture gas can be described by the equation below.

$$\frac{\partial \rho_g Y_i}{\partial t} + \frac{\partial}{\partial x_j}\left(\rho_g u_j Y_i\right) = \frac{\partial}{\partial x_j}\left(D_i \rho_g \frac{\partial Y_i}{\partial x_j} \right) + S_i \tag{4.19}$$

where Y_i represents the mass fraction of each species. D_i is diffusion coefficient. S_i represents the source item due to the phase change of the condensable component.

The conservation equation of droplet number density N_{hom} (m^{-3}) can be calculated by

$$\frac{\partial N_{\text{hom}}}{\partial t}+\frac{\partial}{\partial x_j}\left(N_{\text{hom}}u_j\right)=J \tag{4.20}$$

where J ($\text{m}^{-3}\ \text{s}^{-1}$) represents the nucleation rate of the condensable component.

The conservation equation of the statistical radius of droplet r (m) is expressed as

$$\frac{\partial}{\partial t}(N_{\text{hom}}r)+\frac{\partial}{\partial x_j}\left(N_{\text{hom}}ru_j\right)=N_{\text{hom}}\frac{\text{d}r}{\text{d}t}+Jr_{\text{c}} \tag{4.21}$$

The conservation equation of the wetness fraction β is expressed as

$$\frac{\partial}{\partial t}(\rho\beta)+\frac{\partial}{\partial x_j}\left(\rho\beta u_j\right)=-S_{\text{m}} \tag{4.22}$$

The source terms in Eqs. (4.1)–(4.3) are defined as follows

$$S_{\text{m}}=-\dot{m}_{\text{hom}} \tag{4.23}$$

$$S_{ui}=-\dot{m}_{\text{hom}}u_i \tag{4.23a}$$

$$S_{\text{E}}=-\dot{m}_{\text{hom}}\left(h-h_{\text{lg}}\right) \tag{4.23b}$$

where m_{v} ($\text{kg m}^{-3}\ \text{s}^{-1}$) is the liquid mass per unit volume condensation in unit time, h is the enthalpy, and h_{lg} (kJ kg^{-1}) is the latent heat of condensation. The liquid mass changing rate of homogeneous droplets $\dot{m}_{\text{hom}}$ [10] is given by

$$\dot{m}_{\text{hom}}=J\frac{4\pi r_{\text{c}}^3}{3}+N_{\text{hom}}4\pi r_{\text{hom}}^2 \tag{4.24}$$

4.2.2 Nucleation and Droplet Growth Model

Nucleation rate J ($\text{m}^{-3}\ \text{s}^{-1}$) for moist gas is calculated by Eq. (4.25) [31], which is suited for the water vapor condensation in the presence of carrier gas, and for wet steam takes the form of Eq. (4.26) [28], which is widely used in the wet steam nucleating flow.

$$J = \frac{\upsilon N_{\mathrm{mc}}^2}{S}\sqrt{\frac{2\sigma}{\pi m_{\mathrm{m}}}}\exp\left(-\frac{16\pi}{3}\frac{\upsilon^2\sigma^3}{\left(k_{\mathrm{B}}T_{\mathrm{g}}\right)^3(\ln S)^2}\right) \tag{4.25}$$

where supersaturation ratio S is defined as the ratio of vapor pressure to saturation pressure, $S = p/p_{\mathrm{s}}$. N_{mc} (kg^{-1}) is the number density of molecules of condensable component, υ (m^3) represent the volume of single molecule, σ ($\mathrm{N\ m}^{-1}$) is surface tension. The single water molecule mass $m_{\mathrm{m}} = 2.99 \times 10^{-26}$ kg. Boltzmann's constant $k_{\mathrm{B}} = 1.38 \times 10^{-23}$.

$$J = \frac{q_{\mathrm{c}}}{1+\theta}\left(\frac{2\sigma}{\pi m_{\mathrm{m}}^3}\right)^{1/2}\frac{\rho_{\mathrm{g}}^2}{\rho_l}\exp\left(-\frac{4\pi r_{\mathrm{c}}^2\sigma}{3k_{\mathrm{B}}T_{\mathrm{g}}}\right) \tag{4.26}$$

where condensation coefficient $q_{\mathrm{c}} = 1$. ρ_{l} is the liquid density. The Kantrowitz correction θ is expressed as

$$\theta = 2\frac{\gamma_{\mathrm{g}} - 1}{\gamma_{\mathrm{g}} + 1}\frac{h_{\mathrm{lg}}}{R_{\mathrm{g}}T_{\mathrm{c}}}\left(\frac{h_{\mathrm{lg}}}{R_{\mathrm{g}}T_{\mathrm{g}}} - \frac{1}{2}\right) \tag{4.27}$$

where R_{g} ($\mathrm{J\ kg}^{-1}\ \mathrm{K}^{-1}$) is specific gas constant of condensable component. γ_{g} is the specific heat ratio of water vapor.

The critical radius of the nucleation droplet is calculated by [41]

$$r_{\mathrm{c}} = \frac{2\sigma}{\rho_{\mathrm{l}}R_{\mathrm{g}}T_{\mathrm{g}}\ln S} \tag{4.28}$$

The temperature of droplet surface T_{r} can be obtain by the heat transfer between the droplet and vapor gas from Eq. (4.29), supposing the condensing latent heat will be completely absorbed by the surrounding vapor and carrier gas.

$$4\pi r^2\left(T_{\mathrm{r}} - T_{\mathrm{g}}\right)\sum_{i=1}^{n}\alpha_i = h_{\mathrm{lg}}\frac{dm_{\mathrm{p}}}{dt} = h_{\mathrm{lg}}4\pi r_{\mathrm{hom}}^2\rho_{\mathrm{l}}\frac{dr_{\mathrm{hom}}}{dt} \tag{4.29}$$

where T_{g} is the temperature of gas surface and m_{p} (kg) is a droplet condensing. Thus, the temperature of droplet surface T_{r} [17] is calculated by Eq. (4.30).

$$T_{\mathrm{r}} = T_{\mathrm{s}}(p) - \Delta T\frac{r_{\mathrm{c}}}{r_{\mathrm{hom}}} \tag{4.30}$$

The rate of droplet growth r_{hom} is written as Eq. (4.31).

$$G(r) = \frac{dr_{\mathrm{hom}}}{dt} = \frac{\sum_{i=1}^{n}\alpha_{k,\mathrm{gp}}}{\rho_{\mathrm{l}}h_{\mathrm{fg}}}\left(T_{\mathrm{r}} - T_{\mathrm{g}}\right) \tag{4.31}$$

where $\alpha_{k,\mathrm{gp}}$ (W m^{-2} K^{-1}) is the coefficient of heat transfer between the gas phase and water droplet. Considering the flux matching between condensation nodules and transition layers, Gyarmathy combined the Fick diffusivity and the Fourier thermal conductivity to calculate the growth rate $\frac{\mathrm{d}r_{\mathrm{hom}}}{\mathrm{d}t}$ under two limiting conditions, namely Kn << 1 and Kn >> 1 belonging to the free flow region and continuous flow region [18]. The Gyarmathy model of water droplet growth can be obtained in Eq. (4.32).

$$\alpha_k = \frac{\lambda}{r_{\mathrm{hom}}} \frac{1}{1 + \left(\frac{2\sqrt{8\pi}}{1.5\,\mathrm{Pr}}\right) \times \frac{\gamma}{1+\gamma} K_{\mathrm{n}}} \tag{4.32}$$

where Knudsen number $K_{\mathrm{n}} = \overline{L}/2r$, is a dimensionless number defined as the ratio of the molecular mean free path length to a representative physical length scale. mean free length path of gas molecule $\overline{L}$ (m) is written in general form as follows:

$$\overline{L} = 1.5\mu\sqrt{R_{\mathrm{g}} T_{\mathrm{g}}} \tag{4.33}$$

For condensation flow of water vapor, if let Pr = 1.18, $\gamma = 1.3$. Substituting into Eq. (4.32), the rate of droplet growth r_{hom} can be rewritten by

$$\frac{\mathrm{d}r_{\mathrm{hom}}}{\mathrm{d}t} = \frac{\lambda (T_{\mathrm{s}} - T_{\mathrm{g}})\left(1 - \frac{r_{\mathrm{c}}}{r}\right)}{\rho_{\mathrm{l}} h_{\mathrm{lg}}[1 + 3.18 K_{\mathrm{n}}]} \tag{4.34}$$

The droplet growth model under arbitrary K_{n} is studied based on the Gyarmathy model and Langmuir's three-layer theoretical model of water droplets and then modified at 10–30 kPa low pressure [39]. Young proposed a correction factor β_{c} in the Knudsen layer to overcome the model's deficiencies. The Young model is defined as:

$$\frac{\mathrm{d}r_{\mathrm{hom}}}{\mathrm{d}t} = \frac{\lambda (T_{\mathrm{s}} - T_{\mathrm{g}})\left(1 - \frac{r_{\mathrm{c}}}{r}\right)}{\rho_{\mathrm{l}} h_{\mathrm{lg}}\left[\frac{1}{1+2\beta_{\mathrm{c}} K_{\mathrm{n}}} + 3.78(1 - \nu)\frac{K_{\mathrm{n}}}{\mathrm{Pr}}\right]} \tag{4.35}$$

where a semi-empirical correction coefficient ν is calculated by

$$\nu = \frac{R_{\mathrm{g}} T_{\mathrm{s}}}{h_{\mathrm{lg}}}\left[\alpha - 0.5 - \frac{2 - q_{\mathrm{c}}}{2q_{\mathrm{c}}} \frac{\gamma + 1}{2\gamma} \frac{c_{\mathrm{p}} T_{\mathrm{s}}}{h_{\mathrm{lg}}}\right] \tag{4.36}$$

Besides, the liquid surface tension σ (N/m) is expressed by [27]

$$\sigma = \left(85.27 + 75.612 T_{\mathrm{R}} - 256.889 T_{\mathrm{R}}^2 + 95.928 T_{\mathrm{R}}^3\right) \times 10^{-3} \tag{4.37}$$

where the reduced temperature $T_{\mathrm{R}} = T_{\mathrm{g}}/T_{\mathrm{c}}$. T_{c} is the critical temperature of water vapor, = 647.3 K. The dew point temperature T_{d} is deduced by [26]

$$\begin{aligned} T_\text{d} &= \tfrac{2663.5}{(12.537-\log p_\text{v})} - 273.15, && T_\text{d} \le -60\,^\circ\text{C} \\ T_\text{d} &= -60.45 + 7.0322\ln p_\text{v} + 0.37(\ln p_\text{v})^2, && -60\,^\circ\text{C} \le T_\text{d} \le 0\,^\circ\text{C} \\ T_\text{d} &= -35.957 - 1.8726\ln p_\text{v} + 1.1689(\ln p_\text{v})^2, && T_\text{d} > 0\,^\circ\text{C} \end{aligned} \tag{4.38}$$

The latent heat h_{lg} of water vapor at the arbitrary temperature can be computed by [43]

$$h_{\text{lg}} = h_{\text{lg}}^{0}\left(\frac{1 - T_\text{g}/647.3}{1 - 0.5767}\right)^{0.38} \tag{4.39}$$

where the latent heat h_{lg}^{0} at the normal boiling point (373.16 K) is 2262.2 kJ kg^{-1}.

4.2.3 Entropy Transport Equations

The total specific entropy of the mixture s (J kg^{-1} K^{-1}) is computed by

$$s = s_{\text{in}}^{*} + \Delta s = s_{\text{in}}^{*} + \Delta s_\text{g} - \Delta s_\text{l} \tag{4.40}$$

where Δs_g and Δs_l represent the entropy change of the gas phase and entropy transfer into the gas phase due to latent heat release.

The specific entropy change of the gas phase can be determined as follows:

$$\begin{aligned} s - s_{\text{in}}^{*} = s_{\text{tran}} + s_{\text{gen}} &= s_{\text{tran}} + s_{\text{gen,D}} \\ &+ s_{\text{gen,C}} + s_{\text{gen,L}} + s_{\text{gen,A}} \end{aligned} \tag{4.41}$$

where s_{tran} represents the entropy transfer. The entropy generation s_{gen} includes four parts which are viscous entropy generation $s_{\text{gen,D}}$, heat conduction entropy generation $s_{\text{gen,C}}$, phase change entropy generation $s_{\text{gen,L}}$ and aerodynamic losses $s_{\text{gen,A}}$. The transport equation for entropy is a first-order linear partial differential equation (PDEs) and satisfies the principle of superposition which can be calculated by

$$\rho\frac{D(s_\phi)}{Dt} = \frac{\partial \rho s_\phi}{\partial t} + \frac{\partial}{\partial x_i}\left(\rho s_\phi \overline{u}_i\right) = \text{EN}_\phi \tag{4.42}$$

where the source terms of entropy transport equations of each parts are $\text{EN}_{\text{tran}} = \frac{\partial}{\partial x_i}\left(\frac{\lambda_{\text{eff}}}{T}\frac{\partial T}{\partial x_i}\right)$, $\text{EN}_{\text{gen,D}} = \frac{\Phi}{T}$, $\text{EN}_{\text{gen,C}} = \frac{\Phi_\Theta}{T^2}$, in which $\text{EN}_{\text{gen,D}} = \text{EN}_{\text{gen},\overline{\text{D}}} + \text{EN}_{\text{gen,D}'}$ and $\text{EN}_{\text{gen,C}} = \text{EN}_{\text{gen},\overline{\text{C}}} + \text{EN}_{\text{gen,C}'}$ considering the turbulence effect.

The viscous dissipation $\text{EN}_{\text{gen,D}}$ has two groups of terms [23], one with mean quantities and the other with fluctuating quantities where $Y_{\text{Gen},\overline{\text{D}}}$ is defined as

$$\mathrm{EN}_{\mathrm{Gen},\overline{\mathrm{D}}} = \frac{2\mu \mathrm{SR}_{i,j}\mathrm{SR}_{i,j}}{\overline{T}} = \frac{\mu}{\overline{\overline{T}}}\cdot\left[2\left\{\left(\frac{\partial\overline{u}}{\partial x}\right)^2+\left(\frac{\partial\overline{v}}{\partial y}\right)^2+\left(\frac{\partial\overline{w}}{\partial z}\right)^2\right\}\right.$$
$$\left.+\left(\frac{\partial\overline{u}}{\partial y}+\frac{\partial\overline{v}}{\partial x}\right)^2+\left(\frac{\partial\overline{u}}{\partial z}+\frac{\partial\overline{w}}{\partial x}\right)^2+\left(\frac{\partial\overline{v}}{\partial z}+\frac{\partial\overline{w}}{\partial y}\right)^2\right] \tag{4.43}$$

where the mean strain rate $\mathrm{SR}_{i,j} = \frac{1}{2}\left(\frac{\partial\overline{u}_j}{\partial x_i}+\frac{\partial\overline{u}_i}{\partial x_j}\right)$. According to Wilcox model [7], the specific dissipation rate of $\omega = \frac{\varepsilon}{(\beta^* k)}$ with $\beta^* = C_\mu = 0.09$ for SST $k-\omega$ model. Then, by using the turbulent dissipation model together [23], $\mathrm{EN}_{\mathrm{Gen},\mathrm{D}'}$ is determined as follows:

$$\mathrm{EN}_{\mathrm{Gen},\mathrm{D}'} = \frac{\rho\beta^* k\omega}{\overline{T}} \tag{4.44}$$

Entropy production derived from heat conduction also includes two parts, $\mathrm{EN}_{\mathrm{gen,D}} = \left(\frac{\Phi_\Theta}{T^2}\right) = \mathrm{EN}_{\mathrm{gen},\overline{\mathrm{C}}} + \mathrm{EN}_{\mathrm{Gen,C}'}$: The heat transfer entropy production due to time mean temperature gradients $\mathrm{EN}_{\mathrm{Gen,C}}$ is expressed as

$$\mathrm{EN}_{\mathrm{Gen},\overline{\mathrm{C}}} = \frac{\lambda}{\overline{\overline{T}}^2}\left[\left(\frac{\partial\overline{T}}{\partial x}\right)^2+\left(\frac{\partial\overline{T}}{\partial y}\right)^2+\left(\frac{\partial\overline{T}}{\partial z}\right)^2\right] \tag{4.45}$$

Based on the turbulence model [23], the heat transfer entropy generation due to fluctuating temperature gradients $S_{\mathrm{Gen,C}'}$ can be expressed as:

$$\begin{aligned}\mathrm{EN}_{\mathrm{Gen,C}'} &= \frac{\alpha_\mathrm{t}}{\alpha}\mathrm{EN}_{\mathrm{Gen,C}'}\\ &= \frac{\alpha_\mathrm{t}}{\alpha}\frac{\lambda}{\overline{\overline{T}}^2}\left[\left(\frac{\partial\overline{T}}{\partial x}\right)^2+\left(\frac{\partial\overline{T}}{\partial y}\right)^2+\left(\frac{\partial\overline{T}}{\partial z}\right)^2\right]\end{aligned} \tag{4.46}$$

where thermal diffusivity $\alpha = \frac{\lambda}{\rho c_\mathrm{p}}$. The turbulence thermal diffusivity $\alpha_\mathrm{t} = \frac{\nu_\mathrm{t}}{\mathrm{Pr}_\mathrm{t}}$, where Pr_t is 0.85.

The entropy generation $\mathrm{EN}_{\mathrm{gen,L}}$ due to phase change derived from the temperature difference between liquid and vapor phase is simplified to [22]

$$\mathrm{EN}_{\mathrm{gen,L}} = \Gamma h_{\mathrm{lv}}\left(\frac{1}{T_\mathrm{v}}-\frac{1}{T_\mathrm{l}}\right) \tag{4.47}$$

Based on the exergy balance equation for open system at steady state, the specific flow exergy e_x is expressed as

$$e_\mathrm{x} = h - h_0 - T_0(s-s_0) + \frac{U^2}{2} \tag{4.48}$$

where the superscript '0' denotes the system at the dead state referring to a reference state where a system is in equilibrium with its environment. h (kJ/kg) is specific enthalpy of fluid.

The exergy destruction e_D is directly related to the entropy generation s_{gen} and environment temperature T_0 which is written as

$$e_D = \Delta e_x = T_0 s_{gen} \tag{4.49}$$

The exergy destruction ratio ζ_D is defined as the ratio of irreversible exergy destruction to net exergy of inlet fluid [36], which is determined by

$$\zeta_D = \frac{e_D}{e_{x,in}} = \frac{T_0 \Delta s}{h_{in}^* - h_0 - T_0\left(s_{in}^* - s_0\right)} \tag{4.50}$$

where superscript * denotes the stagnation condition.

4.3 Validation by Experiment

To validate the prediction accuracy of the wet steam model for non-equilibrium condensation of water vapor in supersonic flows, the experimental case of G37-B of the 2/M nozzle [20] was employed. The geometries and dimensions of the 2/M nozzle are shown in Fig. 4.1a. The throat radius height at the 2/M nozzle is set to 10.00 mm. The radius heights at the inlet and outlet of the 2/M nozzle are 19.99 and 17.91 mm, respectively. For the boundary conditions, the total pressure and temperature at the inlet are 13.83 MPa and 674.42 K. The outlet condition is set to the supersonic flow avoiding the aerodynamic shock interactions with condensation shock.

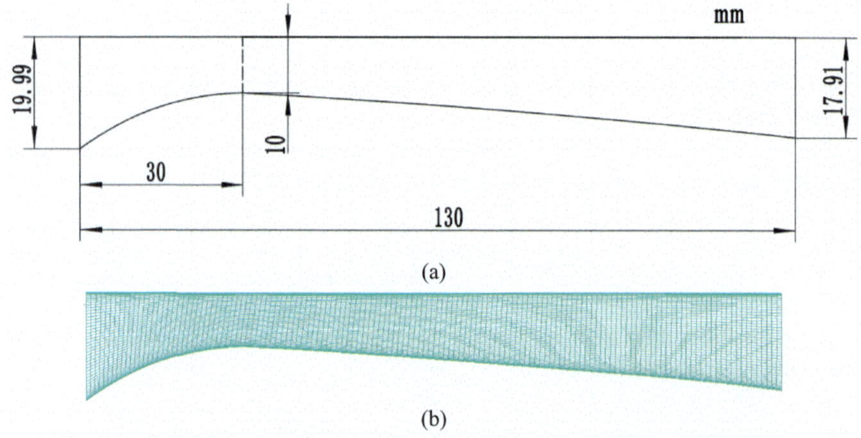

Fig. 4.1 Geometries and computational mesh of the 2/M nozzle: **a** geometries; **b** mesh

Figure 4.1b shows the computational mesh which is refined in the near-wall region to ensure that $y^+ < 1$ at the near-wall cell for better prediction accuracy of SST $k - \omega$ turbulence model. For the grid independence test, the effect of the grid resolution and density on the flow field and condensation characteristics is discussed in detail. In this study, a coarse mesh with 22,400 cells, a medium mesh with 56,400 cells and a fine mesh with 122,400 cells are considered.

The comparison of nucleation rate among these three cases with coarse, medium and fine meshes in the supersonic nozzle are illustrated in Fig. 4.2. The nucleation rate profiles demonstrate that all three resolutions of the quadrilateral mesh predict the consistent condensation behavior for nucleation process. Nonetheless, the medium and fine meshes provide a more intricate depiction of boundary of the nucleation region owing to their increased grid resolution. Therefore, considering both numerical precision and computational expenses, the model validation employs the medium mesh comprising 56,400 cells.

To valid the CFD model by using Hedbäck's experiments, a numerical simulation of single-phase flow excluding phase change, serves as a foundation to clearly illustrate how the condensation process affects flow field through heat and mass transfer between vapor and liquid phases. Comparison results between the numerical simulation and experimental data [20] are depicted in Fig. 4.3, including the contours and profiles of static pressure and droplet size for dry case and condensing case.

It is evident that the proposed wet steam model can effectively capture the initiation of the condensation process and pressure jump deviated from the frozen isentropic flow. Additionally, the numerical results for droplet size align closely with experimental data of Hedbäck. It is indicated that a developed wet steam model can be suitable for predicting non-equilibrium condensation of water vapor in supersonic flows, focusing on static pressure and droplet size.

On the other hand, the prediction accuracy of the CFD model for moist gas flow was also validated by experimental data of Wyslouzil's nozzle [40] and Ding's S2 nozzle [12].

The geometry of Wyslouzil's nozzle was given in Fig. 4.4. The convergent and divergent sections of this nozzle measure 38 mm and 95 mm in length, respectively, and the throat width of the nozzle is 5 mm. Following grid independence testing, the grid was established with 300 nodes in the x-direction and 60 nodes in the y-direction. For boundary conditions, the inlet total temperature and pressure of inlet are 287 K and 60 kPa, respectively. The partial pressure of water vapor at the nozzle inlet are 0.26 kPa, 0.5 kPa and 1 kPa, respectively. The comparison of near wall pressure distribution between the simulation of condensation model and experimental data was shown in Fig. 4.5. The relative differences of p/p_0 between numerical and experimental data are 3.01%, 2.81% and 2.90% receptively. Besides, the differences of temperature between numerical and experimental data are 0.736, 0.237 and 0.952 K which show that the pressure and temperature jumps can be captured accurately.

It should be pointed out that Wyslouzil nozzle is a two-dimensional rectangular nozzle, thus, another three-dimensional circular nozzle S2 [12] is utilized to further validate the accuracy of the established CFD model. The nozzle S2 is a 3D toroidal-throat arc nozzle which curvature radius is 870 mm and the throat of S2 is 10 mm,

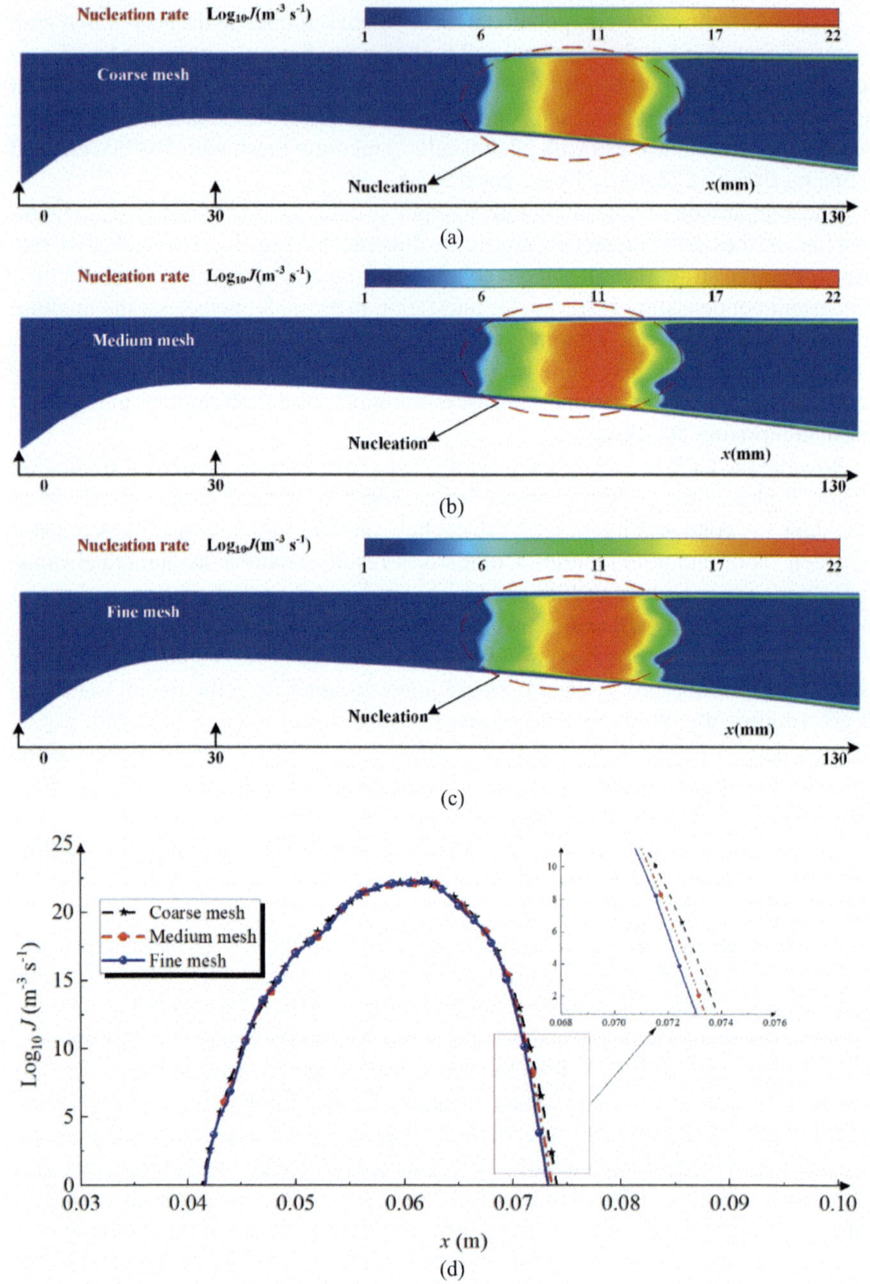

Fig. 4.2 Nucleation rate in the 2/M nozzle with coarse, medium and fine meshes: **a** contour of coarse mesh; **b** contour of medium mesh; **c** contour of fine mesh; **d** profiles of nucleation rate along the axis

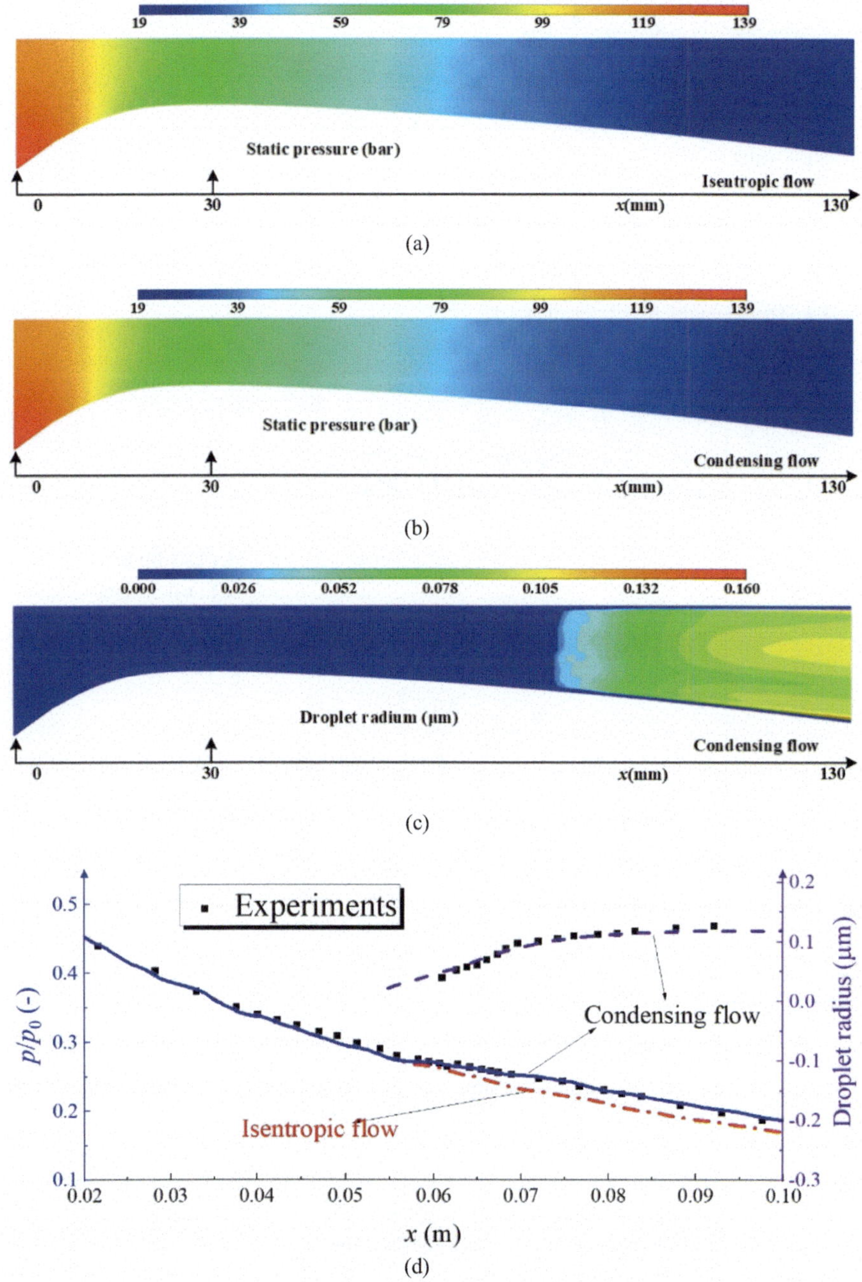

Fig. 4.3 Numerical results and experimental data in the 2/M nozzle: **a** contour of static pressure in dry flows; **b** contour of static pressure in condensing flows; **c** contour of droplet radius in condensing flows; **d** profile of static and droplet radius of dry flow and condensing flow

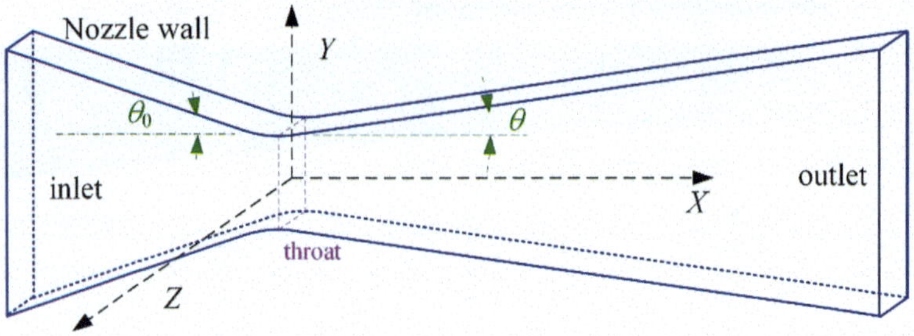

Fig. 4.4 The geometry of Wyslouzil nozzle

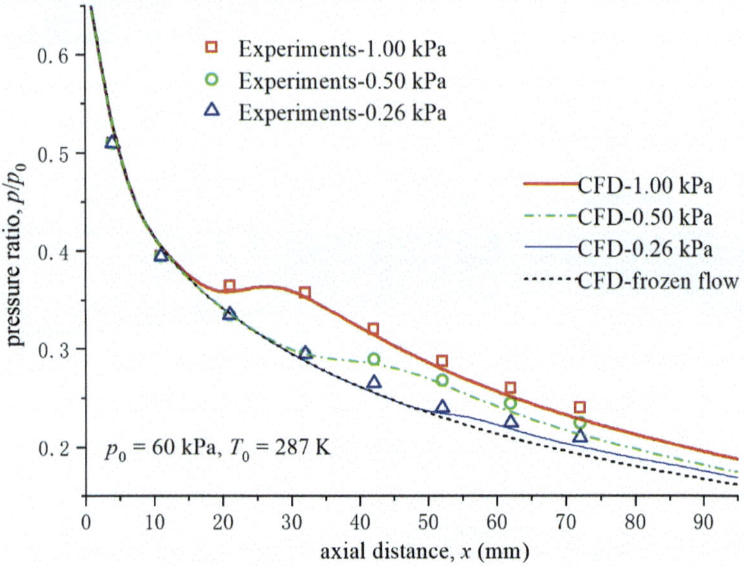

Fig. 4.5 Comparison of pressure profiles between CFD and experiments of moist nitrogen flow in Wyslouzil's nozzle

as shown in Fig. 4.6. The inlet pressure is 400 kPa and inlet temperature is 323 K with inlet relative humidity of 10%, 30%, 70% and 97%, respectively.

The grid consists of structured hexahedral elements, refined near the walls to ensure $y^+ < 1$. Grid independence tests were conducted using Grid Convergence Index (GCI) analysis. GCI is a method for estimating mesh refinement errors based on generalized extrapolation theory, as described by

$$\mathrm{GCI} = \frac{F_s|\varepsilon|}{r^p - 1} \times 100\% \tag{4.51}$$

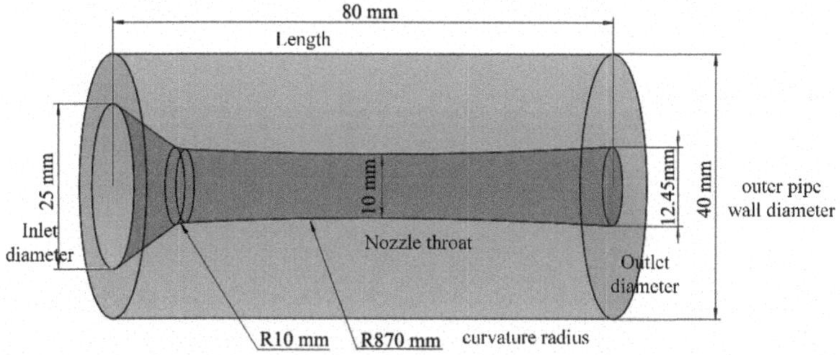

Fig. 4.6 The geometry and dimensions of 3D toroidal-throat arc nozzle with the curvature radius is 870 mm and the throat of 10 mm

where F_s is the safety factor with an empirical value of 3, ε represents the relative error between two grids, r is the refinement factor ratio, and p denotes the algorithm accuracy order. A small GCI indicates minimal difference between solution results of the two grids. The grid configurations include Grid 1 with 16,200 cells, Grid 2 with 13,000 cells, and Grid 3 with 10,000 cells, respectively. According to Table 4.1, based on the outlet pressure as the test parameter, 13,000 grid cells achieve grid-independent solutions.

Figure 4.7 illustrates a close agreement in the pressure distribution near the wall between CFD simulations and experimental data. It means the simulation model can effectively predict the steady-state condensation phenomena in transonic moist air flow.

To validate the unsteady condensation flow, The experiments of nozzle Y167, Y160, Y170 [37], A1 [2] and S2 [12] were utilized. Through the test of grid-independent, the nodes number of nozzles Y167, Y160 and Y170 in the directions of x and y were performed to 500 × 120 and the nodes number of nozzle A1 in the directions of x and y were performed to 600 × 120. In the experiment of condensation in circular arc nozzle, the condition of the inlet stagnation pressure from 40 to 400 kPa, the inlet stagnation temperature from 275 to 313 K and the inlet relative humidity, which defined as the ratio of the actual water vapor density in the air to the saturated water vapor density at the same temperature, was below 250%.

Table 4.1 Test results of grid convergence index

		Grids 1–2 (1-fine, 2-medium)		Grids 2–3 (2-medium, 3-coarse)	
F_s	p	$\varepsilon_{1.2}$ (%)	$GCI_{1.2}$ (%)	$\varepsilon_{1.2}$ (%)	$GCI_{1.2}$ (%)
3	3	0.17	0.87	0.65	2.74

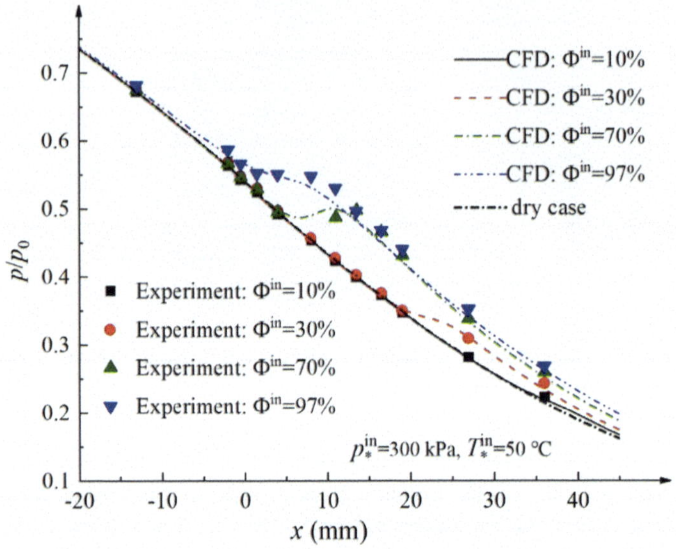

Fig. 4.7 Comparison of CFD and experiments of pressure distribution of nozzle S2 for the moist air flow at inlet pressure of 400 kPa and temperature of 323 K with relative humidity of 10%, 30%, 70% and 97%, respectively

During the grid independence testing, the mesh sizes for nozzles Y167, Y160, and Y170 were set to 500 × 120 nodes in both the x and y directions, while nozzle A1 was meshed with 600 × 120 nodes in the x and y directions. In the experimental study of condensation, inlet stagnation pressure ranges from 40 to 400 kPa, inlet stagnation temperature ranges from 275 to 313 K, and the inlet relative humidity is maintained below 250%.

Table 4.2 presents the frequency comparisons between the present study and the experimental data. Fluctuations in pressure are indicative of variations in internal energy. Figure 4.8 depicts a comparison of CFD results with experimental data regarding the frequency of pressure fluctuations. The differences between CFD simulations and experimental results are consistently below 10% across the frequency range of 800–4000 Hz. Therefore, this simulation model effectively predicts the unsteady feature of the non-equilibrium condensation, as well as the mass and heat transfer processes.

4.4 Non-equilibrium Condensation with Shock Waves

In this section, the effect of the inlet subcooling and relative humidity on the condensation behavior coupling with the shock wave interaction is investigated further to evaluate the performance of the supersonic separation. The nucleation and growth

Table 4.2 Comparing the frequencies of CFD model with the experiments in supersonic nozzles

Num	Nozzle	P_0 (kPa)	T_0 (K)	Φ_0 (%)	f: experiment	f: CFD
1	Y167	88	286	100	1030	1057
2	Y167	88	286	136	1710	1865
3	Y167	115	290	88	954	837
4	Y160	94	287	111	1630	1781
5	Y160	120	290	137	3260	3257
6	Y160	154	294	150	4357	4065
7	Y170	153	294	140	4294	3731
8	A1	100	288	82	905	902.5
9	A1	100	290	80	914	915
10	A1	100	290	95	1000	1115
11	A1	100	290	90	1014	1054
12	S2	300	313	81	2877	2605
13	S2	360	313	81	2592	2685
14	S2	400	313	81	2563	2731
15	S2	360	303	87	2680	2873
16	S2	250	303	80	2492	2477
17	S2	250	303	84	2659	2607
18	S2	250	303	88	2798	2829
19	S2	250	303	95	3435	3300

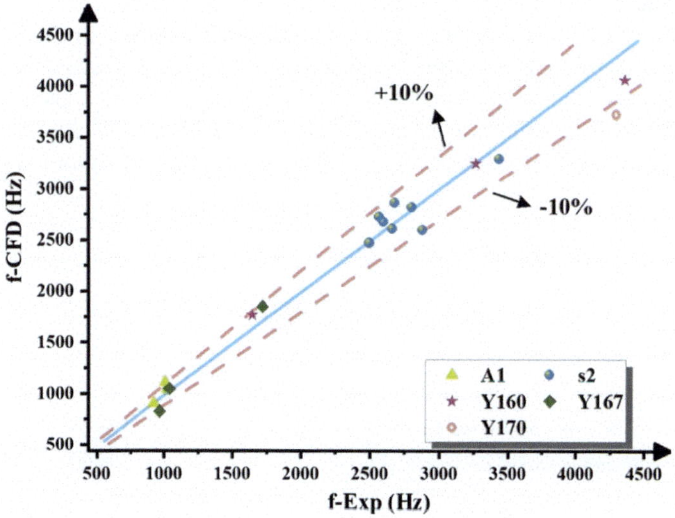

Fig. 4.8 Comparison of frequency between CFD and experiment of pressure fluctuation at wall with frequency range of 800–4000 Hz

process will be restrained sharply due to the aerodynamic shock wave. The incomplete non-equilibrium condensation leads to re-evaporation of condensed droplets, preventing effective water vapor removal, which negatively impacts the supersonic separator's operation and performance. For moist air flow, the flow behaviors of gas and liquid phases in steady and unsteady condensing flow were investigated in detail.

4.4.1 Wet Steam Flow

Two-dimensional simulations are conducted without considering the static vanes, as the primary focus is the non-equilibrium condensation phenomenon. Table 4.3 provides the detailed dimensions of supersonic nozzle. The diverging section of the nozzle is a length of 200 mm. The throat and outlet diameters of the converging–diverging nozzle are 17.40 mm and 21.40 mm, respectively. The lengths of the converging section and the diffuser are 186.60 mm and 300 mm, respectively. Additionally, a constant pipe measuring 200 mm in length is installed both upstream and downstream of the nozzle.

A structured meshes are shown in Fig. 4.9. A grid independence study is performed to evaluate the impact of grid density on predicting non-equilibrium condensation. Three different grid sizes are tested: a coarse mesh with 20,000 cells, a medium mesh with 55,580 cells, and a fine mesh with 287,051 cells. After the independence test, the medium mesh of 55,580 cells is employed for simulating non-equilibrium condensation in supersonic flow considering the numerical accuracy and computational cost, taking into account numerical accuracy and computational cost.

Figure 4.10 presents the static pressure and Mach numbers in the supersonic separation with the non-equilibrium condensation model. The gas accelerates in the converging section of the nozzle, leading to a reduction in static pressure. Upon reaching the critical condition at the nozzle throat, where the Mach number is one, the fluid further expands into supersonic flow. The degree of supersaturation gradually

Table 4.3 Dimensions of the supersonic nozzle

Geometrical parameters	Value (mm)
Inlet diameter of the converging–diverging nozzle (D_{in})	100.00
Throat diameter of the converging–diverging nozzle (D_{th})	17.40
Outlet diameter of the converging–diverging nozzle (D_{out})	21.40
Length of the converging part of the nozzle (L_{nc})	186.60
Length of the diverging part of the nozzle (L_{nd})	200.00
Outlet diameter of the diffuser (D_{di})	60.00
Length of the diffuser (L_{di})	300.00
Length of the constant tube (L_{ct})	200.00

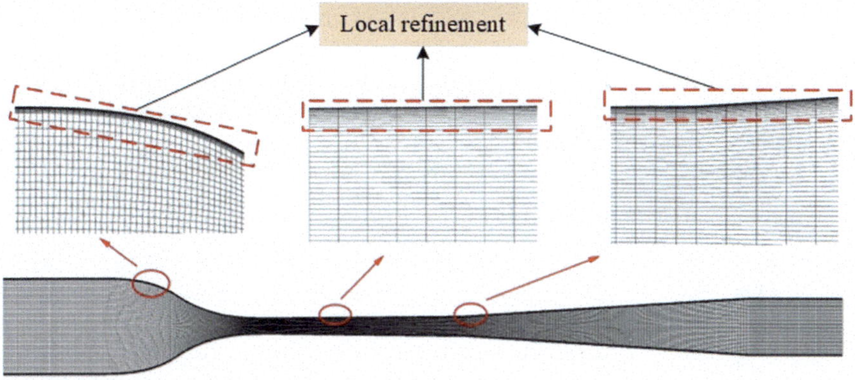

Fig. 4.9 Structured mesh of a supersonic nozzle

increases until the saturation point is achieved, and nucleation does not occur immediately. The vapor phase then transitions into a non-equilibrium state, causing the degree of supersaturation to sharply rise to a peak of 4.28. This non-equilibrium state induces vapor nucleation, which begins to condense at a degree of supersaturation of 2.54 at Wilson Point.

The distributions of subcooling, static pressure, and entropy along the central line of the supersonic nozzle are illustrated in Fig. 4.11 for different inlet subcooling values. It is evident that higher subcooling leads to a greater degree of supersaturation at the nozzle inlet. With lower inlet subcooling, the vapor will expand further along the nozzle before the non-equilibrium condensation occurs. As the inlet subcooling increases, the onset of non-equilibrium condensation, namely Wilson point shifts closer to the nozzle throat. Analyzing the static pressure and entropy profiles of results, it can be seen that in the case of the highest inlet subcooling, vapor condensation occurs immediately after reaching the Wilson point at the nozzle throat. Additionally, the static pressure rises significantly at higher inlet subcooling values, indicating increased latent heat release during the non-equilibrium condensation process, as shown by the entropy distribution at various inlet subcooling in Fig. 4.11c.

Figure 4.12 illustrates the condensation parameters, including nucleation rate, liquid mass generation rate, and liquid fraction, under various inlet subcooling values. With lower inlet subcooling, the lower temperature is obtained in the divergent part of the Laval nozzle before the appearance of the nucleation. There is a higher degree of supersaturation for non-equilibrium condensation, causing a delayed onset of nucleation. It is evident that vapor with higher inlet subcooling begins condensation at a lower supersaturation degree. The rate of liquid mass generation decreases with lower inlet subcooling. In Fig. 4.12b, the liquid mass generation rate drops sharply after reaching the peak at the Wilson point, due to the release of latent heat. The results also show more liquid condenses at higher inlet subcooling during non-equilibrium condensation processes.

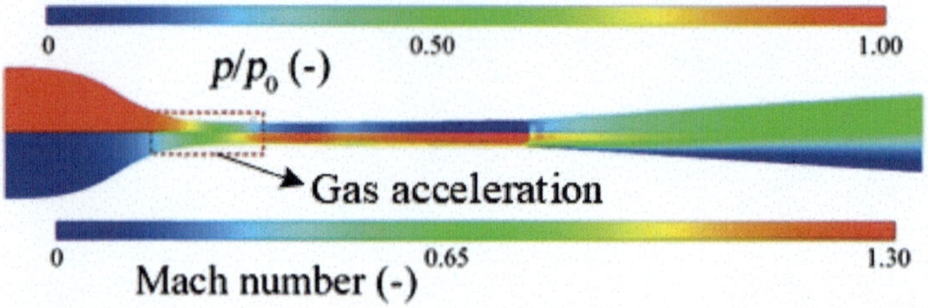

(a)

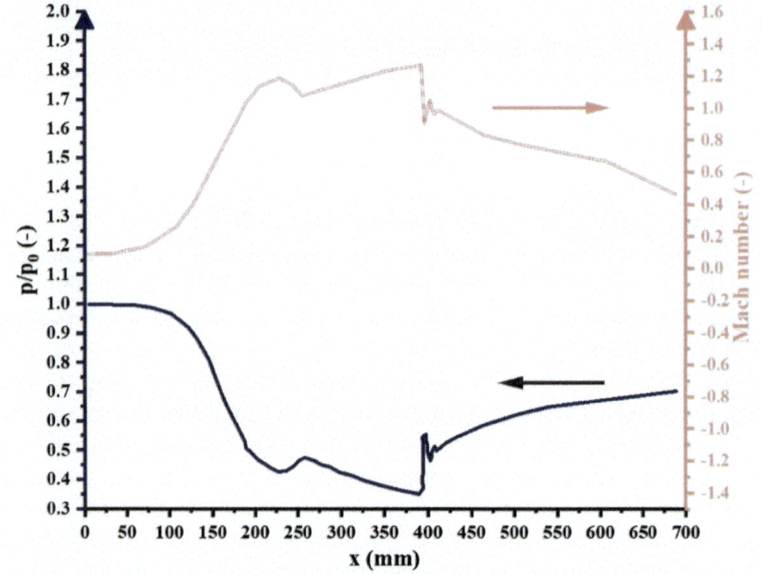

(b)

Fig. 4.10 The static pressure and Mach number distributions in the supersonic nozzle: **a** contours of pressure and Mach number, **b** profiles of pressure and Mach number at the axis

The position of the aerodynamic shock wave changes significantly with different inlet subcooling values, as shown in Fig. 4.11b. A decrease in inlet subcooling shifts the shock wave position upstream of the nozzle. For example, at an inlet subcooling of − 11 K, the aerodynamic shock wave occurs at $x = 382$ mm, but it moves upstream to $x = 371$ mm when the inlet subcooling of water vapor decreases to − 26 K. after the shockwave, the condensed droplets will reevaporates due to the increased pressure and temperature downstream of the shock wave. As a result, the water vapor is not effectively removed by the supersonic separator. Therefore, it is indicated that, increasing the inlet subcooling in natural gas processing can enhance the dehydration performance of the supersonic separator by improving both liquid generation and aerodynamic shock wave position.

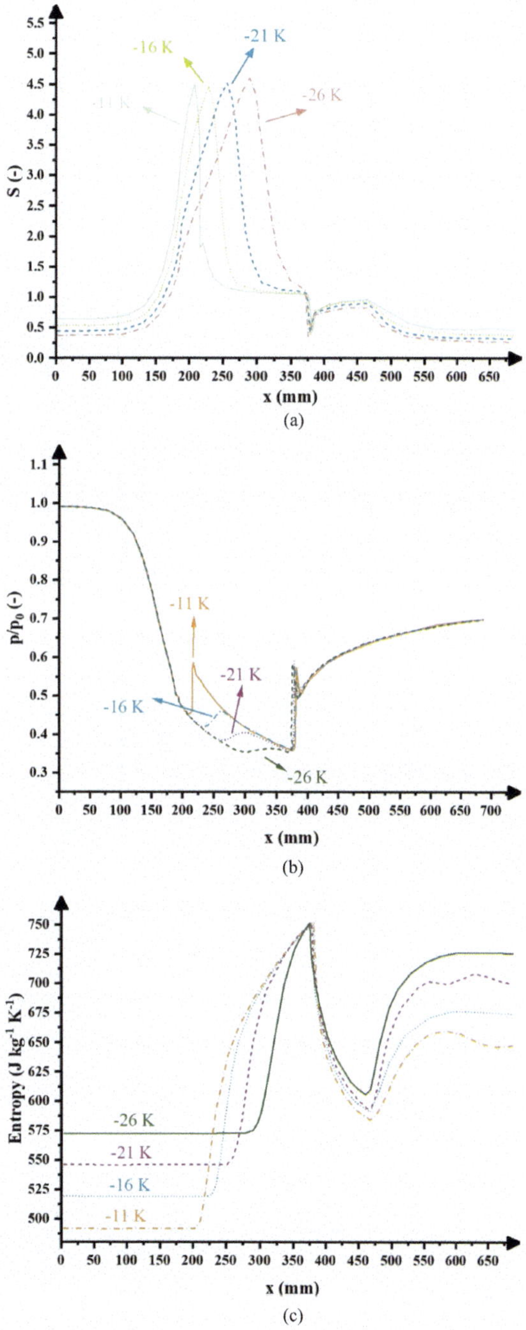

Fig. 4.11 The flow and entropy distributions of the supersonic nozzle at various inlet subcooling values in the supersonic nozzle: **a** degree of supersaturation; **b** static pressure; **c** entropy

Fig. 4.12 The flow parameters of liquid condensing droplets at various inlet subcooling values in the nozzle: **a** nucleation rate; **b** mass generation rate; **c** liquid fraction

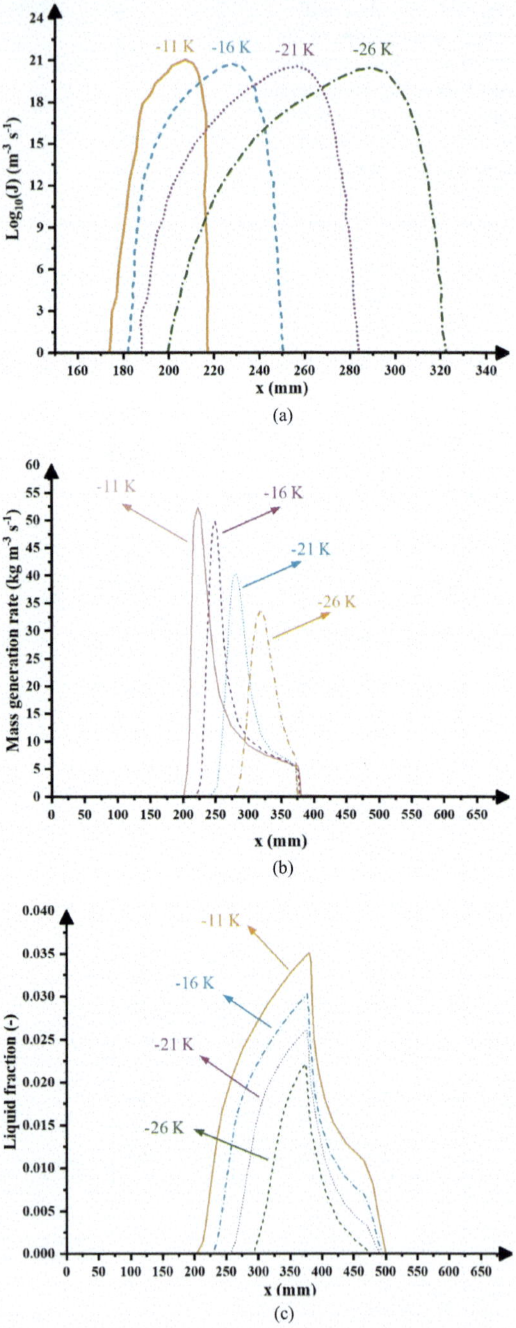

In addition, the impact of the aerodynamic shock wave on non-equilibrium condensation phenomenon in supersonic flows is examined at a low inlet subcooling of – 34 K and an inlet saturation of 0.28, as shown in Fig. 4.13. It depicts the contours and profiles of Mach number, degree of supersaturation, and nucleation rate inside of supersonic nozzle. The low inlet subcooling moves the non-equilibrium condensation region to the diffuser entrance, where a shock wave is present. It is observed that this aerodynamic shock wave overlaps with the area of intense nucleation. Downstream of the shock wave, the subsonic flows are subsequently generated and the static pressure and temperature will increase. Thus, in Fig. 4.13a, the degree of supersaturation fluctuates due to the strong aerodynamic shock wave, falling below unity further downstream of the supersonic nozzle. It means that two-phase flow moves toward thermodynamic equilibrium. Additionally, the nucleation rate drops sharply because of the aerodynamic shock wave, mirroring the profiles of Mach number and degree of supersaturation. This indicates that the nucleation process is restrained, and non-equilibrium condensation is incomplete. Under these conditions, condensed droplets evaporate again, preventing water vapor removal from the natural gas, which is undesirable for the normal operation of the supersonic separator.

4.4.2 *Moist Air Flow*

The study focused on examining the influence of carrier gas pressure on the point at which condensation begins in supersonic nozzles. Two rectangular nozzles, designed similarly to the ISO 9300 nozzle with $R_c = 2d$, $L = 4d$, and $\theta = 3°$, were employed, differing only in throat diameter ($d = 5$ mm and 50 mm). Condensation onset, known as the Wilson point, was determined by using CFD. Throughout the investigation, T_0 remained constant at 300 K, and the inlet vapor pressure, p_{v0}, was a constant of 3.53 kPa, corresponding to the saturation state with $\Phi_0 = 100\%$. The pressure of the carrier gas p_a was systematically varied from 0 to 500 kPa for investigating the effect of carrier gas pressure on the non-equilibrium condensation.

For a nozzle with a throat width of 5.0 mm, the distributions of nucleation ratio and wetness fraction Y at the nozzle center are depicted in Fig. 4.14. The inlet total pressure is given by $p_0 = p_{a0} + p_{v0}$. As the carrier gas pressure p_{a0} increases, the nucleation position shifts closer to the throat, and the wetness fraction β gradually rises. In contrast, when the throat diameter $d = 50$ mm, the distributions of condensation parameters are shown in Fig. 4.15. It's evident that the wetness fraction increases and the nucleation position moves closer to the throat as the throat diameter d increases. Notably, at a specific carrier gas pressure of $p_{a0} = 1.47$ kPa (resulting in $p_0 = 5.0$ kPa), the condensation onset occurs around $x/d = 1.0$ for $d =$ 50 mm, whereas condensation completely disappears for $d = 5$ mm. This indicates that vapor condensation is more likely to occur in larger nozzles.

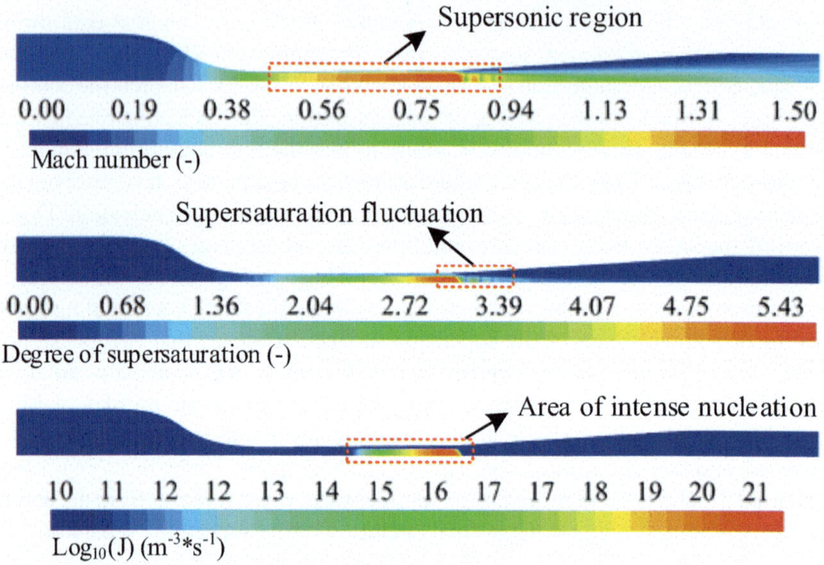

(a)

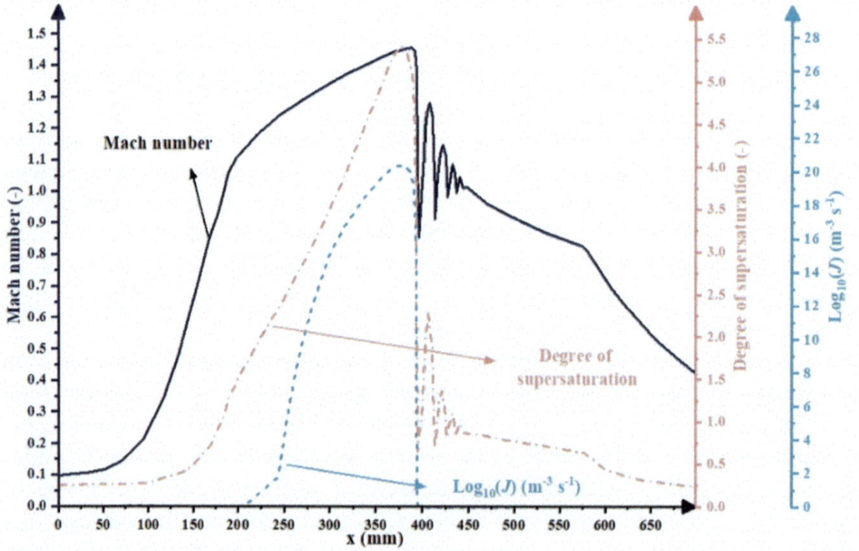

(b)

Fig. 4.13 The interaction between condensation process and shock waves in supersonic flows: **a** contours of Mach number; degree of supersaturation and nucleation rate; **b** profiles of Mach number; **c** degree of supersaturation and nucleation rate

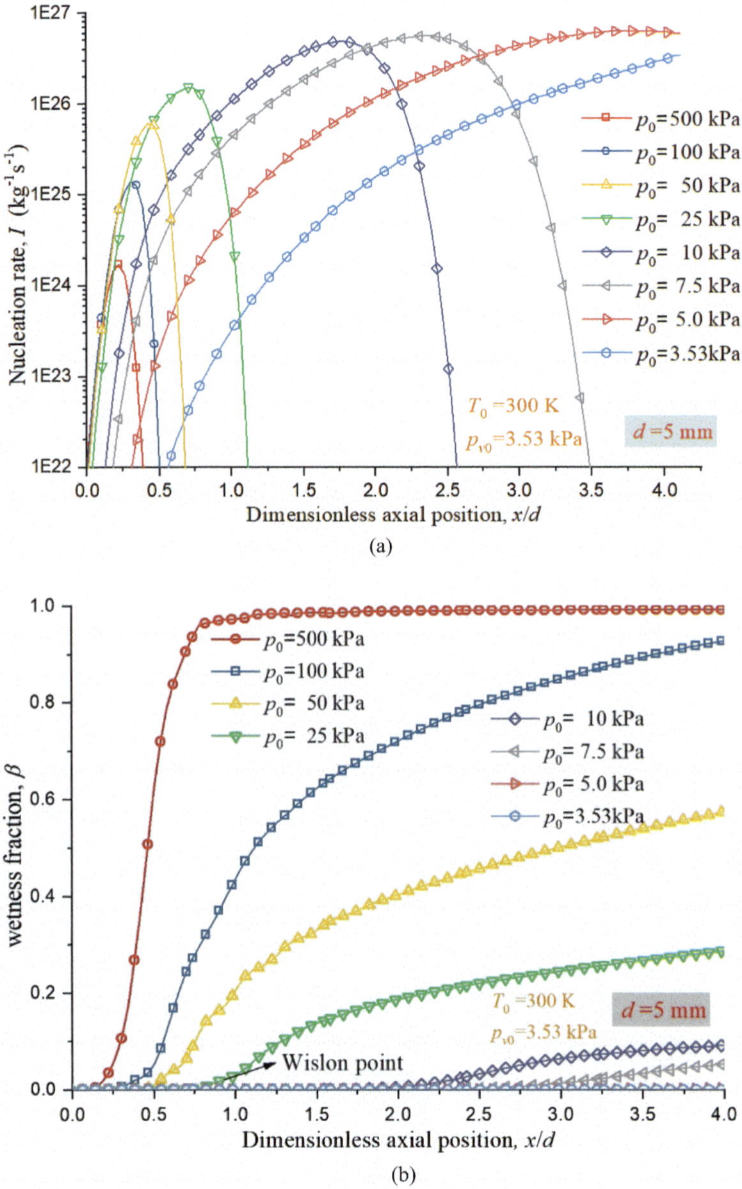

Fig. 4.14 Profiles of the condensation parameters at the center axis when $d = 5$ mm, $T_0 = 300$ K and $p_{v0} = 3530$ Pa: **a** nucleation rate; **b** wetness fraction

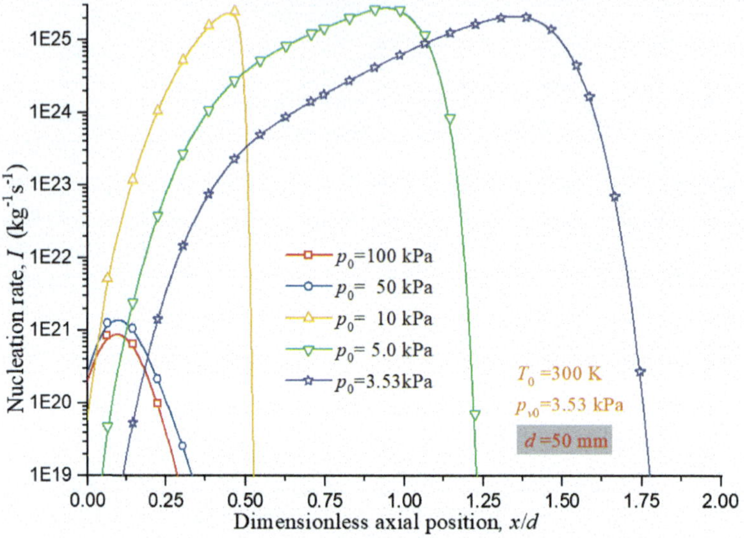

(a)

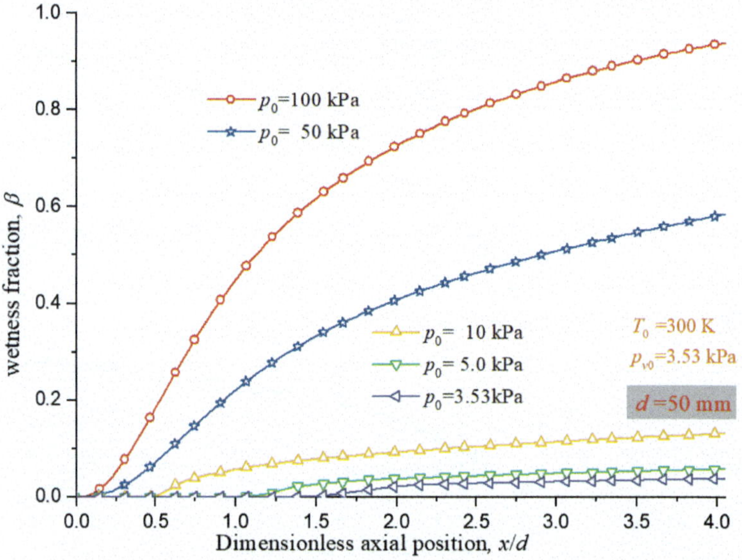

(b)

Fig. 4.15 Profiles of condensation parameters at nozzle center axis when $d = 50$ mm, $T_0 = 300$ K and $p_{v0} = 3530$ Pa: **a** nucleation rate; **b** wetness fraction

Comparison of pressure ratio p/p_0 between the condensing and dry flows with different carrier gas pressures and nozzle sizes are illustrated in Fig. 4.16. As the carrier gas pressure p_{a0} increases or the nozzle size becomes large, the nucleation position, namely pressure jump moves upstream of the supersonic nozzle.

In addition to steady flow, there will be unsteady states with symmetric and asymmetric oscillations when the relative humidity is high. The pressure fluctuation intensity (PF) and frequency (f) of near-wall pressure fluctuations at the throat ($x =$ 0 mm) exhibit distinct patterns during the transition from steady to unsteady states. As shown in Fig. 4.17, the frequency decreases slightly before rising, coinciding with an increase in pressure fluctuation intensity. The frequency of pressure fluctuations tends to increase from mode 3 to mode 1. The PF reaches the maximum value at $\Phi_0 = 110\%$, and then decrease along the increase of relative humidity. Thus, values of relative humidity $\Phi_0 = 72\%$, 90%, and 200% are selected as the typical values for modes 3, 2, and 1, respectively.

Figure 4.18 depicts the intensity (PF) and frequency (f) of pressure fluctuation at the throat ($x = 0$ mm) during asymmetric oscillation of nozzle A1. When the relative humidity in range from 72 to 85%, a bifurcation in frequency is evident. As Φ_0 increases to 85%, the frequency abruptly transitions from the lower branch to the upper branch. Conversely, decreasing Φ_0 along the upper branch until reaches 72%, the frequency shifts from the upper branch back to the lower branch.

Figure 4.19 illustrates Mach number contours for three modes of symmetric oscillation in nozzle Y167. The transition from mode 3 to mode 1 shows a shift of the nucleation area from downstream to upstream of the throat, accompanied by an expansion in the nucleation area's range. The Mach number contours vary significantly across these modes. In contrast, Fig. 4.20 presents Mach number contours for symmetric and asymmetric oscillations in nozzle A1. Comparing with the description in Fig. 4.18, it is evident that the Mach number contours for asymmetric oscillation in differ significantly from those of symmetric oscillation.

The pressure gradient along the axis of nozzle Y167 for symmetric oscillation modes 1–3 is depicted in Fig. 4.21. In modes 1, 2, and 3, the pressure gradient exhibits periodic variations. In mode 3, a condensation shock forms downstream of the throat and moves upstream without passing through the throat, reappearing only after a new shock builds up. Mode 2 also features a condensation shock forming downstream of the throat, passing through the throat, and reaching upstream of the throat. In mode 1, the condensation shock originates upstream of the throat and moves toward the inlet direction.

The frequency of pressure oscillation f in unsteady condensing flow is associated with the gas pressure and temperature of carrier gas, relative humidity of water vapor and the characteristic length of supersonic nozzle. A classical empirical formula [37] for dimensional frequency f^0 is expressed as

$$f^0 \equiv \frac{fl}{a_{cr}} = 4.007\Phi_0 w_0^{0.53} \tag{4.52}$$

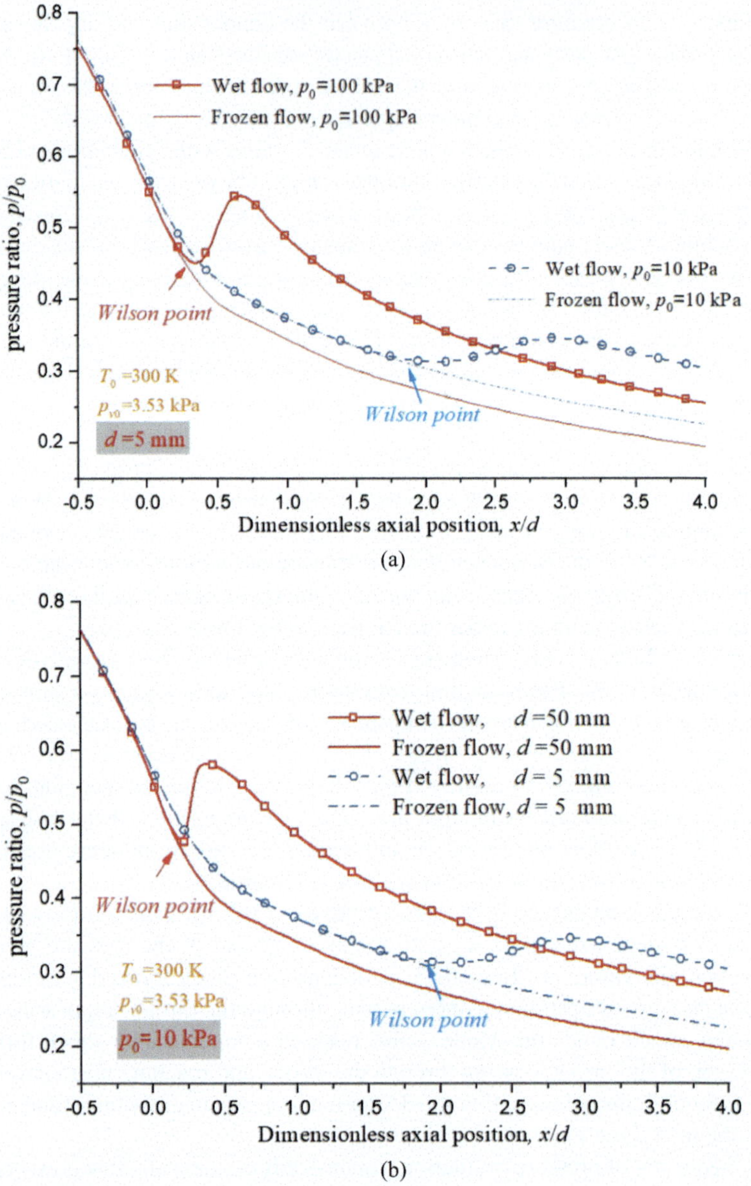

Fig. 4.16 Comparison of pressure ratio p/p_0 between the condensing flow and dry frozen flow: **a** different carrier gas pressures; **b** different nozzle sizes

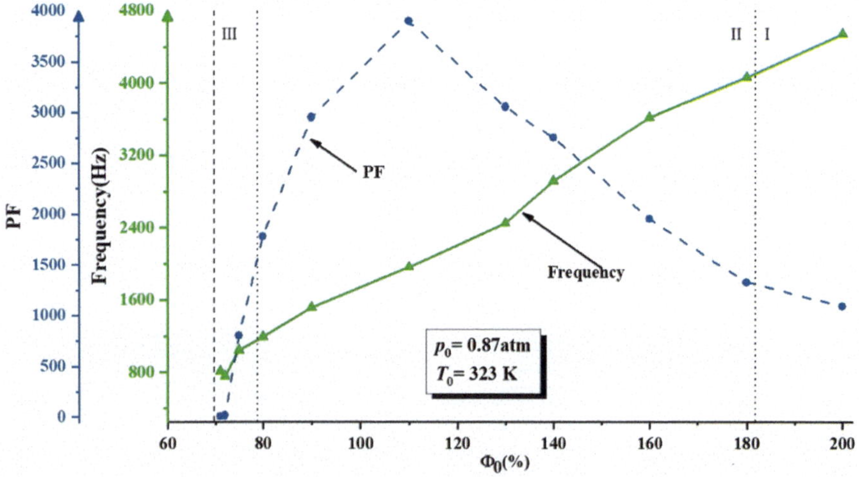

Fig. 4.17 The intensity and frequency of the near-wall pressure fluctuation at throat ($x = 0$ mm) for symmetric oscillation of nozzle Y167

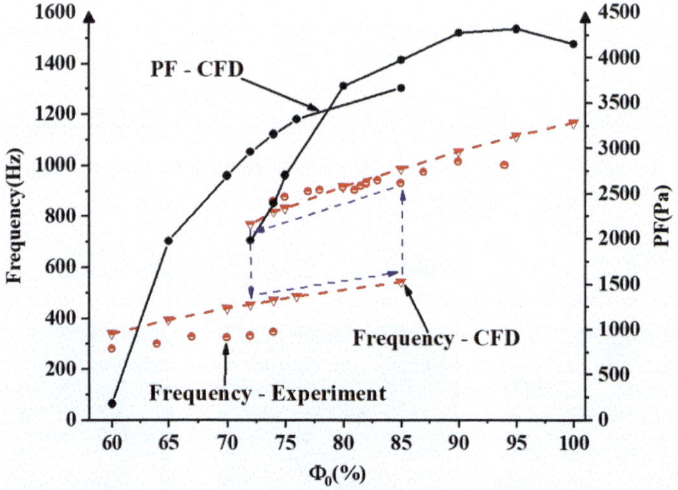

Fig. 4.18 The intensity and frequency of pressure fluctuation at throat for asymmetric bifurcation oscillation of nozzle A1

where Φ_0 is inlet relative humidity, w_0 is the inlet mass fraction of water vapor. Besides, the sonic speed at critical state $a_{cr} = \left(2\gamma R_g T^*/(\gamma + 1)\right)^{0.5}$. However, two variables Φ_0 and w_0 in Eq. (4.52) are not mutually independent.

Reduced by Buckingham's theorem, the dimensionless frequency f^0 is calculated by Eq. (4.53), where the characteristic length $l = (R^* H^*)^{0.5}$, R^* is the throat radius of curvature, H^* is the throat diameter.

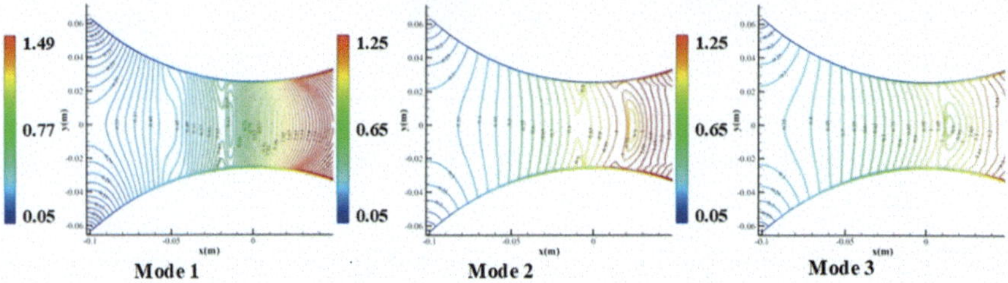

Fig. 4.19 Mach number contours for three modes of symmetric oscillation in nozzle Y167

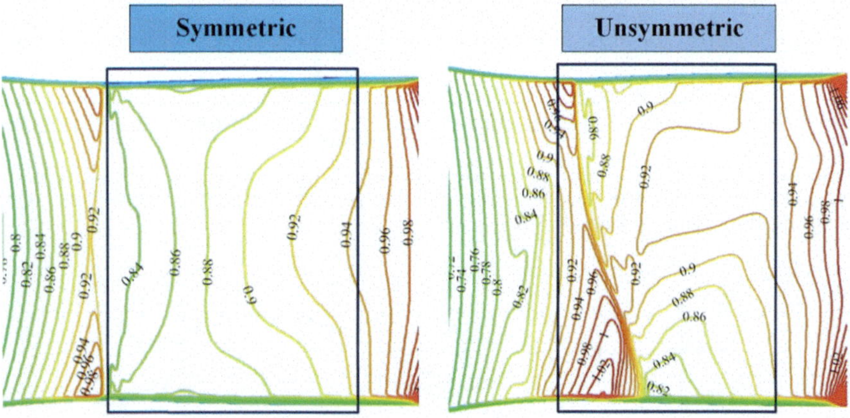

Fig. 4.20 Mach number contours of symmetric and asymmetric bifurcation oscillations in nozzle A1 at $p_0 = 101$ kPa, $T_0 = 290$ K and $\Phi_0 = 80\%$

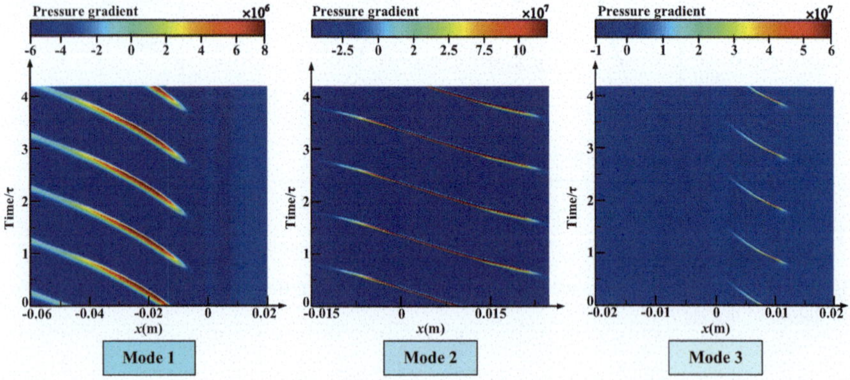

Fig. 4.21 The pressure gradient along the central axis of nozzle Y167 for symmetric oscillating modes 1–3

$$f^0 \equiv \frac{fl}{a_{\mathrm{cr}}} = G_1\left(\frac{P_{\mathrm{v0}}}{P_{\infty 0}}, \frac{P_{\mathrm{v0}}}{P_0}, \frac{h_{\mathrm{lg}}}{c_{\mathrm{p}}T_0}, \frac{H^*}{l}, \frac{\tau_{\mathrm{c}}a^*}{l}, \gamma\right) \tag{4.53}$$

Besides, τ_c is the characteristic time of condensation. The subscript v and ∞ denotes vapor and flat surface of liquid.

The first two terms $P_{v0}/P_{\infty 0}$ and P_{v0}/P_0 are respectively proportional to the inlet relative humidity Φ_0 and mass fraction of water vapor w_0 which is a function of Φ_0 in Eq. (4.54), thus variable w_0 is replaced with saturated mass fraction of water vapor.

$$w_0 = f(\Phi_0, w_{s0}) \tag{4.54}$$

When two independent variables are selected, a more precise semi-empirical formula can be obtained by fitting the experimental data. Thus, f^0 can be expressed as

$$f^0 = G_2\left(\Phi_0, w_{s0}, \frac{h_{\mathrm{lg}}}{c_{\mathrm{p}}T_0}, \frac{H^*}{l}, \frac{\tau_{\mathrm{c}}a^*}{l}, \gamma\right) \tag{4.55}$$

Given the narrow range of inlet temperatures, $h_{lg}/(c_pT_0)$ is considered constant. For the similar nozzle geometries H^*/l, the parameter $\tau_c a^*/l$ is also the same. Finally, a simple semi-empirical function is deduced as follow

$$f^0 = G_3(\Phi_0, w_{s0}) \tag{4.56}$$

Based on above results of experiment and mathematical model, the final identified semi-empirical equation is [12]

$$f^0 = a_0(b_1\Phi_0 + b_0)\left(c_2 w_{s0}^{-2} + c_1 w_{s0}^{-1} + c_0\right) \tag{4.57}$$

where $a_0 = 1.32$, $b_1 = 1.79$, $b_0 = -0.73$, $c_2 = -8.58 \times 10^{-6}$, $c_1 = 0.0013$, $c_0 = 0.70$.

The frequency and intensity of pressure fluctuation are depicted in Figs. 4.22, 4.23 and 4.24. The experimental results are compared with those from the mathematical model. The fitted curves in these figures also illustrate the trend of changes in the frequency and intensity of pressure fluctuations. Table 4.4 presents the ranges of humidity, temperature, and carrier gas pressure, along with corresponding values for frequency (f) and intensity (PF).

At an inlet condition of $P_0 = 250$ kPa and $T_0 = 30$ °C, the frequency of oscillations increases as the humidity parameter Φ_0 rises as shown in Fig. 4.22. Unsteady flow phenomena manifest when Φ_0 exceeds 60%. As Φ_0 increases, the oscillation intensity PF initially escalates rapidly before generally tapering off. This trend stems from the enlarged condensation, causing the condensation shock position to shift upstream with rising humidity. The increased release of latent heat and reduced critical heat

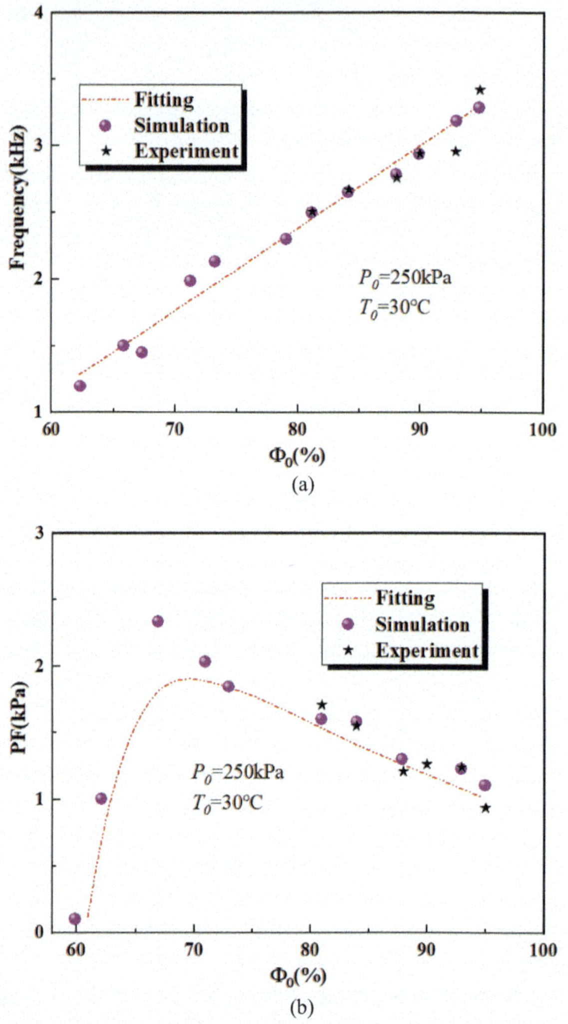

Fig. 4.22 Frequency and PF variations with different inlet humidity Φ_0 in nozzle S2: **a** frequency of unsteady flow; **b** pressure fluctuation intensity PF

addition collectively amplify the oscillation, thereby elevating the self-excited oscillation frequency. However, in conditions of higher humidity, the oscillating condensation shock migrates from the divergent to the convergent parts, weakening the supersaturation in the condensation zone and consequently diminishing oscillation intensity.

When $P_0 = 360$ kPa and $\Phi_0 = 87\%$, the critical temperature threshold for unsteady flow is 22 °C in Fig. 4.23. As temperature rises, the oscillation frequency of unsteady follow also increases, particularly noticeable in the lower temperature range. The intensity parameter PF exhibits variation across the entire temperature range approaching zero at low temperature and still rising at high temperature.

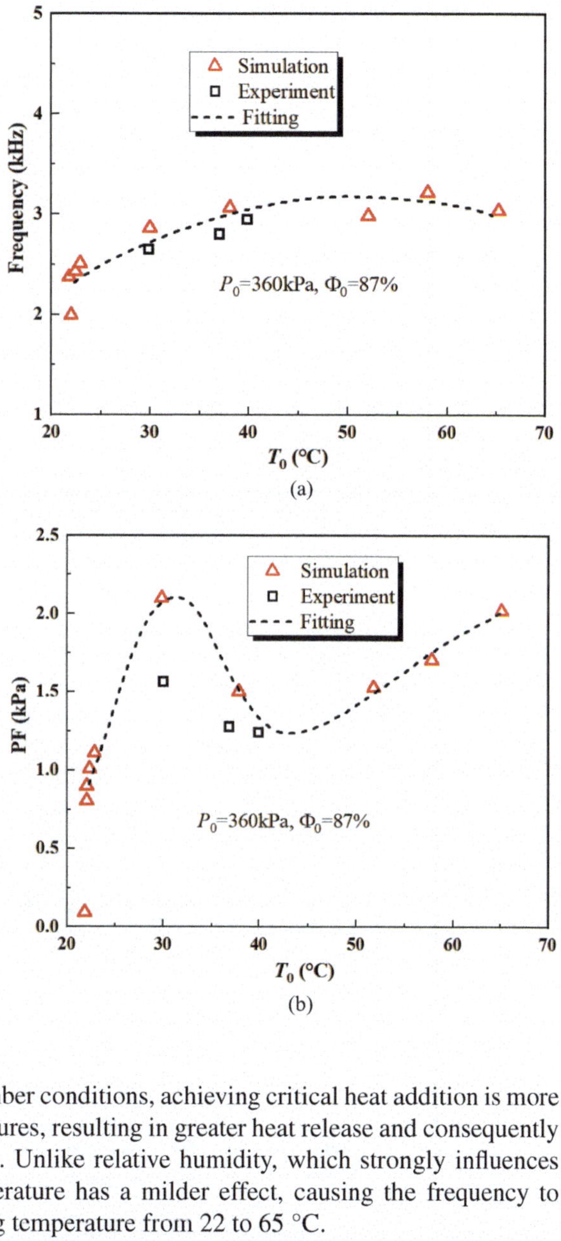

Fig. 4.23 Frequency and PF variations with different inlet temperature T_0 in nozzle S2: **a** frequency; **b** pressure fluctuation intensity PF

Under the same Mach number conditions, achieving critical heat addition is more achievable at higher temperatures, resulting in greater heat release and consequently higher-frequency oscillations. Unlike relative humidity, which strongly influences condensation position, temperature has a milder effect, causing the frequency to change a little with increasing temperature from 22 to 65 °C.

As observed in Fig. 4.24, at $T_0 = 40$ °C and $\Phi_0 = 81\%$, the unsteady flow occurs only when $P_0 \leq 620$ kPa, with the inlet pressure of carrier gas rises, both the frequency and the intensity PF of pressure initially increase before declining at higher pressures. At lower pressures, the release rate of latent heat accelerates with increasing carrier gas pressure, thereby boosting both frequency and intensity.

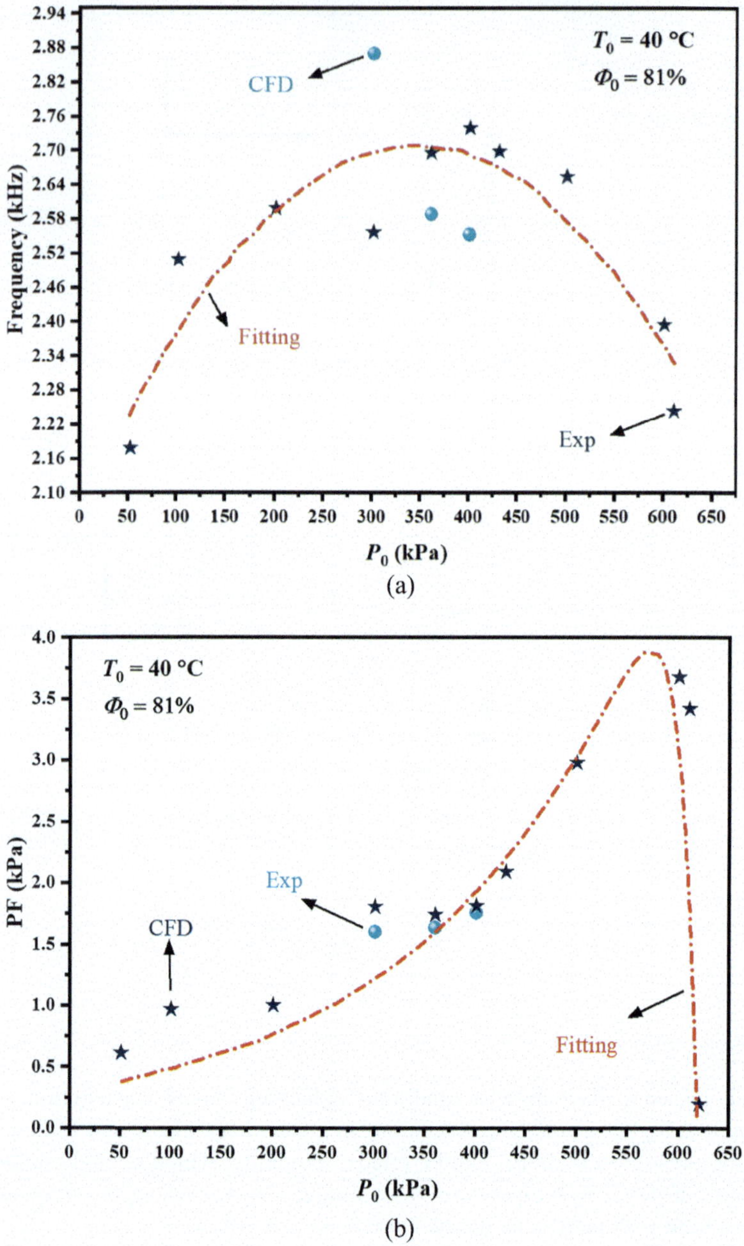

Fig. 4.24 Frequency and PF variations with different inlet pressure P_0 in nozzle S2: **a** frequency; **b** pressure fluctuation intensity PF

Table 4.4 The ranges of inlet conditions, frequency f and intensity PF for nozzle S2

Constants	$P_0 = 250$ kPa, $T_0 = 30$ °C	$P_0 = 360$ kPa, $\Phi_0 = 87\%$	$T_0 = 40$°C, $\Phi_0 = 81\%$
Variables	$\Phi_0 = 60\%$ ~ 95%	$T_0 = 22$ °C ~ 65 °C	$P_0 = 50$ kPa ~ 620 kPa
f	1.2 kHz ~ 3.4 kHz	2.0 kHz ~ 3.2 kHz	2.2 kHz ~ 2.9 kHz
PF	0.1 kPa ~ 2.3 kPa	0.1 kPa ~ 2.1 kPa	0.6 kPa ~ 3.7 kPa

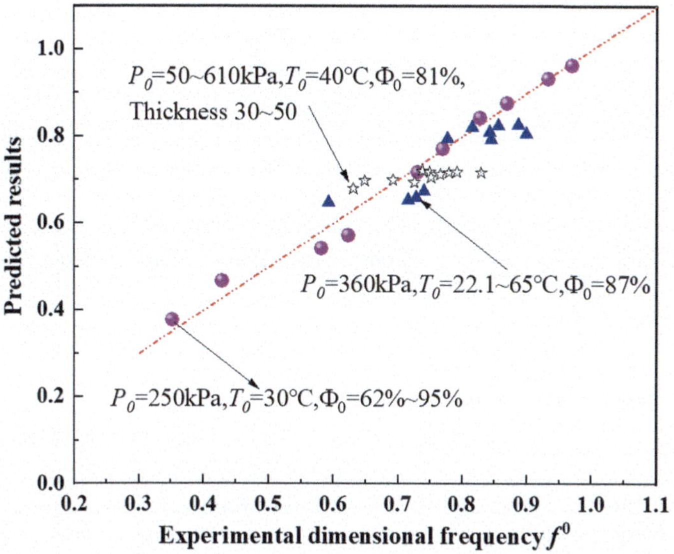

Fig. 4.25 The dimensional frequency f^0 between experimental and predicted values

In the high-pressure regime, the release rate stabilizes while critical heat addition continues to escalate rapidly, resulting in a decrease in oscillation frequency and intensity. Ultimately, with further increases in inlet pressure, the unsteady flow transitions to a stable state.

The predicted result of Eq. (4.57) is illustrated in Fig. 4.25. The average error is 5.51% and the maximum error is 13.53%, the sum of squares error is 0.08 and R^2 is 0.86. It means the semi-empirical formula can predict the frequency of unsteady flow very well.

4.5 Analytic Method for Wilson Point of Condensation

An analytical approach through mathematical derivation or closed-form expressions for calculating the Wilson point is built to acquire a broader perspective on the nucleation phenomenon in nozzle flow. It offers the advantage of quickly obtaining precise results of Wilson point. This approach not only provides insights into the underlying mechanisms of a problem but also allows for efficient computation without the need for extensive computational resources and times. This technique can be extended to determine the Wilson point in supersonic nozzle in supersonic separator for guiding flow mechanism research and structural optimization design.

4.5.1 Assumptions and Models

Assumptions are made that: (1) the flow is one-dimensional, steady, and inviscid; (2) the vapor behaves as an ideal gas; (3) there is no slip or dissipation in interphase velocities; (4) the specific heat capacity c_p of the liquid remains constant; (5) no aerodynamic shock wave occurs before the condensation; (6) the liquid droplet is uniformly dispersed, and the volume of the liquid phase is neglected.

Firstly, the governing equations with Clausius–Clapeyron equation are proposed. For steady flow, the Lagrangian momentum and energy equations in gas–liquid two-phase mixture is expressed by

$$\rho_{\text{mix}}\frac{\mathrm{D}u}{\mathrm{D}t}+\frac{\partial p}{\partial x}=0,\quad \rho_{\text{mix}}\frac{\mathrm{D}h_{\text{mix}}}{\mathrm{D}t}=\frac{\mathrm{D}p}{\mathrm{D}t} \tag{4.58}$$

where, the mixture phase density $\rho_{\text{mix}} = \rho/(1-Y)$ because the volume of the liquid phase can be neglected. $\mathrm{WF_v}$ is wetness fraction of water vapor. $\mathrm{D}/\mathrm{D}t$ is the Lagrangian time derivative and $\mathrm{d}/\mathrm{d}x$ is Eulerian time derivative. The mixture of enthalpy of mixture h_{mix} is

$$h_{\text{mix}}=c_pT-\mathrm{WF_v}\left(h_{\text{lg}}-c_p\Delta T\right) \tag{4.59}$$

The relationship between saturation temperature and pressure follows the Clausius–Clapeyron equation, expressed as follows:

$$\frac{1}{T_s(p)}\frac{\mathrm{d}T_s(p)}{\mathrm{d}t}=\frac{R_vT_s(p)}{h_{\text{lg}}}\frac{1}{p}\frac{\mathrm{d}p}{\mathrm{d}t} \tag{4.60}$$

Near the Wilson point, h_{lg}, M and specific heat c_p remain nearly constant in Rayleigh flow, while the temperature, pressure, and their derivatives related to nucleation. The flow parameter relationship between nucleation and the isentropic process is as follows [13].

$$T=T_i\left(1+\frac{\gamma M^2-1}{M^2-1}\frac{h_{\text{lg}}\mathrm{WF_v}}{c_pT_i}\right),\quad p=p_i\left(1+\frac{\gamma M_i^2}{M_i^2-1}\frac{h_{\text{lg}}\mathrm{WF_v}}{c_pT_i}\right) \tag{4.61}$$

For steady flow, Gyarmathy and Meyer proposed that the location of the Wilson point depends on the expansion rate k_p, which is determined by the nozzle geometry as follows [21]

$$k_{\mathrm{p}} = \frac{-1}{p}\frac{\mathrm{D}p}{\mathrm{D}t} = \frac{-1}{p}u\frac{\mathrm{d}p}{\mathrm{d}x} \tag{4.62}$$

Taking into account the latent heat release $h_{\mathrm{lg}}Y$ per unit mass of vapor, the expansion rate k_{p} and its first-order derivative can be calculated using the expansion rate coefficients f_{p} and f_{p1} respectively.

$$f_{\mathrm{p}} = \frac{k_{\mathrm{p}}}{k_{\mathrm{ip}}} = 1 - \frac{\dot{\mathrm{WF}}_{\mathrm{v}}}{k_{\mathrm{ip}}}\frac{\gamma M_{\mathrm{iw}}^2}{M_{\mathrm{iw}}^2 - 1}\frac{h_{\mathrm{lg}}}{c_{\mathrm{p}}T_{\mathrm{iw}}} \tag{4.63}$$

$$f_{\mathrm{p1}} = \frac{\dot{k}_{\mathrm{p}}}{\ddot{k}_{\mathrm{ip}}} = 1 - \frac{\ddot{\mathrm{WF}}_{\mathrm{v}}}{\dot{k}_{\mathrm{ip}}}\frac{\gamma M_{\mathrm{iw}}^2}{M_{\mathrm{iw}}^2 - 1}\frac{h_{\mathrm{lg}}}{c_{\mathrm{p}}T_{\mathrm{iw}}} \tag{4.63a}$$

where k_{ip} is the expansion rate of isentropic flow.

Combining Eqs. (4.58)–(4.62), and neglecting the insignificant terms,

$$\frac{\Delta \dot{T}_{\mathrm{w}}}{T} + L \times \dot{\mathrm{WF}}_{\mathrm{w}} - \gamma_{\mathrm{s}} f_{\mathrm{p}} k_{\mathrm{ipw}} = 0, \tag{4.64}$$

where the over-head dot represents Lagrangian time derivation d/dt. The subscript 'w' denotes Wilson point.

$$L = \frac{h_{\mathrm{lg}} - c_{\mathrm{p}}\Delta T}{c_{\mathrm{p}}T} \tag{4.64a}$$

$$\gamma_{\mathrm{s}} = \frac{\gamma - 1}{\gamma} - \frac{R_{\mathrm{v}}T_{\mathrm{s}}}{h_{\mathrm{lg}}}\frac{T_{\mathrm{s}}}{T} \tag{4.64b}$$

where the parameter L and γ_{s} is considered a constant evaluated at the Wilson point, typically having values of 4.0 and 0.2 respectively. Since the first derivative of subcooling $\Delta\, \dot{\mathrm{WF}}_{\mathrm{w}} = 0$ at the Wilson point, Eq. (4.64) gives,

$$\dot{\mathrm{WF}}_{\mathrm{w}} = \frac{\gamma_{\mathrm{s}} f_{\mathrm{p}}}{L} k_{\mathrm{ipw}}, \quad \ddot{\mathrm{WF}}_{\mathrm{w}} = \frac{\gamma_{\mathrm{s}} f_{\mathrm{p1}}}{L} \dot{k}_{\mathrm{ipw}} \tag{4.65}$$

$$\begin{aligned} \frac{\Delta\, \dot{\mathrm{WF}}_{w}}{T_{w}} &= \gamma_{s} f_{p} k_{ipw} - L\, \dot{\mathrm{WF}}_{w} = 0 \\ \frac{\Delta\, \ddot{T}_{w}}{T_{w}} &= \gamma_{s} f_{p1} \dot{k}_{ipw} - L\, \ddot{Y}_{w} \end{aligned} \tag{4.65a}$$

The nucleation rate is calculated by a truncated Taylor expansion which ignores the higher order term.

$$I(\tau) = I_{\mathrm{w}} \exp\left(-\tau^2/\tau_{\mathrm{n}}^2\right) \tag{4.66}$$

where

$$I_{\mathrm{w}} = I_{0\mathrm{w}} \exp\left(-\frac{\Theta_{\mathrm{w}}^2}{\Delta T_{\mathrm{w}}^2}\right) \tag{4.66a}$$

$$\tau_{\mathrm{n}} = \frac{\Delta T_{\mathrm{w}}}{\Theta_{\mathrm{w}}}\left(\frac{\Delta T_{\mathrm{w}}}{-\Delta \ddot{T}_{\mathrm{w}}}\right)^{0.5} \tag{4.66b}$$

Before the Wilson point, the nucleation rate follows a half-Gaussian curve. According to Clausius–Clapeyron equation, the critical radius of Eq. (4.28) can be converted into

$$r_{\mathrm{c}} = \frac{2\sigma T_{\mathrm{s}}}{\rho_{\mathrm{L}} h_{\mathrm{lg}} \Delta T} \tag{4.67}$$

The droplet radius r, depends on ΔT_{w} and $\Delta \ddot{T}_{\mathrm{w}}$, can be determined. Before the Wilson point, during the growth period, the droplet size is significantly smaller than the mean free path of vapor molecules, making growth rate $\dot{r}$ is independent of r. It is assumed that the Wilson point occurs under free-molecule conditions, which is generally applicable for low-pressure steam situations. The droplet growth rate at the Wilson point is described by Young and can be expressed as follows:

$$\dot{r} = C_{\mathrm{r}} \Delta T, \quad C_{\mathrm{r}} = \frac{p}{(2\pi R_{\mathrm{v}} T_{\mathrm{s}})^{0.5} \rho_{\mathrm{L}}}\left(\frac{\gamma + 1}{2\gamma}\right)\left(\frac{c_{\mathrm{p}}}{h_{\mathrm{lg}}}\right) \tag{4.68}$$

There are two approximation methods for droplet radius and their growth laws are shown in Fig. 4.26. First one assumes the initial droplet radius is zero and the droplet growth rate remains constant until the Wilson point [22]. The other method assumes the initial droplet radius is constant and always equals the critical radius r_{c} at the Wilson point [13], with the growth rate similar to the first method. Given that the critical nucleation radius is not substantially smaller than the nucleation radius at the Wilson point, the second approach appears more justified. Thus, Dobbins' theory of droplet growth is selected.

Based on the approximate process described above, the droplet radius r is a function of two-time variables: the current time t_{c} and the nucleation time t. At the Wilson point, the current time $t_{\mathrm{c}} = t_{\mathrm{w}}$. The nucleation time t can be transformed to $\tau = t_{\mathrm{w}} - t$. The growth coefficient C_{r} is approximated by a truncated Taylor expansion at the Wilson point, $C_{\mathrm{r}} = C_{\mathrm{rw}}(1 - k_{\mathrm{rw}}\tau)$. Therefore, the expression for the droplet radius at the Wilson point is as follows:

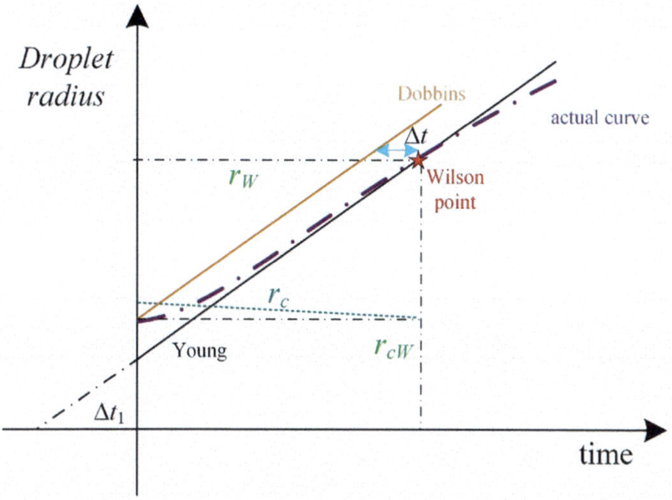

Fig. 4.26 Two approximations for the droplet growth theory presented by Young and Dobbins. The critical radius r_c will decreases over time

$$
\begin{aligned}
r_w(\tau) &\equiv r_w(t_w, t_w - \tau) \\
&= r_c + C_{rw}\Delta T_w\left(\tau - \frac{1}{2}k_{rw}\tau^2 + \frac{1}{6}\frac{\Delta \ddot{T}_w}{\Delta T_w}\tau^3\right)
\end{aligned}
\tag{4.69}
$$

where

$$
k_{rw} = -k_{pw}\left\{1 + \left(\frac{R_v T_s}{h_{fg}}\right)_w \left[\frac{1}{2} - \frac{D(\ln h_{fg})}{D(\ln T_s)}\right]_w\right\}
\tag{4.69a}
$$

Neglecting higher-order pre-exponential terms in the Gamma function, Eq. (4.69) is simplified to expression $r_w(\tau) = r_c + C_{rw}\Delta T_w\tau$. At the Wilson point, the wetness fraction and its derivatives are determined by the following computations

$$
\mathrm{WF}_w = \frac{4}{3}\pi\rho_L \int_0^\infty I_w \exp\left(-\frac{\tau^2}{\tau_n^2}\right)\cdot(r_c + C_{rw}\Delta T_w \tau)^3 d\tau
\tag{4.70}
$$

$$
\dot{\mathrm{WF}}_w = \rho_L \frac{4}{3}\pi\left[I_w r_w^3(t_w, t_w) + \int_0^\infty I(\tau) 3 r_w^2(\tau)\dot{r}_w(\tau) d\tau\right]
\tag{4.70a}
$$

$$
\ddot{\mathrm{WF}}_w = \rho_L \frac{4}{3}\pi\left[\int_0^{t_c} \frac{I(t)\partial^2 r^3(t_c, t)}{\partial t^2} dt + \frac{I(t_c)\partial r_w^3(t_w, t_w)}{\partial t_c}\right]
\tag{4.70b}
$$

Define the nucleation time coefficients $\zeta = \frac{r_c}{C_{rw}\Delta T_w \tau_n}$ and Eqs. (4.70)–(4.70a) can be further transformed into,

$$\mathrm{WF_w} = \frac{2}{3}\pi\rho_L I_w (C_{rw}\Delta T_w)^3 \tau_n^4 [1 + g_1(\zeta)] \tag{4.71}$$

$$\dot{\mathrm{WF}}_w = \frac{2}{3}\pi\rho_L I_w (C_{rw}\Delta T_w)^3 \tau_n^4 [1 + g_1(\zeta)] \tag{4.71a}$$

$$\ddot{Y}_w = 4\pi\rho_L I_w (C_{rw}\Delta T_w)^3 \tau_n^2 [1 + g_1(\zeta)] - k_{rw}\dot{Y}_w \tag{4.71b}$$

where

$$g_1(\zeta) = \frac{3\sqrt{\pi}}{2}\zeta + 3\zeta^2 + \sqrt{\pi}\zeta^3 \tag{4.72}$$

$$g_2(\zeta) = \frac{4\zeta}{\sqrt{\pi}} + 2\zeta^2 + \frac{4}{3\sqrt{\pi}}\zeta^3 \tag{4.72a}$$

$$g_3(\zeta) = \sqrt{\pi}\zeta + \zeta^2 \tag{4.72b}$$

When the gas includes non-condensable gases under low-pressure conditions, only the growth coefficient C_r is influenced by the carrier gas. Therefore, the modified growth coefficient C_{rW} at the Wilson point is expressed by

$$C_{rw} = \frac{p}{(2\pi R_v T_s)^{0.5}\rho_L}\left(\frac{\gamma+1}{2\gamma}\right)\left(\frac{c_p}{h_{lg}}\right) + \frac{p_a}{(2\pi R_a T_s)^{0.5}\rho_L}\left(\frac{\gamma_a+1}{2\gamma_a}\right)\left(\frac{c_{pa}}{h_{lg}}\right) \tag{4.73}$$

where R_a, γ_a, and c_{pa} is the gas constant, specific heat ratio and specific heat of the carrier gas respectively.

In isentropic flow, the relation of characteristic Mach number λ and the flow area A at critical conditions can be described by implicit function as

$$A = \left(\frac{\gamma+1}{2}\right)^{\frac{-1}{\gamma-1}} \lambda_i^{-1}\left(1 - \frac{\gamma-1}{\gamma+1}\lambda_i^2\right)^{\frac{-1}{\gamma-1}} \tag{4.74}$$

Alternatively, the flow area A varies with both the nozzle's geometric configuration and its axial position x. For instance, in the Barschdorff nozzle, this relationship is exemplified by

$$A = \left[1 + 2R\left(1 - \sqrt{1 - \left(\frac{x}{R}\right)^2}\right)\right]^2 \tag{4.75}$$

where R represents the dimensionless radius of wall curvature normalized by d.

From Eqs. (4.74)–(4.75), it can derive the following relationships

$$\frac{\mathrm{d}(p)}{\mathrm{d}} = \left[1 + 2R\left(1 - \sqrt{1 - \left(\frac{x}{R}\right)^2}\right)\right]^2 \tag{4.76}$$

$$\frac{d\lambda_i}{dA} = \left(\frac{\gamma+1}{2}\right)^{\frac{1}{(\gamma-1)}} \frac{\lambda_i^2}{\lambda_i^2 - 1}(1 - \frac{\gamma-1}{\gamma+1}\lambda_i^2)^{\frac{\gamma}{\gamma-1}} \tag{4.77}$$

Substitute Eq. (4.76) into Eq. (4.77), the expansion rate k_{ip} can be calculated by

$$k_{\text{ip}} = \frac{2\gamma c_{\text{icr}}}{d}\left(\frac{\gamma+1}{2}\right)^{\frac{2-\gamma}{\gamma-1}} \frac{\lambda_i^4}{\lambda_i^2 - 1}\left(1 - \frac{\gamma-1}{\gamma+1}\lambda_i^2\right)^{\frac{1}{\gamma-1}} \frac{\mathrm{d}A}{\mathrm{d}x} \tag{4.78}$$

4.5.2 General Solution of Equations

Therefore, based on the aforementioned analysis, we can summarize the four significant equations related to the Wilson point as follows:

$$\frac{\Delta \dot{T}_{\text{w}}}{T_{\text{w}}} = \gamma_{\text{s}} f_{\text{p}} k_{\text{ipw}} - L\dot{\text{WF}}_{\text{w}} = 0 \tag{4.79}$$

$$\frac{\Delta \ddot{T}_{\text{w}}}{T_{\text{w}}} = \gamma_{\text{s}} f_{\text{p1}} \dot{k}_{\text{ipw}} - L\ddot{\text{WF}}_{\text{w}} = -\frac{\Delta T_{\text{w}}^3}{\Theta_{\text{w}}^2 T_{\text{w}} \tau_{\text{n}}^2} \tag{4.79a}$$

$$\dot{\text{WF}}_{\text{w}} = \pi^{\frac{3}{2}} \rho_{\text{L}} I_{\text{w}} (C_{\text{rw}} \Delta T_{\text{w}})^3 \tau_{\text{n}}^3 [1 + g_2(\zeta)] \tag{4.79b}$$

$$\ddot{\text{WF}}_{\text{w}} = 4\pi \rho_{\text{L}} I_{\text{w}} (C_{\text{rw}} \Delta T_{\text{w}})^3 \tau_{\text{n}}^2 [1 + g_3(\zeta)] \tag{4.79c}$$

Combine the third and fourth equations into the first and second equations extensively, get

$$\gamma_{\text{s}} f_{\text{p}} k_{\text{ipw}} = L\pi^{\frac{3}{2}} \rho_{\text{L}} J_{\text{w}} (C_{\text{rw}} \Delta T_{\text{w}})^3 \tau_{\text{n}}^3 [1 + g_2(\zeta)] \tag{4.80}$$

$$\gamma_{\text{s}} f_{\text{p1w}} \dot{k}_{\text{ipw}} - 4L\pi \rho_{\text{L}} I_{\text{w}} (C_{\text{rw}} \Delta T_{\text{w}})^3 \tau_{\text{n}}^2 [1 + g_3(\zeta)] = -\frac{\Delta T_{\text{w}}^3}{\Theta_{\text{w}}^2 T_{\text{w}} \tau_{\text{n}}^2} \tag{4.80a}$$

According to Eqs. (4.79)–(4.80a), the x_w, λ_w, p_{iw}, T_{iw}, $\Delta \ddot{T}_w$, f_p, f_{p1}, ΔT_w, k_{ipw}, $\dot{k}_{ipw}$, τ_n, $\dot{WF}_w$, $\ddot{WF}_w$ is be calculated. Given the inlet stagnation temperature, inlet stagnation pressure, and nozzle geometry, the solution can be determined through an iterative procedure, despite the complexity of this nonlinear algebraic equation. Once ΔT_w and τ_n have been determined, the total droplet number per unit mass of mixture can be calculated by

$$N_w = \int_0^\infty I(\tau)d\tau = I_w \int_0^\infty \exp\left(-\frac{\tau^2}{\tau_n^2}\right)d\tau = \frac{1}{2}\pi^{0.5} I_w \tau_n \tag{4.81}$$

According to Eqs. (4.71)–(4.71b), the mass-mean droplet radius is given by

$$(r_{NM})_w = \frac{C_{rw}\Delta T_w \tau_n}{\pi^{\frac{1}{6}}}\left[1 + g_1(\zeta)\right]^{\frac{1}{3}} \tag{4.82}$$

Sauter mean droplet radius r_{32} can be calculated by

$$\begin{aligned}(r_{32})_w &= \frac{\int_0^\infty I(\tau) r_w^3(\tau)d\tau}{\int_0^\infty I(\tau) r_w^2(\tau)d\tau} \\ &= \frac{2C_{rw}\Delta T_w \tau_n}{\sqrt{\pi}} \frac{1 + \frac{3\sqrt{\pi}\zeta}{2} + 3\zeta^2 + \sqrt{\pi}\zeta^3}{1 + \frac{4\zeta}{\sqrt{\pi}} + 2\zeta^2} \\ &= 1.128\left(1 + 0.402\zeta + 0.093\zeta^2\right) C_{rw}\Delta T_w \tau_n\end{aligned} \tag{4.83}$$

The droplet size distribution function $f(r_w, r_w(\tau))$ at the Wilson point also can be obtained, as follows,

$$f(r_w, r_w(\tau)) = \frac{I_w}{C_{rw}\Delta T_w N_w} e^{-\left[\frac{1}{\zeta+1}\left(\zeta + \frac{r_w(\tau)}{r_w}\right)\right]^2} \tag{4.84}$$

As the parameter ζ approaches 0, the distribution of droplets is approximated as a half-Gaussian distribution.

4.6 Comparison with Analytic Method, CFD and Experiments

Table 4.5 demonstrates a strong agreement between the exact numerical method and the algebraic solution across three cases. The relative deviations of the dimensionless cross-sectional area A_0 between calculated and tested data at the Wilson point are 1.12%, 0.24%, and 0.52%. Since nucleation status is closely tied to A_0 rather than the coordinate x_0, these deviations highlight the analytical method's ability

Table 4.5 Comparison of analytical solutions at the Wilson point with numerical simulation and experiments data: $T_0 = 287$ K, $p_{a0} + p_0 = 60.0$ kPa, $c_p = 1880$ J kg^{-1} K^{-1}, $\frac{dA}{dx} = 0.077538$, $\frac{d^2A}{dx^2} = 0$

Parameters	$p_0 = 0.26$ kPa			$p_0 = 0.50$ kPa			$p_0 = 1.00$ kPa		
	Analytical	CFD	Expt.	Analytical	CFD	Expt.	Analytical	CFD	Expt.
x^0 (–)	3.669	3.82	3.858	2.32	2.25	2.284	1.186	1.300	1.260
x (mm)	46.6	48.5	49.0	29.5	28.6	29.0	15.06	16.50	16.0
A^0 (–)	1.2845	1.2961	1.2992	1.1801	1.1746	1.1771	1.0919	1.1007	1.0977
p_W (Pa)	62.21	61.03	60.74	144.5	146.1	145.4	353.5	345.5	348.2
T_W (K)	206.3	205.4	\	215.5	216.1	\	225.8	224.6	\
ΔT_W (K)	41.7	42.4	\	41.3	40.9	\	40.9	41.8	\
λ	1.468	1.476	\	1.382	1.377	\	1.279	1.291	\
u (m/s)	568.0	571.2	\	534.8	532.6	\	494.8	499.7	\
k_{ipW} (/s)	5.69×10^3	5.62×10^3	\	6.56×10^3	6.63×10^3	\	8.05×10^3	7.82×10^3	
f_{pW}	0.7283	0.7302	\	0.7043	0.7025	\	0.6602	0.6669	
ζ	0.0987	\	\	0.1077	\	\	0.1101	\	\
τ_n (s)	6.6×10^{-6}	\	\	5.5×10^{-6}	\	\	4.4×10^{-6}	\	\
r_{32} (nm)	4.40	9.45	\	4.32	7.21	\	4.19	6.54	\
N_W (kg^{-1})	2.1×10^{18}	1.7×10^{18}	\	2.2×10^{18}	1.4×10^{18}	\	2.3×10^{18}	1.1×10^{18}	\
Y_W (%)	0.026	0.067	\	0.025	0.066	\	0.024	0.064	\

Expt. Experiment

to provide accurate forecasts. Additionally, the maximum deviations of subcooling ΔT_W between algebraic solution and CFD result are within 0.7 K, 0.4 K, and 0.9 K respectively, while velocity deviations are only 3.2 m/s, 2.2 m/s, and 4.9 m/s. Therefore, this approximate model demonstrates satisfactory agreement with the numerical solution.

Take $R^0 = 10$, $p_0 = 78{,}500$ Pa and $T_0 = 380.55$ K for example, the relationship between the above parameters and nozzle diameter d at Wilson point is illustrated in Fig. 4.27. With the increase of d, the axial position x of Wilson point moves toward the nozzle outlet but the dimensional throat flow area A^0 decreases. Meanwhile, the values of k_p, ΔT and M all show a decreasing trend.

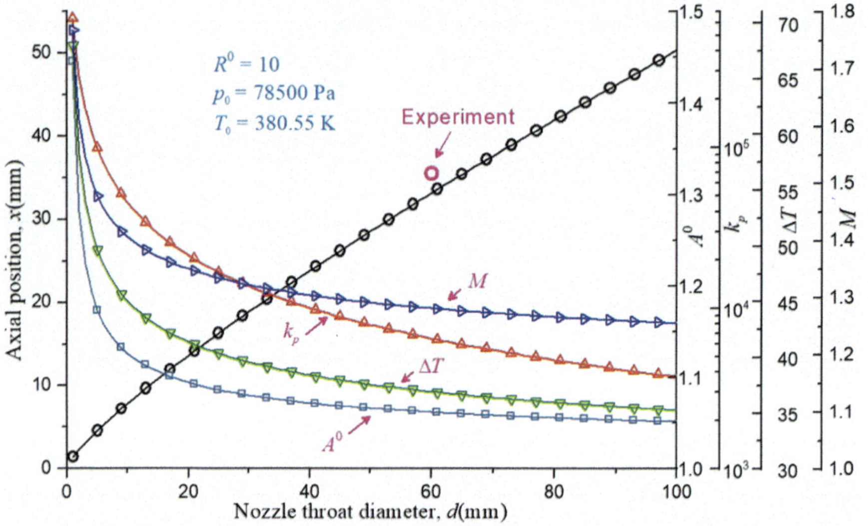

Fig. 4.27 The relationships of axial position x, non-dimensional area A_0, expansion rate coefficient k_p, subcooling ΔT and Mach number M at Wilson point modifying in accordance with the nozzle diameter d, for $R^0 = 10$, $p_0 = 78{,}500$ Pa and $T_0 = 380.55$ K

References

1. Abadi SNR, Kouhikamali R, Atashkari K (2015) Two-fluid model for simulation of supersonic flow of wet steam within high-pressure nozzles. Int J Therm Sci 96:173–182
2. Adam S, Schnerr GH (1997) Instabilities and bifurcation of non-equilibrium two-phase flows. J Fluid Mech 348:1–28
3. Aliabadi MAF, Jahangiri A, Khazaee I, Lakzian E (2020) Investigating the effect of water nano-droplets injection into the convergent-divergent nozzle inlet on the wet steam flow using entropy generation analysis. Int J Therm Sci 149:106181
4. Alnoush W, Castier M (2019) Shortcut modeling of natural gas supersonic separation. J Nat Gas Sci Eng 65:284–300
5. Bian J, Cao X, Yang W, Guo D, Xiang C (2020) Prediction of supersonic condensation process of methane gas considering real gas effects. Appl Therm Eng 164:114508
6. Bolaños-Acosta AF, Restrepo JC, Simões-Moreira JR (2021) Two semi-analytical approaches for solving condensation shocks in supersonic nozzle flows. Int J Heat Mass Transf 173:121212
7. Bredberg J (2001) On two-equation eddy-viscosity models. Department of Thermo and Fluid Dynamics, Chalmers University of Technology, Göteborg, Sweden
8. Delale CF, Schnerr GH, Zierep J (1993) Asymptotic solution of transonic nozzle flows with homogeneous condensation. I. Subcritical flows. Phys Fluids A Fluid Dyn 5(11):2969–2981
9. Ding H, Wang C, Zhao Y (2014) An analytical method for Wilson point in nozzle flow with homogeneous nucleating. Int J Heat Mass Transf 73:586–594
10. Ding H, Wang C, Wang G, Chen C (2015) Analytic equations for the Wilson point in high-pressure steam flow through a nozzle. Int J Heat Mass Transf 91:961–968
11. Ding H, Li Y, Lakzian E, Wen C, Wang C (2019) Entropy generation and exergy destruction in condensing steam flow through turbine blade with surface roughness. Energy Convers Manage 196:1089–1104
12. Ding H, Li Y, Wang C, Wen C, Tian Y (2020) Feature extraction of oscillating flow with vapor condensation of moist air in a sonic nozzle. IEEE Trans Instrum Meas 69(9):6465–6477

13. Dobbins RA (1983) A theory of the Wilson line for steam at low pressures. J Fluids Eng 105(4):414–422
14. Dykas S, Wróblewski W (2011) Single-and two-fluid models for steam condensing flow modeling. Int J Multiph Flow 37(9):1245–1253
15. Dykas S, Majkut M, Smołka K, Strozik M (2017) Comprehensive investigations into thermal and flow phenomena occurring in the atmospheric air two-phase flow through nozzles. Int J Heat Mass Transf 114:1072–1085
16. Dykas S, Majkut M, Smołka K (2020) Influence of air humidity on transonic flows with weak shock waves. J Therm Sci 29:1551–1557
17. Gerber AG, Kermani MJ (2004) A pressure based Eulerian–Eulerian multi-phase model for non-equilibrium condensation in transonic steam flow. Int J Heat Mass Transf 47(10):2217–2231
18. Gyarmathy G (1982) The spherical droplet in gaseous carrier streams: review and synthesis. Multiph Sci Technol 1:99–279
19. Hamidi S, Kermani MJ (2015) Numerical study of non-equilibrium condensation and shock waves in transonic moist-air and steam flows. Aerosp Sci Technol 46:188–196
20. Hedbäck AJ (1982) Theorie der spontanen kondensation in düsen und turbine. ETH Zurich
21. Huang L, Young JB (1996) An analytical solution for the Wilson point in homogeneously nucleating flows. Proc R Soc Lond Ser A Math Phys Eng Sci 452:1459–1473
22. Kermani MJ, Gerber AG (2003) A general formula for the evaluation of thermodynamic and aerodynamic losses in nucleating steam flow. Int J Heat Mass Transf 46:3265–3278
23. Kock F, Herwig H (2005) Entropy production calculation for turbulent shear flows and their implementation in CFD codes. Int J Heat Fluid Flow 26(4):672–680
24. Liang D, Ji Z, Li Y, Zhou Z (2024) The effect of the inlet steam superheat degree on the non-equilibrium condensation in steam turbine cascade. Phys Fluids 36(4)
25. Lv X, Bai B (2014) A multi-fluid model for non-equilibrium condensation in gaseous carrier flows. Appl Therm Eng 65(1–2):24–33
26. Marti J, Mauersberger K (1993) A survey and new measurements of ice vapor pressure at temperatures between 170 and 250 K. Geophys Res Lett 20(5):363–366
27. Mccallum M, Hunt R (1999) The flow of wet steam in a one-dimensional nozzle. Int J Numer Methods Eng 44(12):1807–1821
28. McDonald JE (1963) Homogeneous nucleation of water vapor condensation. I. Thermodynamic aspects. Am J Phys 30:870–877
29. Menter FR (2009) Review of the shear-stress transport turbulence model experience from an industrial perspective. Int J Comput Fluid Dyn 23(4):305–316
30. Menter F, Esch T (2001) Elements of industrial heat transfer predictions. 16th Brazilian congress of mechanical engineering (COBEM) 2001
31. Oxtoby DW (1992) Homogeneous nucleation: theory and experiment. J Phys Condens Matter 4:7627–7650
32. Pillai A, Prasad BVSSS (2018) Effect of wall surface roughness on condensation shock. Int J Therm Sci 132:435–445
33. Rhee J, Im J, Kim J, Song SJ (2019) Humidity effects on the aerodynamic performance of a transonic compressor cascade. Int J Heat Mass Transf 140:743–751
34. Shamsi M, Obaid AA, Vaziri M, Mousavian S, Hekmatian A, Bonyadi M (2024) A comprehensive comparison of the turbo-expander, Joule–Thomson, and combination of mechanical refrigeration and Joule–Thomson processes for natural gas liquids production. Energy 295:131032
35. Stripf M, Schulz A, Bauer HJ, Wittig S (2009) Extended models for transitional rough wall boundary layers with heat transfer—part I: model formulations. J Turbomach 131(3):031016
36. Vučković GD, Stojiljković MM, Vukić MV (2015) First and second level of exergy destruction splitting in advanced exergy analysis for an existing boiler. Energy Convers Manage 104:8–16
37. Wegener PP, Cagliostro DJ (1973) Periodic nozzle flow with heat addition. Combust Sci Technol 6(5):269–277

38. Wen C, Karvounis N, Walther JH, Ding H, Yang Y (2020) Non-equilibrium condensation of water vapour in supersonic flows with shock waves. Int J Heat Mass Transf 149:119109
39. White AJ, Young JB, Walters PT (1996) Experimental validation of condensing flow theory for a stationary cascade of steam turbine blades. Philos Trans R Soc A 354:59–88
40. Wyslouzil BE, Heath CH (2000) Binary condensation in a supersonic nozzle. J Chem Phys 113(17):7317–7329
41. Young JB (1982) The spontaneous condensation of steam in supersonic nozzles. Phys Chem Hydrodyn 3(1):57–82
42. Zhang G, Yang Y, Chen J, Jin Z, Majkut M, Smołka K, Dykas S (2023) Effect of relative humidity on the nozzle performance in non-equilibrium condensing flows for improving the compressed air energy storage technology. Energy 280:128240
43. Zhong X (1984) An improved generalized Watson equation for prediction of latent heat of vaporization. Chem Eng Commun 29:257–269

Chapter 5
Polydisperse Droplet Spectrum and Exergy Analysis

5.1 Introduction

Numerous studies have focused on the spectrum of droplets, where droplet radius significantly influences heat transfer and phase change calculations between vapor and liquid phases [22]. Typically, two primary methods are employed to model the liquid droplet phase in wet steam flow [24]. One method is the monodispersed model (Mono), which is straightforward but fails to account for droplet size distribution effects. Hence, the polydisperse model is more suitable for studying wet steam flow within turbine blades.

Various techniques exist for modeling the polydisperse liquid phase, including methods such as a finite number of droplet groups using a Lagrangian approach, fixed-volume bins using a Eulerian approach, spectrum pruning, the method of moments (MOM), and the Quadrature Method of Moments (QMOM) [21]. White and Young [23] discretized droplet distributions into multiple groups. Gerber and Kermani [9] studied the fixed-volume bins method, employing conservation fixed-volume integration to discretize liquid and gas phase conservation equations. Aliabadi et al. [3] introduced a method dividing the nucleation zone into twenty parts to form 20 droplet groups.

However, determining the appropriate number of droplet groups or fixed-volume bins and nucleation time step or spatial interval poses challenges. White and Hounslow [22] and Gerber and Kermani [9] explored MOM and QMOM for polydispersed droplets, predicting droplet distributions along nozzle lengths. Hughes et al. [11] investigated spectrum pruning and QMOM to reduce the number of droplet groups. Afzalifar et al. [2] found MOM solves several transport equations of droplet sizes, reducing computational time, with QMOM offering superior prediction of droplet size distribution effects.

H. Ding et al., *Supersonic Separators for Sustainable Gas Purification*, Green Energy and Technology, https://doi.org/10.1007/978-3-031-85213-8_5

Moreover, MOM and QMOM allow reconstruction of the droplet spectrum, specifically the droplet number density distribution, through several moments. Hulburt and Kartz [12] assumed an empirical form for particle size spectrum, reconstructing it using low-order moments. Diemer and Olson [6] proposed a nonlinear regression method for spectrum reconstruction based on prior shapes. John et al. [14] presented a discrete method for reconstructing droplet distribution using moments. De Souza et al. [5] improved PDF reconstruction using spline functions and an adaptive algorithm to quickly reconstruct droplet distribution without prior knowledge. The Quadrature Method of Moments (QMOM) is applied to solve numerically the transport equations of the probability density function coupled with conserved source terms incompressible Navier–Stokes equations for the liquid phase [13]. In the Quadrature Method of Moments (QMOM), the population balance equation undergoes transformation into a series of transport equations governing the lower-order statistical moments of the size distribution function. Efficient evaluation of integral source terms is achieved through Gaussian quadrature, ensuring computational accuracy [16]. This method's advantage lies in its minimal reliance on a small number of scalar fields, enhancing its suitability for seamless integration into Computational Fluid Dynamics (CFD) codes.

5.2 Polydisperse Model Using Quadrature Method of Moments (QMOM)

It assumed spherical, polydisperse droplets as illustrated in Fig. 5.1 and employed QMOM to calculated the integral of droplet number density distribution with its first $2m$ moments (μ_k, for $k = 0$ to $2m-1$, where $2m$ is the total number of moments). The negligible velocity slip between vapor and droplets (approximately 0.1 m s^{-1}) was justified by the submicron size of condensation-induced droplets, with maximum slip around 1 m s^{-1} [8]. Consequently, the influence of these droplets on turbulence was deemed insignificant. The mixture density, ρ_{mix}, is given by $\xi_v \rho_v + \xi_l \rho_l$, where ξ_l represents the volume fraction of liquid fraction $\xi_l = V_l/V$, and $\xi_v + \xi_l$ equals 1. The wetness fraction, β, denotes the liquid mass fraction, $\beta = \xi \rho_l/\rho_{\text{mix}}$.

5.2.1 *Quadrature Method of Moments (QMOM)*

Hulburt and Katz [12] pioneered the method of moments (MOM) for modeling droplet number density functions to accurately describe the droplet growth and nucleation dynamics. McGraw [17] introduced the Quadrature Method of Moments (QMOM) specifically tailored for the polydisperse wet steam flow, employing a limited set of droplet size moments. QMOM is noted for its exceptional accuracy and computational efficiency, as it does not necessitate a predefined growth law or

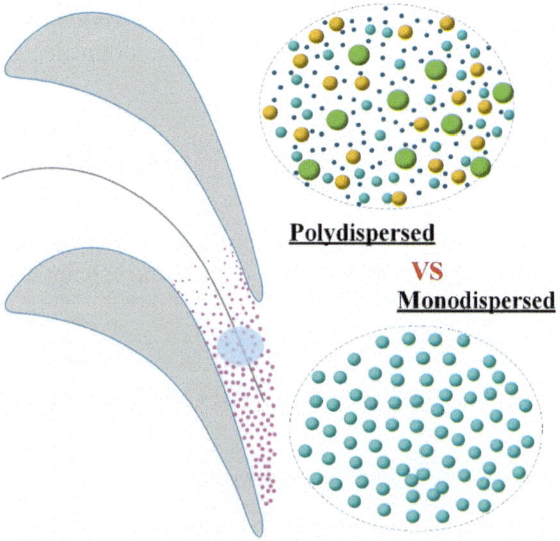

Fig. 5.1 Assumptions of polydisperse and monodispersed droplets for gas–liquid two-phase flows with homogeneous condensation

assume a specific mathematical form for the droplet number distribution. The droplet number density function $F(r, x_i, t)$ (m^{-1} kg^{-1}) is represented as

$$\frac{\partial \rho_m F}{\partial t} + \frac{\partial}{\partial x_j}(\rho_m F u_j) + \frac{\partial}{\partial r}(\rho_m F G) = J\delta(r - r_c) \tag{5.1}$$

where $F dr$ denotes the number of droplets in the interval $[r, r + dr]$ per unit mass. The growth rate, $G(r)$ varies with droplet size. The symbol j is used in tensor notation, u_j denotes the mean velocity component. δ represents the Dirac delta function.

The kth order moment of number density can be calculated by

$$\mu_k = \int_0^\infty r^k F(r)\mathrm{d}r, \quad k = 0, 1, 2, \ldots \tag{5.2}$$

where μ_0 (kg^{-1}) represents the number of droplets per unit mass of the mixture. μ_1 ($m\ kg^{-1}$) is the radius accumulation of droplets per unit mass. μ_2 ($m^2\ kg^{-1}$) is the total surface area of droplets per unit mass. μ_3 ($m^3\ kg^{-1}$) represents the total volume of droplets per unit mass of the mixture. The number-average radius, r_{10}, is given by μ_1/μ_0. The volume-surface-average radius, or Sauter radius, r_{32}, is given by μ_3/μ_2.

Thus, the wetness fraction β is calculated by

$$\beta = \int_0^\infty \frac{4\pi}{3}\pi\rho_l r^3 F(r)\mathrm{d}r = \frac{4\pi}{3}\rho_l\mu_3 \tag{5.3}$$

Combined with Eqs. (5.1)–(5.2), the governing equations for the evolution of the moments of droplet number density are formulated as follows:

$$\frac{\partial \rho_m \mu_k}{\partial t} + \frac{\partial}{\partial x_j}\left(\rho_m \mu_k u_j\right) = k \int_0^{\infty} r^{k-1} \rho_m G F \mathrm{d}r + J r_{\mathrm{c}}^k \tag{5.4}$$

The integral in right side of Eq. (5.4) can be obtained by a sum using weights w_i (kg^{-1}) and abscissa r_i according to the m-point Gaussian quadrature by McGraw [17], as follows:

$$\frac{\partial \rho_m \mu_k}{\partial t} + \frac{\partial}{\partial x_j}\left(\rho_m \mu_k u_j\right) = k \rho_m \sum_{i=1}^{m} r_i^{k-1} G(r_i) w_i + J r_{\mathrm{c}}^k \tag{5.5}$$

where the moments can be rewritten as

$$\mu_k = \int_0^{\infty} r^k F(r) \mathrm{d}r = \sum_{i=1}^{m} r_i^k w_i \tag{5.6}$$

The product-difference (PD) algorithm, as applied in this study with $m = 3$, utilizes the first 6 moments for $k = 0$–5 to compute 3 quadrature weights w_i and corresponding abscissas r_i. This is achieved through the construction of a symmetric tridiagonal matrix, specified in the PD algorithm [10].

$$J_3 = \begin{bmatrix} a_1 & b_1 & 0 \\ b_1 & a_2 & b_2 \\ 0 & b_2 & a_3 \end{bmatrix} \tag{5.7}$$

where

$$a_1 = \mu_1, \; b_1^2 = \mu_2 - \mu_1^2 \tag{5.7a}$$

$$a_2 = \frac{\left(\mu_1^3 - 2\mu_1\mu_2 + \mu_3\right)}{\left(\mu_2 - \mu_1^2\right)} \tag{5.7b}$$

$$b_2^2 = \frac{\left(-\mu_2^3 + 2\mu_1\mu_2\mu_3 - \mu_3^2 - \mu_1^2\mu_4 + \mu_2\mu_4\right)}{\left(\mu_2 - \mu_1^2\right)^2} \tag{5.7c}$$

$$a_3 = \frac{\mu_1\left(-\mu_2^3 + 2\mu_1\mu_2\mu_3 - \mu_3^2 - \mu_1^2\mu_4 + \mu_2\mu_4\right)}{\left(\mu_1\mu_3 - \mu_2^2\right)\left(\mu_2 - \mu_1^2\right)} + \frac{\left(\mu_2 - \mu_1^2\right)\left(-\mu_3^3 + 2\mu_2\mu_3\mu_4 - \mu_1\mu_4^2 - \mu_2\mu_5^2 + \mu_1\mu_3\mu_5\right)}{\left(-\mu_2^3 + 2\mu_1\mu_2\mu_3 - \mu_3^2 - \mu_1^2\mu_4 + \mu_2\mu_4\right)\left(\mu_1\mu_3 - \mu_2^2\right)} \tag{5.7d}$$

Then, the values of abscissas r_i are the eigenvalues of J_3. The weights w_i are calculated by the eigenvectors w_{i1}, as follows:

$$w_i = \mu_0 \upsilon_{i1}^2 \tag{5.8}$$

The mass change rate $\dot{m}_\mathrm{v}$ (kg m^{-3} s^{-1}) is rewritten by

$$\begin{aligned}\dot{m}_\mathrm{v} &= \frac{4}{3}\pi\rho_\mathrm{l} J r_\mathrm{c}^3 + 4\pi\rho_\mathrm{l}\rho_m \int_0^\infty r^2 GF \mathrm{d}r \\ &= \frac{4}{3}\pi\rho_\mathrm{l} J r_\mathrm{c}^3 + 4\pi\rho_\mathrm{l}\rho_\mathrm{m} \sum_{i=1}^{m} G(r_i) r_i^2 w_i \end{aligned} \tag{5.9}$$

where growth rate $G(r)$ is calculated by Eq. (4.31).

5.2.2 *Exergy Destruction and Erosion Rate*

In a polydisperse model, the entropy generation $s_\mathrm{gen,L}$ resulting from phase change, encompassing condensation and evaporation, relies on the temperature differential between the liquid and vapor phases, as described in

$$\frac{\partial \rho_m s_\mathrm{gen,L}}{\partial t} + \frac{\partial}{\partial x_i}\left(\rho_m s_\mathrm{gen,L} u_i\right) = Y_\mathrm{L} \tag{5.10}$$

where the source term, the local entropy generation rate Y_L, is computed as

$$\begin{aligned} Y_\mathrm{L} &= 4\pi\rho_m h_\mathrm{lg} \int_0^\infty \rho_\mathrm{l} r^2 GF \left(\frac{1}{T_\mathrm{v}} - \frac{1}{T_\mathrm{l}}\right) \mathrm{d}r \\ &= 4\pi\rho_m h_\mathrm{lg} \sum_{i=1}^{m} \rho_\mathrm{l} G(r_i) r_i^2 w_i \left(\frac{1}{T_\mathrm{v}} - \frac{1}{T_\mathrm{l}(r_i)}\right) \end{aligned} \tag{5.11}$$

For analyzing the condensation flow in turbine, the efficiency ratio ξ_η can be expressed as

$$\xi_\eta = \frac{\eta_\mathrm{wet}}{\eta_\mathrm{dry}} = \frac{1 - \zeta_\mathrm{D,wet}}{1 - \zeta_\mathrm{D,dry}} = 1 - a_\mathrm{B}\beta_\mathrm{m} \tag{5.12}$$

where a_B represents the Baumann factor. β_m is half of the mass-weighted average outlet wetness fraction $0.5Y_\mathrm{out}$, which considers partial dry expansion with non-equilibrium condensation in this single stage [20]. η_wet and η_dry denote the wet and dry total isentropic efficiencies, respectively. The isentropic efficiency is determined using the method outlined in Abadi et al. [1].

$$\eta_{\text{dry}} = \frac{h_{\text{in}} - h_{\text{out}}}{h_{\text{in}} - h_{\text{out,is}}} \tag{5.13}$$

where the subscript "is" denotes the isentropic condition, and h represents the specific enthalpy of vapor.

Additionally, the impact of moving droplets colliding with turbine blades leads to blade erosion, affecting the safety and reliability of turbine operation [18]. The damage caused by a single 500 μm droplet is significantly greater than that caused by 125 droplets each with a diameter of 100 μm. Research indicates that droplet radius is the primary factor influencing turbine erosion. Blade erosion, as described by Lee et al. [15], is formulated as

$$r = k_{\text{e}} \left(\frac{m'_{\text{p}}}{m_{\text{ref}}} \right) \left(\frac{u_{\text{p}}}{u_{\text{ref}}} \right)^{5.1} \left(\frac{r_{\text{p}}}{r_{\text{ref}}} \right)^{b} 10^{\gamma_{\text{c}} H_{\text{v}}} \tag{5.14}$$

where k_{e} is the erosion constant, m'_{p} (kg mm^2 h^{-1}) is the droplet mass flow flux. and u_{p} and r_{p} denote the droplet velocity and radius, respectively. The exponent of the power function b ranges between 2 and 4.5, and γ_{c} and H_{v} represent constants for hardness and Vickers hardness value, respectively. Governed by Lee's model, erosion rate Er correlated with the water droplet radius and wetness fraction can be calculated by

$$\text{Er} \propto \beta r^b \tag{5.15}$$

where the typical value of parameter $b = 2$, the normalized local erosion rate $\hat{\text{E}}\text{r}$ is defined as

$$\hat{\text{E}}\text{r}_{\text{avg}} = \frac{\beta r^2}{\beta_{\text{ref}} r_{\text{ref}}^2} \tag{5.16}$$

where the reference parameter β_{ref} and r_{ref} are equal to 0.01 and 1e^{-7} m.

In the context of polydisperse droplets, the normalized local erosion rate $\hat{\text{E}}\text{r}_{\text{QMOM}}$ is written by

$$\hat{\text{E}}\text{r}_{\text{QMOM}} = \frac{\int_0^\infty \frac{4\pi}{3} \pi \rho_{\text{l}} r^3 r^2 F(r) \text{d}r}{\beta_{\text{ref}} r_{\text{ref}}^2} \tag{5.17}$$

Therefore, the mass-weighted average of the normalized local erosion rate is given by

$$\hat{\mathrm{E}}\mathrm{r}_{\mathrm{QMOM}} = \frac{4\pi}{3}\frac{\rho_1\mu_5}{\left(\beta_{\mathrm{ref}} r_{\mathrm{ref}}^2\right)} \tag{5.18}$$

where the mass-weighted average of a quantity $\overline{\phi}$ is expressed as

$$\overline{\phi} = \frac{\int \phi\rho\vec{u}_{\mathrm{loc}} \cdot d\vec{u}_{\mathrm{fa}}}{\int \rho\vec{u}_{\mathrm{loc}} \cdot d\vec{u}_{\mathrm{fa}}} = \frac{\sum_{i=1}^{n_{\mathrm{fa}}} \phi_i\rho_i\left|\vec{u}_{\mathrm{loc},i} \cdot \vec{u}_{\mathrm{fa},i}\right|}{\sum_{i=1}^{n_{\mathrm{fa}}} \rho_i\left|\vec{u}_{\mathrm{loc},i} \cdot \vec{u}_{\mathrm{fa},i}\right|} \tag{5.19}$$

where n_{fa} indicates the cumulative count of facets. $\vec{u}_{\mathrm{fa}}$ and $\vec{u}_{\mathrm{loc}}$ denote the vectors of facet area and local velocity.

5.3 Reconstruction Algorithm for Distribution

Actually, the droplet spectra, namely droplet size distribution $f(r)$ can also be reconstructed if a finite number of moments are known. The usual method is to use a priori knowledge concerning the shape of probability density function (PDF), such as Gauss distribution, log–normal distribution, and gamma distribution. The Gaussian normal distribution is expressed as

$$f(r) = \frac{1}{\sigma_{\mathrm{g}}\sqrt{2\pi}} \exp\left(-\frac{(r - r_{10})^2}{2\sigma_{\mathrm{g}}^2}\right) \tag{5.20}$$

where the mean radius $r_{10} = \mu_1/\mu_0$, and standard deviation σ_{g} is calculated by

$$\sigma_{\mathrm{g}} = \sqrt{\frac{\mu_2}{\mu_0} - \left(\frac{\mu_1}{\mu_0}\right)^2} = r_{10}\sqrt{\frac{\mu_0\mu_2}{\mu_1^2} - 1} = r_{10}C_{\mathrm{V}} \tag{5.20a}$$

where C_{V} is coefficient of variation.

A log–normal (or lognormal) distribution is a continuous probability distribution of a random variable whose logarithm is normally distributed, as follows:

$$f(r) = \frac{1}{r\sigma_{\log}\sqrt{2\pi}\exp\left(-\frac{(\ln r - \ln r_{\log})^2}{2\sigma_{\log}^2}\right)} \tag{5.21}$$

where the median $r_{\log}$ is given by

$$r_{\log} = r_{10}\exp\left(-0.5\sigma_{\log}^2\right) \tag{5.21a}$$

$\sigma_{\log}$ is shape parameter, namely the standard deviation of the log–normal function is equal to

$$\sigma_{\log} = \sqrt{\ln \frac{\mu_0 \mu_2}{\mu_1^2}} \tag{5.21b}$$

$\sigma_{\log}$ is shape parameter (standard deviation of the log–normal function). The mode (the most probable value) of log–normal distribution is equal to $r_{10} \exp\left(-1.5\sigma_{\log}^2\right)$ which is smaller than mean radius r_{10}.

The gamma distribution is another widely used distribution which can be calculated by

$$f(r) = \frac{1}{b^a \Gamma(a)} r^{a-1} \exp\left(-\frac{r}{b}\right), \quad a > 0 \tag{5.22}$$

where the parameters a and b denote the shape and scale parameters respectively, defined as

$$a = \frac{r_{10}^2}{\sigma_{\mathrm{g}}^2} = \frac{1}{C_{\mathrm{V}}^2} \tag{5.22a}$$

$$b = \frac{\sigma_{\mathrm{g}}^2}{r_{10}} \tag{5.22b}$$

It is noteworthy that $a > 1$ in the current investigation. The gamma function $\Gamma(a)$ is defined as

$$\Gamma(a) = \int_0^\infty z^{a-1} \exp(-z) \mathrm{d}z \tag{5.22c}$$

The mode radius, where the probability density function reaches its peak, is expressed by $(a-1)b = r_{10} - \sigma_{\mathrm{g}}^2 / r_{10}$. However, determining the median lacks a straightforward closed-form solution.

The available moments are solely employed to fit the parameters of the probability density function (PDF). However, there is a constraint that the PDF must exhibit a straightforward form. To address this challenge, a reconstruction algorithm employing low-order splines can be employed [14]. This method approximates the PDF shape using piecewise polynomial functions, eliminating the necessity for any prior assumptions about the shape of the PDF.

Let $\{r_1 < r_2 < r_3 < \cdots < r_{n+1}\}$ denote a set of specified knots with $a = r_1 < r_2 < r_3 < \cdots < r_{n+1} < b$. The piecewise polynomial spline $S^{(l)}(r)$ has a degree l on range of $[a, b]$. $S_i(r)$ is the spline defined over subinterval $[r_i\ r_{i+1}]$, $i = 1, \ldots, n$. It possesses $n + 1$ knots, resulting in n intervals and $(l + 1)n$ unknowns. Then ansatz for piecewise polynomials is expressed by

$$f(r) \approx S_i(r) = \sum_{j=0}^{l} S_{ij}(r - r_i)^j, \quad r \in [r_i, r_{i+1}], \quad i = 1, \ldots, n \tag{5.23}$$

By setting the $(l + 1)n$ free coefficients S_{ij}, the entire spline is established. Assuming $f(r)$ is zero beyond $[a, b]$, the left boundary conditions consist of $l - 1$ conditions. These conditions require the first through $l - 1$ order derivatives to be zero at point a.

$$f\begin{pmatrix} S_{10} \\ S_{11} \\ S_{12} \\ \vdots \\ S_{1(l-1)} \end{pmatrix}_{l\times 1} = \begin{pmatrix} 0 \\ 0 \\ 0 \\ \vdots \\ 0 \end{pmatrix}_{l\times 1} \tag{5.24}$$

The right boundary conditions consist of l conditions are

$$\begin{pmatrix} 1 & x_{n+1} - x_n & (x_{n+1} - x_n)^2 & (x_{n+1} - x_n)^3 & \cdots & (x_{n+1} - x_n)^l \\ 0 & 1 & 2(x_{n+1} - x_n) & 3(x_{n+1} - x_n)^2 & \cdots & l(x_{n+1} - x_n)^{l-1} \\ 0 & 0 & 2 & 6(x_{n+1} - x_n) & \cdots & l(l-1)(x_{n+1} - x_n)^{l-2} \\ \vdots & \vdots & \vdots & \vdots & \ddots & \vdots \\ 0 & 0 & 0 & 0 & \cdots & l!(x_{n+1} - x_n) \end{pmatrix}_{l\times(l+1)}$$

$$\times \begin{pmatrix} S_{n0} \\ S_{n1} \\ S_{n2} \\ S_{n3} \\ \vdots \\ S_{nl} \end{pmatrix}_{(l+1)\times 1} = \begin{pmatrix} 0 \\ 0 \\ 0 \\ 0 \\ \vdots \\ 0 \end{pmatrix}_{l\times 1} \tag{5.25}$$

Additional $l(n - 1)$ conditions ensure the continuity of $S_i(r)$ across adjacent intervals. These conditions govern the first through l order derivatives as follows:

$$\begin{pmatrix} 1 & x_{i+1} - x_i & (x_{i+1} - x_i)^2 & \cdots & (x_{i+1} - x_i)^l & -1 & 0 & \cdots & 0 \\ 0 & 1 & 2(x_{i+1} - x_i) & \cdots & l(x_{i+1} - x_i)^{l-1} & 0 & -1 & \cdots & 0 \\ \vdots & \vdots & \vdots & \ddots & \vdots & \vdots & \vdots & \ddots & 0 \\ 0 & 0 & 0 & \cdots & l!(x_{i+1} - x_i) & 0 & 0 & 0 & -1 \end{pmatrix}_{l\times(2l+1)}$$

$$\times \begin{pmatrix} S_{i0} \\ S_{i1} \\ S_{i2} \\ \vdots \\ S_{il} \\ S_{(i+1)0} \\ S_{(i+1)1} \\ \vdots \\ S_{(i+1)(l+1)} \end{pmatrix}_{(2l+1)\times 1} = \begin{pmatrix} 0 \\ 0 \\ \vdots \\ 0 \end{pmatrix}_{l\times 1}, \quad i = 1, \ldots, n \tag{5.26}$$

Equations (5.24)–(5.26) yield $l + l + l(n-1) = l(n+1)$ conditions in total. Therefore, to determine the $(l+1)n$ unknowns required for calculation, additional $n - l$ conditions should still be supplied. These can be provided by the $n - l$ known moments.

$$\begin{aligned} \int_a^b r^k S^{(l)}(r)\mathrm{d}r &= \sum_{i=1}^{n} \int_{x_i}^{x_{i+1}} r^k S_i(r)\mathrm{d}r \\ &= \sum_{i=1}^{n} \sum_{j=0}^{l} S_{ij} \int_{x_i}^{x_{i+1}} r^k (r - r_i)^j \mathrm{d}r, \quad k = 0, 1 \ldots, (n - l - 1) \\ &= \sum_{i=1}^{n} \Big[X_1 S_{i0} + (X_2 - x_i X_1) S_{i1} + \big(X_3 - 2x_i X_2 + x_i^2 X_1\big) S_{i2} \\ &\quad + \cdots + \sum_{j=0}^{l} (-1)^j C_l^j x_i^j X_{l-j} \Big] = \mu_k \end{aligned} \tag{5.27}$$

where C_l^j $(j = 0, 1, \ldots, l)$ represents the binomial coefficient.

$$X_j = \frac{x_{i+1}^{k+j} - x_i^{k+j}}{k + j} \tag{5.28}$$

Thus, $(l + 1)n \times (l + 1)n$ linear system of equations $\boldsymbol{XS} = \boldsymbol{Y}$ has been established and $(l + 1)n$ coefficients S_{ij} is obtained reconstruct $f(r)$ in Eq. (5.23). In this study, the number of moments $n - l = 6$. For linear spline with $l = 1$, the number intervals $n = 7$. For quadratic spline with $l = 2$, $n = 8$. For cubic spline with $l = 3$, the number intervals $n = 9$.

The challenge lies in identifying the optimal interval $[a, b]$. To enhance stability and expedite the search process, normalized moments implemented in the following:

$$\hat{\mu}_k = \frac{\frac{\mu_k}{\mu_0}}{\left(\frac{\mu_1}{\mu_0}\right)^k} \tag{5.29}$$

where the 0th and 1st moments are consistently normalized to 1, denoted as $\hat{\mu}_0 = \hat{\mu}_1 = 1$. Therefore, Eq. (5.2) is reformulated into a normalized format, depicted as

$$\hat{\mu}_k = \int_0^\infty \hat{r}^k \hat{f}(\hat{r})\mathrm{d}\hat{r} = \sum_{i=1}^m \hat{r}_i^k \hat{w}_i, \quad k = 0, 1, 2, \ldots \tag{5.30}$$

where $\hat{r}_i$ is normalized radius (abscissa), $= r/r_{10}$ and $\hat{w}_i$ is normalized weight, $= w/\mu_0 . \hat{f}(\hat{r})$ is PDF of normalized radius which is equal to $f(r)/r_{10} . f(r)$ is probability density function of radius, namely $f(r) = F(r)/\mu_0$, m^{-1}, thus

$$\hat{f}(\hat{r}) = F(r)\mu_1/\mu_2 \tag{5.31}$$

The mean of $\hat{r}_i$ has been normalized to 1; therefore, the initial value of solution interval can be set as $a = 1 - \xi_1$ and $b = 1 + \xi_2$ where the default value of $\xi_1 = \xi_2 = 10$. For the next step, the values of a^{k+1} and b^{k+1} can be calculated by

$$a^{k+1} = a^k + 0.01\left(1 - a^k\right), \quad \text{if} \int_{\hat{r}_1}^{\hat{r}_2} S_1(\hat{r})\mathrm{d}\hat{r} < 0.01 f(\hat{r}_2 - \hat{r}_1)_{\max} \tag{5.32}$$

$$b^{k+1} = b^k + 0.01\left(1 - b^k\right), \quad \text{if} \int_{\hat{r}_n}^{\hat{r}_{n+1}} S_n(\hat{r})\mathrm{d}\hat{r} < 0.01 f(\hat{r}_{n+1} - \hat{r}_n)_{\max} \tag{5.33}$$

The TSVD method is employed to regularize ill-posed problems by truncating small singular values ($< 1\mathrm{e}^{-5}$) in the decomposition $\boldsymbol{X} = \boldsymbol{U}\boldsymbol{\Lambda}\boldsymbol{V}^{\mathrm{T}}$. Both $\boldsymbol{U}$ and $\boldsymbol{V}$ are orthonormal matrices. $\boldsymbol{\Lambda}$ is a diagonal matrix with singular values of $\boldsymbol{X}$, where $\lambda_1, \lambda_2, \ldots, \lambda_j$ are positive.

When the coefficient of variation C_{V} exceeds 0.5, the solution exhibits a non-physical negative radius. To mitigate this issue, a constraint on the first-order derivative of the left boundary condition is imposed. This results in an under-determined linear system $\boldsymbol{X}_1\boldsymbol{S} = \boldsymbol{Y}$ of dimensions $(l + 1)n - 1 \times (l + 1)n$, which can be effectively addressed using the TSVD method. The pseudo-inverse solution [26] is given by $\boldsymbol{S} = \boldsymbol{V}\boldsymbol{\Lambda}^+\boldsymbol{U}^{\mathrm{T}}\boldsymbol{Y}$, where $\boldsymbol{\Lambda}^+$ denotes the Moore–Penrose generalized inverse of $\boldsymbol{\Lambda}$.

5.4 Model Validation by Experiments

For experimental verification, Table 5.1 presents the boundary conditions at the inlet and outlet for Moore's nozzle B [19] and Bakhtar's turbine blade [4]. Moore nozzle B is characterized by an inlet total pressure $p^*_{\text{in}} = 25$ kPa and an inlet total temperature $T^*_{\text{in}} = 357.6$ K. The Bakhtar turbine blade, on the other hand, has an inlet pressure $p^*_{\text{in}} = 172$ kPa and an inlet temperature $T^*_{\text{in}} = 380.66$ K. The outflow pressure p_{out} is set to 0.48 times p^*_{in}.

The radius spectrum of the submicron droplet is rather difficult to measure. Moore et al. and Bakhtar et al. only measured Sauter radius r_{32} at the outlets of nozzle and blade which has been widely used in the validations of various simulation studies. Thus, the values of Sauter radius r_{32} of CFD are compared with the experimental Sauter radius for the purpose of validation of CFD model.

For wet steam flow through Moore nozzle B, Computational Fluid Dynamics (CFD) simulations compare pressure distribution and droplet size using polydisperse and monodispersed models against experimental findings by Moore [19]. Table 5.2 depicts the nozzle's geometry details, featuring a throat diameter d of 100 mm. A finer mesh, consisting of 450 × 110 grid cells, ensures a solution that is independent of grid size.

In Fig. 5.2, the polydisperse model employing the Quadrature Method of Moments (QMOM) shows a smaller droplet Sauter radius r_{32} compared to the monodisperse model. Specifically, at $x = 370$ mm, r_{32} is 0.04920 μm for QMOM and 0.0645 μm for the monodisperse method.

Experimental measurements using a spectral turbidity instrument yield $r_{32} = 0.05$ μm. The relative errors of the two models are − 1.6% and 29%, respectively. This indicates that the polydisperse model's r_{32} is closer to the experimental value.

Additionally, the static pressure distributions along the axis of Moore nozzle B for both QMOM and monodisperse models are compared with experimental data. The term "dry flow" denotes superheated steam flow without condensation. Computational pressure profiles along the nozzle axis exhibit good agreement with experimental data.

Table 5.1 Boundary conditions of Moore nozzle B and Bakhtar turbine blade

Experiments	p^*_{in} (kPa)	p_{out}	T^*_{in} (K)	$T_{\text{s,in}}$ (K)
Moore nozzle B	25	7 kPa (supersonic)	357.6	338.1
Bakhtar turbine blade	172	$0.48p_{\text{in}}{}^*$	380.66	388.66

Table 5.2 Dimensions of Moore nozzle B

Point	1	2	3	4
x (mm)	− 250	− 200	0	500
y (mm)	56.35	56.35	50.0	72.0

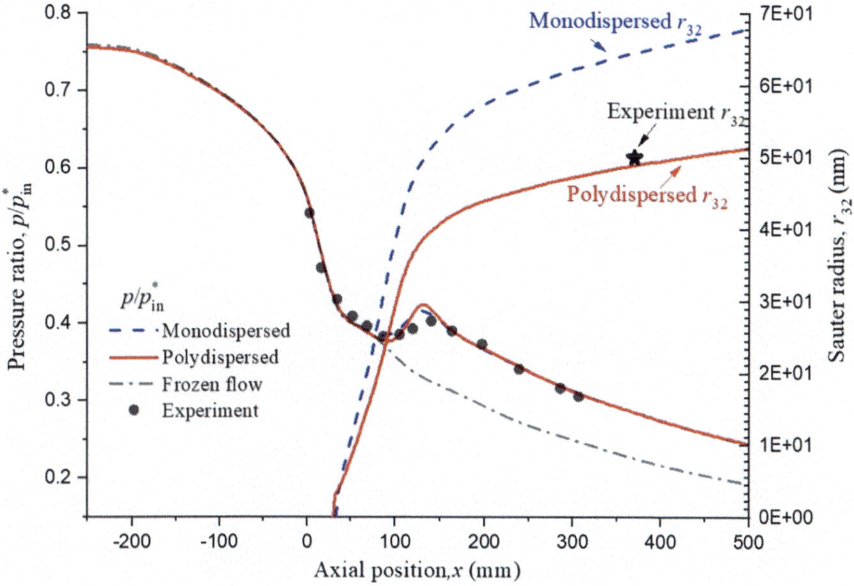

Fig. 5.2 Static pressure distribution and droplet radius of polydisperse and monodisperse models were compared with experimental data at the axis of Moore nozzle B

The Wilson points, where the time derivative of subcooling ΔT_W is zero and the nucleation rate reaches its maximum, are $x_w = 75.32$ mm and 66.12 mm for the polydisperse and monodisperse models, respectively. The theoretical Wilson point predicted by the algebraic formula in Chap. 4 is $x_w = 78.4$ mm, and the experimental value is approximately 75 mm. These findings suggest that the polydisperse model more accurately predicts the Sauter radius.

The polydisperse model was also validated using experimental data from Bakhtar [4], with details provided in Table 5.3 for the turbine blade. The outlet angle of the mid passage is $\varphi_{out} = 22.8°$, and the nominal throat diameter of the passage is 4.914 mm. The near-wall grid was refined with the wall ($y+ \approx 1$) to accurately predict velocity profiles, entropy generation, and entropy transfer. A grid-independent test determined the optimal grid size to be 290 × 120, as illustrated in Fig. 5.3.

Table 5.3 Dimensions of Bakhtar's turbine blade

Length	Chord	Pitch	Axial chord	Inlet flow angle	Grid
76.00 mm	35.76 mm	18.26 mm	25.27 mm	0°	290 × 120

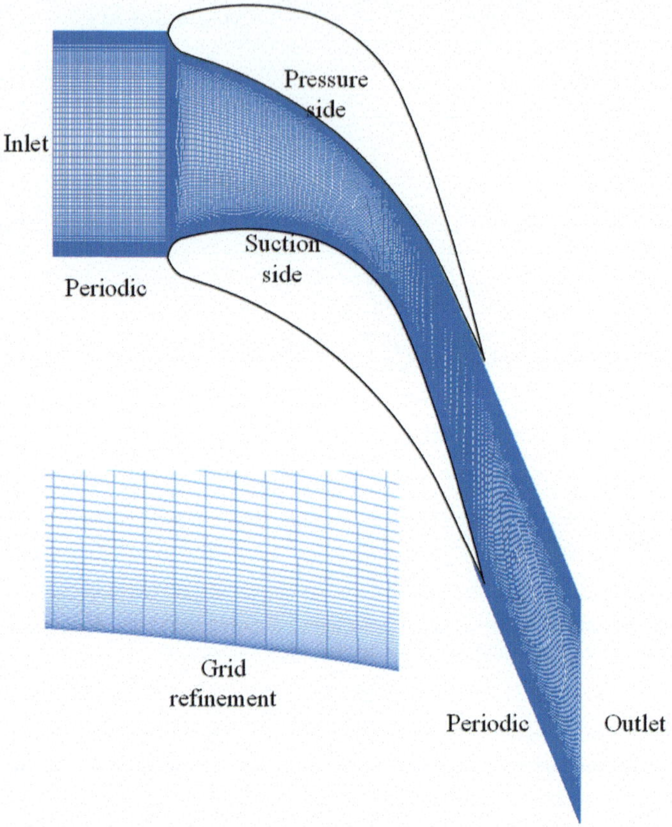

Fig. 5.3 The geometry size and grid of Bakhtar turbine blade

Figure 5.4 depicts static pressure profiles on the pressure and suction sides for the polydisperse model, monodispersed model, and experimental data. On the suction side, two pressure jumps occur due to condensation pseudo-shock from heat release at $x = 18.5$ mm and aerodynamic shock wave at $x = 20.5$ mm. The polydisperse model shows a slight delay in the condensation process compared to the monodispersed model.

Figure 5.5 illustrates the Sauter radius r_{32} at the mid passage of the turbine blade for the polydisperse model, monodispersed model, and experimental results. It indicates that the Sauter radius r_{32} in the polydisperse model is lower than in the monodispersed model and closer to the experimental measurement. At the midline outlet of the passage, the mean radius r_{32} values are 5.962×10^{-8} m, 8.861×10^{-8} m, and 5.70×10^{-8} m for the polydisperse method, monodisperse method, and experimental measurement, respectively. The predicted errors relative to the experimental result are

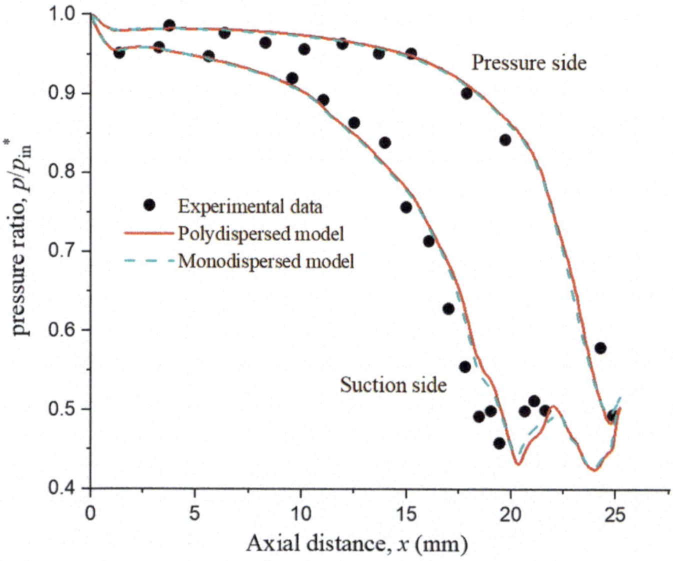

Fig. 5.4 Comparison of static pressure distributions on the pressure and suction sides among the polydisperse model, monodisperse model, and experimental data

4.6% for the polydisperse method and 55.4% for the monodispersed method. These results indicate that the polydisperse method provides better accuracy in predicting non-equilibrium condensation in the turbine.

5.5 Polydisperse Droplet Spectrum of Nozzle Flow

The characterization of droplet size distributions in nozzle flows plays a crucial role in various engineering and scientific applications, particularly in areas such as steam turbine, natural dehydration, and modern thermal power plant. In contrast to the simplified assumptions of monodisperse models, which consider droplets of uniform size, polydisperse models account for a range of droplet sizes, reflecting the more complex and realistic nature of many practical systems.

Polydisperse droplet spectra in nozzle flows exhibit a distribution of droplet sizes, typically described by parameters such as the Sauter mean diameter or the volume or number fraction distributions. This spectrum arises due to the diverse mechanisms governing droplet formation and interaction within the nozzle environment, including nucleation, condensation, coalescence, and breakup phenomena. Understanding and accurately predicting the polydisperse droplet spectrum are essential for optimizing nozzle performance.

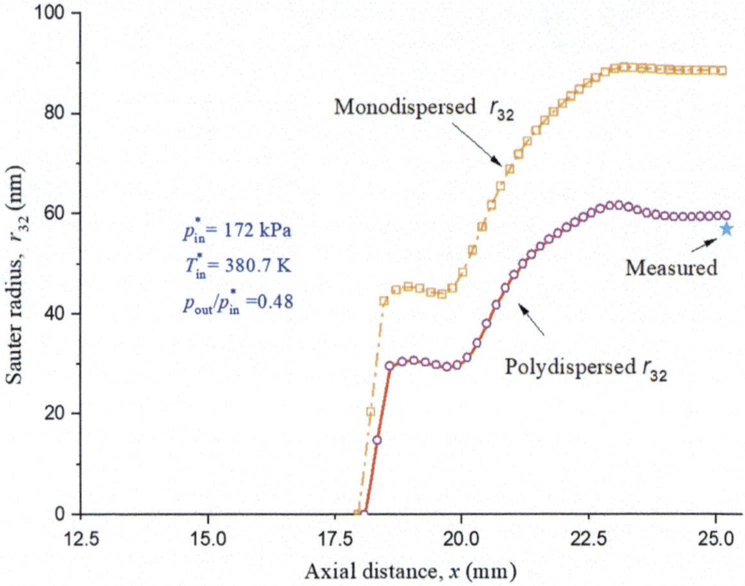

Fig. 5.5 Sauter radius r_{32} of the polydisperse model compared with the monodisperse model and experimental measurements along the mid passage of the turbine blade

5.5.1 Normalized Moments and Droplet Spectrum

In the context of wet steam flow involving condensation and evaporation processes, droplet size serves as a pivotal parameter according to entropy generation and erosion formulas. This study examines the normalized moments and droplet spectrum under identical boundary conditions as Moore nozzle B with the conditions of $p_{\text{in}}^* = 25\,\text{kPa}$, $p_{\text{out}} = 1.0\,\text{kPa}$, and $T_{\text{in}}^* = 357.6\,\text{K}$.

Figure 5.6 illustrates the profiles of normalized moments along the nozzle axis, where the 0th and 1st moments are consistently normalized to 1. Analyzing nucleation rate profiles reveals a close relationship between normalized moments and the nucleation process. Specifically, the values of normalized moments, such as $\hat{\mu}_2$, $\hat{\mu}_3$, $\hat{\mu}_4$, and $\hat{\mu}_5$, are observed to be higher in the nucleation zone compared to the growth zone.

Furthermore, the coefficient of variation C_V is calculated as $C_V = \sqrt{\hat{\mu}_2 - 1}$. A higher C_V indicates greater dispersion of droplet radii around the mean radius. This parameter provides insights into the variability and spread of droplet sizes within the flow, crucial for understanding the dynamics of condensation and evaporation processes in wet steam flows.

In Fig. 5.6, it is evident that the dispersion of droplet sizes in the nucleation zone exceeds that in the growth zone. The maximum values of $\hat{\mu}_2$, $\hat{\mu}_3$, $\hat{\mu}_4$ and $\hat{\mu}_5$ reach peaks of 1.649, 3.977, 13.119, and 54.985 respectively, occurring at different positions along the nozzle axis: $x = 63.55$ mm, 42.55 mm, 40.18 mm, and 39.00 mm.

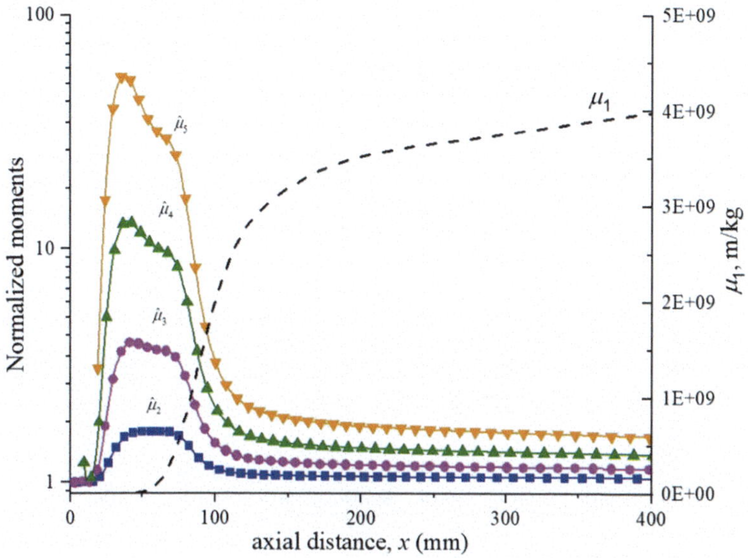

Fig. 5.6 Profiles of normalized moments at the axis of Moore nozzle B

The varying peak positions and trajectories of these normalized moments suggest that the shape of the droplet size distribution $f(r)$, or probability density function, changes along the streamline.

Furthermore, as the flow progresses toward the nozzle outlet, $\hat{\mu}_2$, $\hat{\mu}_3$, $\hat{\mu}_4$, and $\hat{\mu}_5$ tend to stabilize at constant values. This stabilization is attributed to lower subcooling degrees and reduced droplet growth rates in this region of the flow. Understanding these variations in normalized moments provides critical insights into how droplet sizes evolve within wet steam flow.

Table 5.4 provides computed values of six moments, the number-average radius r_{10}, and the coefficient of variation C_V from Point A to Point O along the nozzle axis. The number of droplets per unit mass of mixture μ_0 increases significantly, ranging from 5.57×10^{12} to 8.87×10^{16} kg^{-1}. Concurrently, the mean radius r_{10} expands from 1.28 to 47 nm. The coefficient of variation C_V initially increases and then decreases along the streamline in droplets nucleation and growth process. This trend reflects changes in the dispersion of droplet sizes relative to the mean radius along different points of the nozzle axis. Understanding these variations is crucial for evaluating how droplet characteristics evolve throughout the nozzle flow.

Using the six moments, quadrature weights w_i, and abscissa r_i of the Quadrature Method of Moments (QMOM) for closure requirements is calculated using the PD algorithm. As an example, the moments' values at point L in Table 5.4 are selected. Initially, the moments are normalized using Eq. (5.29). Subsequently, the symmetric tridiagonal matrix of the PD algorithm is employed. Once the Jacobi matrix is determined, the abscissas and weights can be solved, which are related to eigenvalues and eigenvectors.

Table 5.4 Computed values of six moments and the number-average radius at various locations along the axis of Moore nozzle B

Point	x (mm)	μ_0 (kg^{-1})	μ_1 (m kg^{-1})	μ_2 (m^2 kg^{-1})	μ_3 (m^3 kg^{-1})	μ_4 (m^4 kg^{-1})	μ_5 (m^5 kg^{-1})	r_{10} (m)	C_V
A	30	5.57×10^{12}	7153.63	1.26×10^{-5}	3.51×10^{-14}	1.50×10^{-22}	8.59×10^{-31}	1.28×10^{-9}	0.609
B	40	4.26×10^{14}	8.89×10^{5}	2.96×10^{-3}	1.52×10^{-11}	1.06×10^{-19}	9.10×10^{-28}	2.09×10^{-9}	0.774
C	50	4.76×10^{15}	1.68×10^{7}	9.81×10^{-2}	8.09×10^{-10}	8.38×10^{-18}	1.02×10^{-25}	3.55×10^{-9}	0.779
D	60	1.85×10^{16}	9.90×10^{7}	0.86767	1.03×10^{-8}	1.49×10^{-16}	2.47×10^{-24}	5.34×10^{-9}	0.780
E	70	4.62×10^{16}	3.37×10^{8}	4.0084	6.35×10^{-8}	1.21×10^{-15}	2.61×10^{-23}	7.29×10^{-9}	0.794
F	75	6.40×10^{16}	5.33×10^{8}	7.0371	1.22×10^{-7}	2.50×10^{-15}	5.79×10^{-23}	8.32×10^{-9}	0.767
G	80	8.01×10^{16}	8.25×10^{8}	12.619	2.46×10^{-7}	5.61×10^{-15}	1.44×10^{-22}	1.03×10^{-8}	0.698
H	85	8.74×10^{16}	1.16×10^{9}	20.688	4.53×10^{-7}	1.15×10^{-14}	3.23×10^{-22}	1.32×10^{-8}	0.593
I	90	8.86×10^{16}	1.41×10^{9}	28.315	6.76×10^{-7}	1.84×10^{-14}	5.53×10^{-22}	1.59×10^{-8}	0.514
J	100	8.90×10^{16}	1.96×10^{9}	49.911	1.43×10^{-6}	4.52×10^{-14}	1.56×10^{-21}	2.21×10^{-8}	0.388
K	125	8.98×10^{16}	2.85×10^{9}	97.761	3.60×10^{-6}	1.42×10^{-13}	5.93×10^{-21}	3.17×10^{-8}	0.283
L	200	9.03×10^{16}	3.51×10^{9}	143.79	6.22×10^{-6}	2.83×10^{-13}	1.35×10^{-20}	3.88×10^{-8}	0.238
M	300	9.24×10^{16}	3.87×10^{9}	170.70	7.90×10^{-6}	3.83×10^{-13}	1.94×10^{-20}	4.19×10^{-8}	0.228
N	400	8.77×10^{16}	3.97×10^{9}	188.13	9.33×10^{-6}	4.84×10^{-13}	2.61×10^{-20}	4.53×10^{-8}	0.218
O	500	8.87×10^{16}	4.17×10^{9}	204.88	1.05×10^{-5}	5.62×10^{-13}	3.13×10^{-20}	4.70×10^{-8}	0.212

The values of normalized abscissas $\hat{r}_i$ are 0.7281969, 1.0923890, and 1.5649833, corresponding to quadrature abscissas of moments r_i at 2.8265331×10^{-8} m, 4.2401628×10^{-8} m, and 6.0745614×10^{-8} m, respectively, where the mean radius r_{10} is equal to 3.88×10^{-8} m. The values of the normalized weights are 0.3565972, 0.4664764, and 0.1769263, summing up to 1.

The profiles of abscissas r_i and normalized weights w_i determined by the PD algorithm for QMOM at the nozzle axis are depicted in Fig. 5.7 and Table 5.5. It is notable that the weights of QMOM exhibit significant variations within the nucleation zone at ranges of $x = 20 \sim 100$ mm due to the intricate relationships among the six moments. The normalized moment $\hat{w}_1$ decreases from its initial value of 1.0–0.36. Beyond the nucleation zone, all weights of QMOM remain relatively consistent downstream. These results provide essential closure conditions for the governing equations of moments in modeling complex droplet size distributions in nozzle flows.

Figure 5.8 presents a comparison of reconstructed probability density functions $f(r)$ of droplet size between a spline-based algorithm using six moments and priori shape functions derived from three moments. At point C within the initial nucleation zone, the spline reconstruction closely resembles a gamma distribution, depicted in Fig. 5.8a. The mode values of the spline distribution and gamma distribution are 1.11×10^{-9} m and 1.35×10^{-9} m, respectively, whereas the mode values of Gauss and log–normal distributions are 1.69×10^{-9} m and 3.55×10^{-9} m. Notably, due to a deviation coefficient C_V of 0.779, the radius of the Gauss distribution exhibits an unphysical negative value.

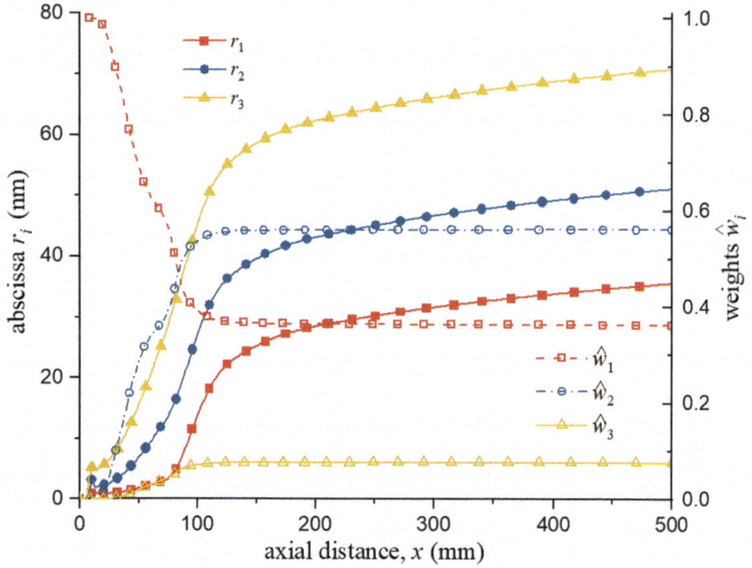

Fig. 5.7 Abscissa distributions r_i and normalized weight distributions w_i computed using the PD algorithm for QMOM along the axis of Moore nozzle B

Table 5.5 Quadrature weights w_i and abscissa r_i calculated by PD algorithm of six normalized moments at point L (200 mm, 0 mm)

i	$\hat{\mu}_i$	$\hat{a}_i$	$\hat{b}_i$	$\hat{r}_i$	$\hat{w}_i$	r_i (m)	w_i (kg^{-1})
1	1	1	0.2376448	0.7281969	0.3565972	2.8265331×10^{-8}	3.2213845×10^{16}
2	1.0564750	1.1343497	0.3221083	1.0923890	0.4664764	4.2401628×10^{-8}	4.2139981×10^{16}
3	1.1770126	1.2512196		1.5649833	0.1769263	6.0745614×10^{-8}	1.5982959×10^{16}
4	1.3792684						
5	1.6950068						

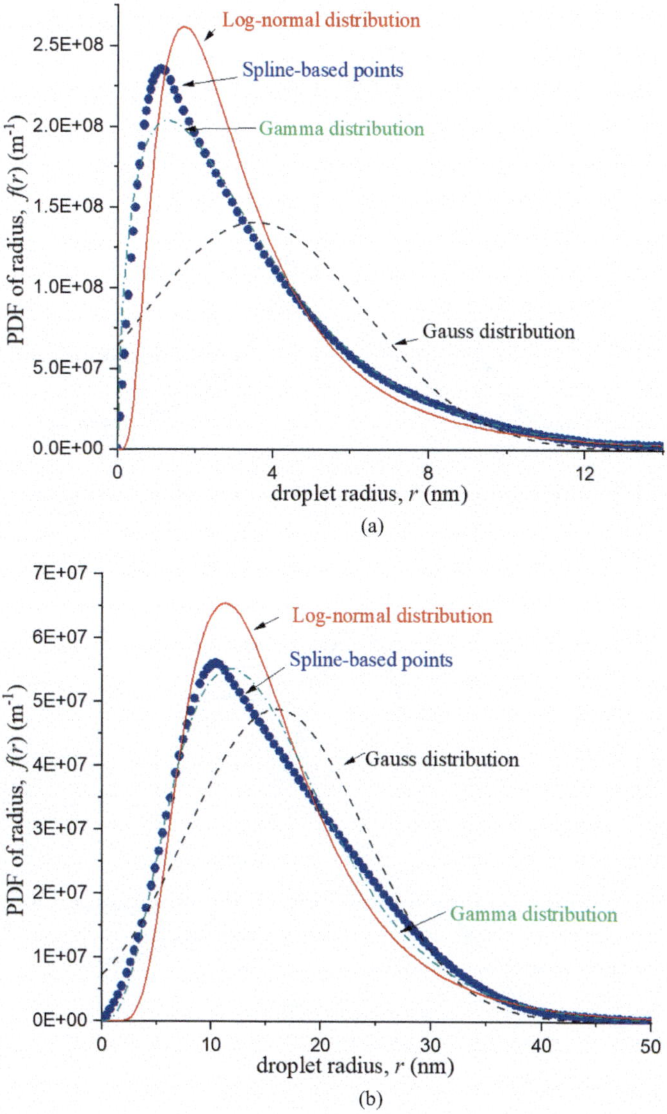

Fig. 5.8 Reconstruction of the probability density function $f(r)$ using a spline-based algorithm and priori shape functions derived from several moments at various locations along the axis of Moore nozzle B: **a** point C (50, 0); **b** point I (90, 0); **c** point L (200, 0); **d** point O (500, 0)

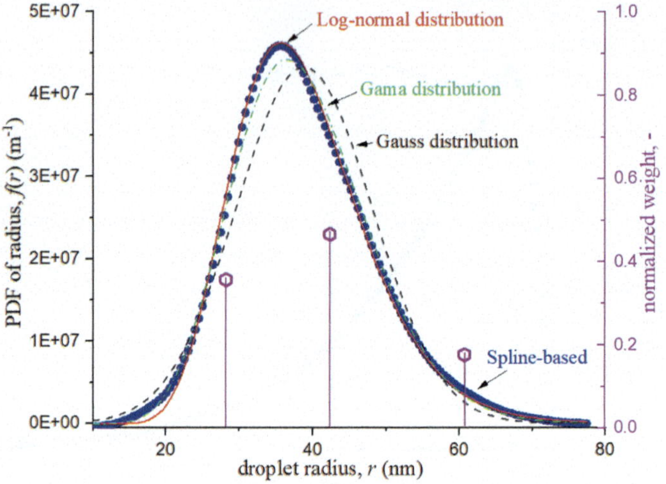

(c)

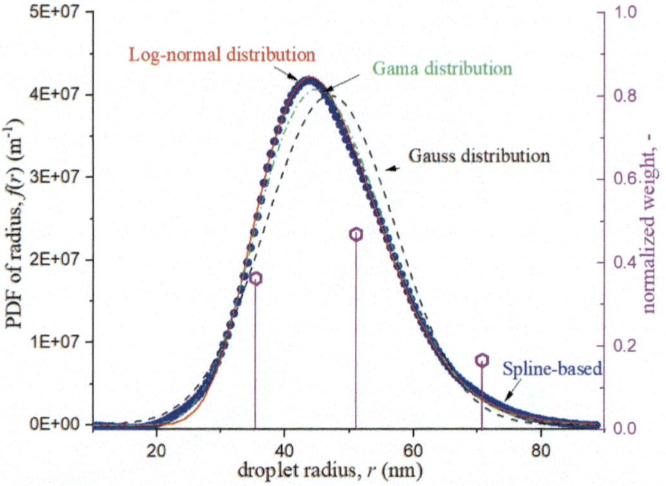

(d)

Fig. 5.8 (continued)

Similarly, the shape of the probability density function $f(r)$ of droplet size at point I in Fig. 5.8b closely resembles that at point C in Fig. 5.8a, indicating that the actual probability distribution in the nucleation zone is more akin to a gamma distribution.

As droplets transition into the growth zone, the droplet size distribution undergoes gradual changes. Figure 5.8c–d presents the reconstructions of the probability density function $f(r)$ using the spline-based algorithm at points L and O. These reconstructions indicate that the spline distribution in the growth zone closely approximates a log–normal distribution.

5.5.2 *Evolution of Droplet Spectrum with Aerodynamic Shock*

Droplet evaporation occurs either in the nozzle or turbine when the droplets encounter a superheated region or pass through an aerodynamic shock wave. Various back pressure ratios were then tested to assess how shock waves impact the QMOM model's ability to predict droplet size distribution. Figure 5.9 displays the static pressure profiles along the nozzle centerline for both subsonic and supersonic outflows in dry and polydisperse wet scenarios. Figure 5.10 presents the corresponding data for the Sauter radius and wetness fraction of the polydisperse model under wet conditions.

In scenarios featuring supersonic outflow, no aerodynamic shock waves are present in the flow field. Instead, a condensation pseudo-shock occurs at point $x = 75.32$ mm (known as the Wilson point), exhibiting a maximum supercooling degree of 35.05 K. At this point, the pressure increases from 0.37 to 0.43. Following the peak nucleation, the supercooling degree rapidly decreases to 2 K, as depicted in Fig. 5.10. The figure also illustrates a sharp rise in wetness fraction near the pseudo-shock, followed by a continuous increase along the flow direction.

In cases of subsonic outflow with aerodynamic shocks, outlet back pressures of $p_{\mathrm{out}} = 12$, 14, and 16 kPa were specified. As seen in Fig. 5.9, as the back pressure increases, the aerodynamic shock moves closer to the nozzle throat. Additionally, it is observed that in wet conditions, the shock location lags behind that of

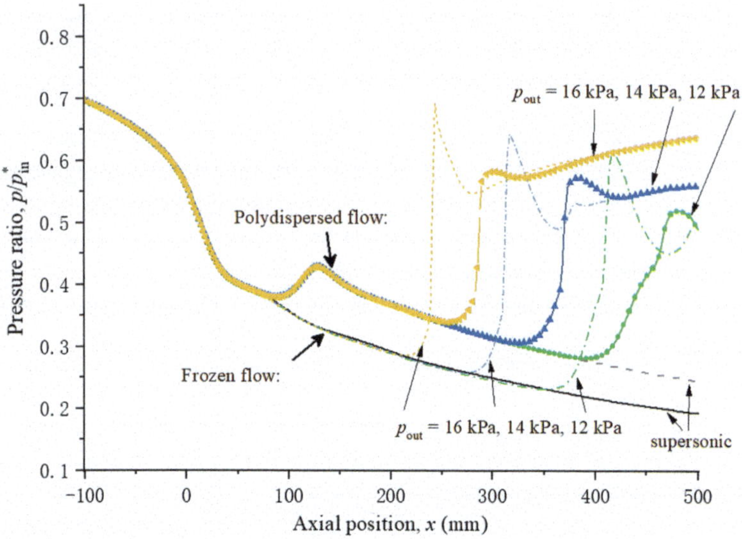

Fig. 5.9 Pressure ratio variation of polydisperse model utilizing QMOM technique at the centerline Moore nozzle B with subsonic and supersonic outlet in both dry and wet conditions

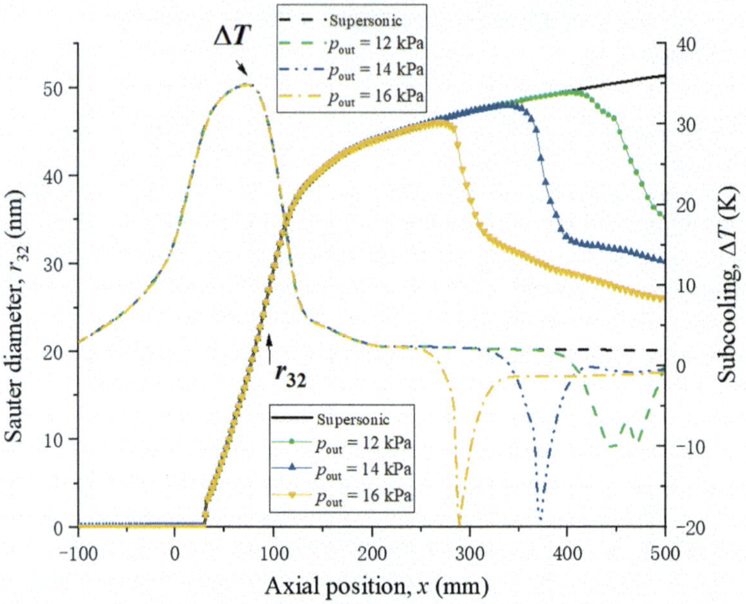

Fig. 5.10 Sauter radius and wetness fraction for the polydispersed model using the QMOM method at the central axis of Moore nozzle B under different back pressure conditions

dry conditions under similar circumstances. Beyond the aerodynamic shock, the supercooling degree abruptly turns negative, indicating the vapor is superheated, as shown in Fig. 5.10. Consequently, droplets will rapidly evaporate in these superheated conditions.

Furthermore, downstream of the aerodynamic shock, a cap-shock pattern is observed with a slight pressure dip, consistent with Papamoschou's experiment by [25]. The shock compression results in a pressure jump followed by a rapid expansion wave, after which the pressure stabilizes on a small plateau. This phenomenon is well-explained by the structure of the shock itself.

Figure 5.11 illustrates the Mach number contours and shock separation in the nozzle under pout = 14 kPa (subsonic outflow). It shows an asymmetric lambda shock and separated flow within a symmetric nozzle configuration. The lambda shape includes the incident shock C1, reflected shock C2, triple point T1, and Mach stem [7]. In the nozzle, two types of shock-induced separations can occur: free shock separation (FSS) and restricted shock separation (RSS). The presence of lambda shocks and separations significantly influences the evaporation rate and wetness fraction.

Figure 5.12 illustrates the reconstructions of the probability density function $f(r)$ of droplet size at the nozzle outlet point O (500, 0) under various back pressure values. As previously discussed, the reconstructed spline shape at the outlet with supersonic outflow closely resembles a log–normal distribution. However, the reconstructed

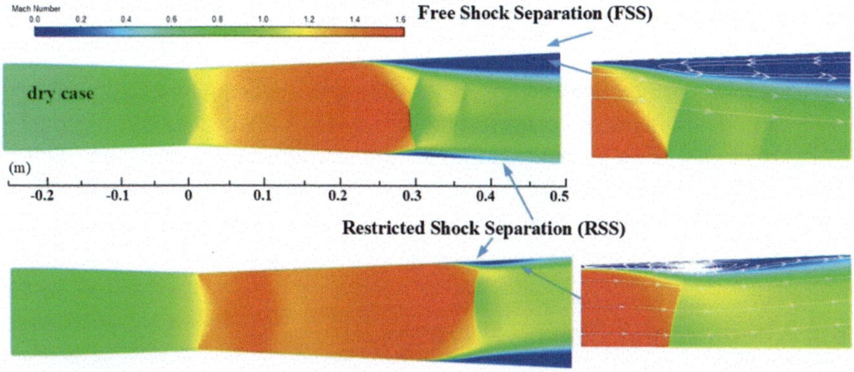

Fig. 5.11 Contours depicting Mach number and shockwave separation in Moore nozzle B under $p_{out} = 14$ kPa (subsonic outflow): compared dry flow with wet flow

shape shifts toward an approximate gamma distribution due to evaporation effects when the back pressures are 12 and 16 kPa.

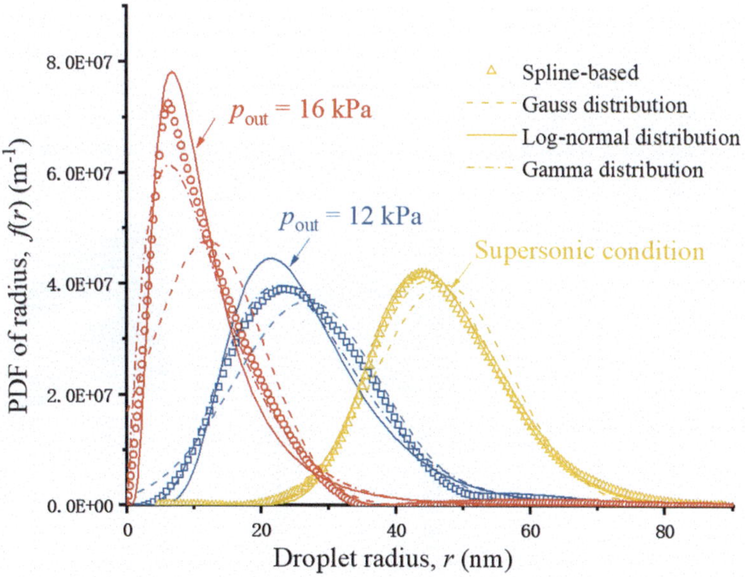

Fig. 5.12 Reconstructed $f(r)$ using a spline-based algorithm and predefined priori functional forms at the outlet point O (500, 0) of the nozzle for different back pressure levels

5.6 Polydisperse Droplet Spectrum of Turbine Flow

This section aims to explore the characteristics and implications of polydisperse droplet spectra within the context of turbine flow systems. By examining how droplet size distributions affect turbine performance and exergy efficiency, this section seeks to provide insights into the complexities of droplet behavior in turbulent flow environments.

5.6.1 Wetness and Droplet Spectrum

The proposed polydisperse model employing the QMOM approach will be used to analyze droplet distribution, exergy loss, and erosion rate in a linear turbine blade. Initially, contours of Mach number, mass transfer rate, and wetness fraction for the polydisperse model were plotted in Figs. 5.13 and 5.14. The inlet conditions are $p_{\text{in}}^{*} = 172$ kPa and $T_{\text{in}}^{*} = 380.66$ K, with a subcooling degree of 8 K, placing the Wilson point in the subsonic region without a condensation pseudo-shock in the nucleation region.

In Fig. 5.13a, a faint oblique shock wave remains supersonic beyond the shock wave near the trailing edge when the back pressure ratio $p_{\text{out}}/p_{\text{in}}^{*} = 0.30$. The pressure side shock wave S_{P} occurs at Ma = 1.43, and the suction side shock wave S_{S} appears at Ma = 1.50. In Fig. 5.13b, c, an evaporation zone with of mass transfer rate m_{v} is equal to − 1850 kg m^{-3} s^{-1} and a secondary condensation zone emerges

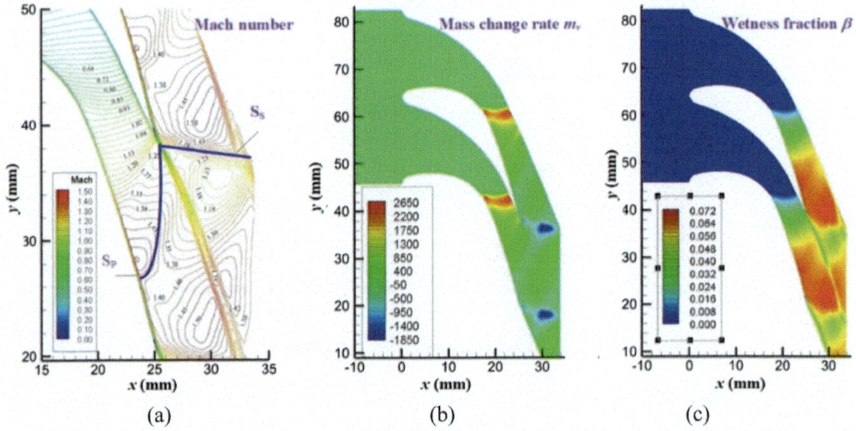

Fig. 5.13 Contours of flow parameters in two-phase flows for polydisperse model using QMOM method on turbine blade at back pressure ratio $p_{\text{out}}/p_{\text{in}}^{*} = 0.30$: **a** Mach number; **b** mass change rate; **c** wetness fraction

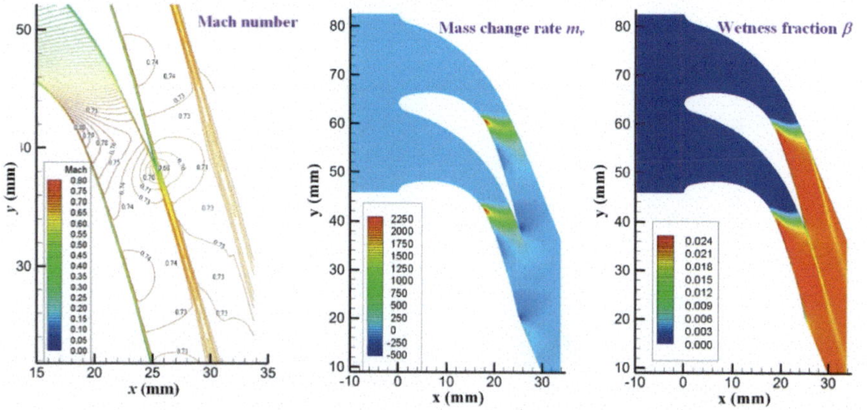

(a) (b) (c)

Fig. 5.14 Contours of flow parameters in two-phase flows for polydisperse model using QMOM method on turbine blade at back pressure ratio $p_{\text{out}}/p_{\text{in}}^* = 0.70$: **a** Mach number; **b** mass change rate; **c** wetness fraction

downstream of the oblique shock. Consequently, four circular isolines are visible along the blade passage, as depicted in Fig. 5.13a. Moreover, it is evident that the wake and flow deflection angle are influenced by the interaction of a reflected shock with the boundary layer.

At a back pressure ratio of $p_{\text{out}}/p_{\text{in}}^* = 0.50$, the flow field is subsonic, as depicted in Fig. 5.14a. The shock wave is absent at the trailing edge. Evaporation is minimal, with the maximum evaporation rate reaching only 500 kg m^{-3} s^{-1} in Fig. 5.14b. Figure 5.14c shows that the wetness fraction remains relatively constant at approximately 0.024, indicating that the weak evaporation and secondary condensation effects are negligible.

Subsequently, the reconstructed probability density function $f(r)$ of droplet size using a spline-based algorithm on the turbine blade was plotted and analyzed. Comparisons between the reconstructed $f(r)$ using the spline-based algorithm and predefined functional forms were made at points A (20, 44), B (21, 42), and C (30, 20) of the turbine blade for back pressure ratios $p_{\text{out}}/p_{\text{in}}^* = 0.30$ and 0.50, shown in Fig. 5.15a, b. Taking point B as an example, the mean radius r_{10} is 3.64×10^{-8} m and the Sauter radius r_{32} is 4.59×10^{-8} m.

The mode of the spline-based probability density function is equal to 3.20×10^{-8} m with a probability density of 3.21×10^7 m^{-1}, while the mode of the gamma distribution is 3.13×10^{-8} m with a probability density of 3.18×10^7 m^{-1}. The results indicate that the reconstructed splines closely approximate the gamma distribution throughout the turbine blade.

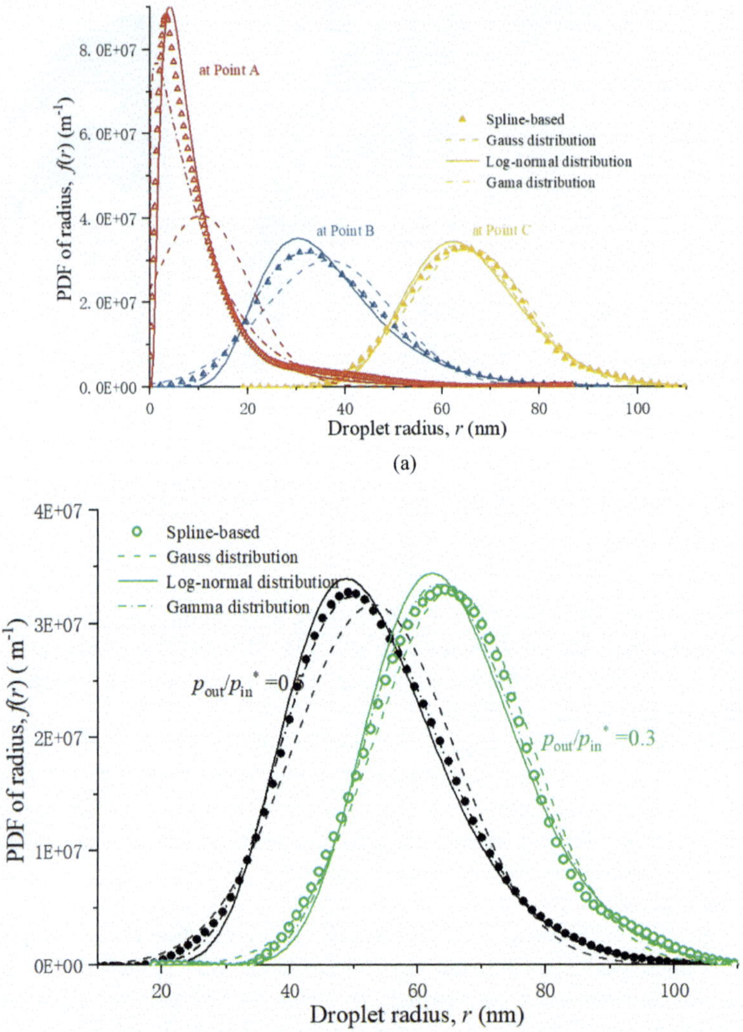

Fig. 5.15 Reconstruction of $f(r)$ using a spline-based algorithm and predefined functional forms based on several moments on the turbine blade: **a** at points A, B, and C under the identical back pressure ratio $p_{out}/p_{in}^* = 0.30$; **b** at point C for the back pressure ratios $p_{out}/p_{in}^* = 0.30$ and 0.50

5.6.2 Exergy Destruction and Erosion Rate

At last, the study utilized a polydisperse model employing the QMOM approach to analyze entropy generation, exergy destruction, Baumann factor, and erosion rate in wet steam turbine blades. The analysis focused on variations in back pressure ratios, wall roughness conditions, and phase change effects.

Figure 5.16 illustrates radar charts detailing the multi-component entropy generation across different back pressures for both smooth and rough blade surfaces. It highlights that entropy generation due to condensation $s_{gen,L}$ and aerodynamic shocks $s_{gen,A}$ significantly dominates in both smooth and rough blade configurations. For instance, in the smooth blade scenario, $s_{gen,L}$ increases from 5.886 to 19.775 J kg^{-1} K^{-1} as the back pressure ratio p_{out}/p^*_{in} ranges from 0.30 to 0.70. The introduction of roughness affects boundary layer characteristics, turbulence viscosity, and thermal diffusivity, influencing these entropy components.

Table 5.6 presents mass-weighted average entropy generation data for dry and wet flows, emphasizing the impact of phase change on entropy production. Notably, entropy generation s_{gen} in the smooth blade increases from 16.428 J kg^{-1} K^{-1} for dry flow to 64.638 J kg^{-1} K^{-1} for wet flow at $p_{out}/p^*_{in} = 0.30$. Comparison between polydisperse and monodispersed models reveals that the QMOM method generally predicts an entropy generation about 8% higher than the Mono method.

Additionally, specific exergy destruction values e_D are detailed in Table 5.6, with a maximum e_D of 25.293 kJ kg^{-1} observed under certain conditions. For LP steam turbines, this translates to a maximum exergy destruction rate e_D of 3.65 MW, considering a specific flow rate of $q_m = 144.5$ kg s^{-1}.

Figure 5.17a displays the mass-weighted averages of exergy destruction ratio ζ_D, which indicates the efficiency losses influenced by back pressure, phase change effects, and wall roughness. The QMOM model shows higher ζ_D values compared to the Mono model, with ζ_D increasing as back pressure decreases. In the smooth blade scenario, ζ_D ranges from 1.04 to 3.39% across different back pressures, reaching a maximum of 4.45% at $p_{out}/p^*_{in} = 0.30$ under rough wall conditions.

Figure 5.17b presents the calculated Baumann factor a_B for wetness loss correlation, ranging from 0.574 to 1.312. The average a_B is 0.728 for the QMOM method and 0.644 for the Mono method in the smooth blade configuration, demonstrating good agreement with empirical data.

Finally, Fig. 5.17c shows the normalized erosion rate based on droplet size and wetness fraction. It indicates that the erosion rate predicted by the polydisperse QMOM model is 58.4–64.4% lower than that predicted by the monodispersed model. This discrepancy is attributed to the more accurate prediction of Sauter radius by the QMOM approach, underscoring its effectiveness in predicting droplet spectrum, energy loss, and erosion rate in wet steam turbines.

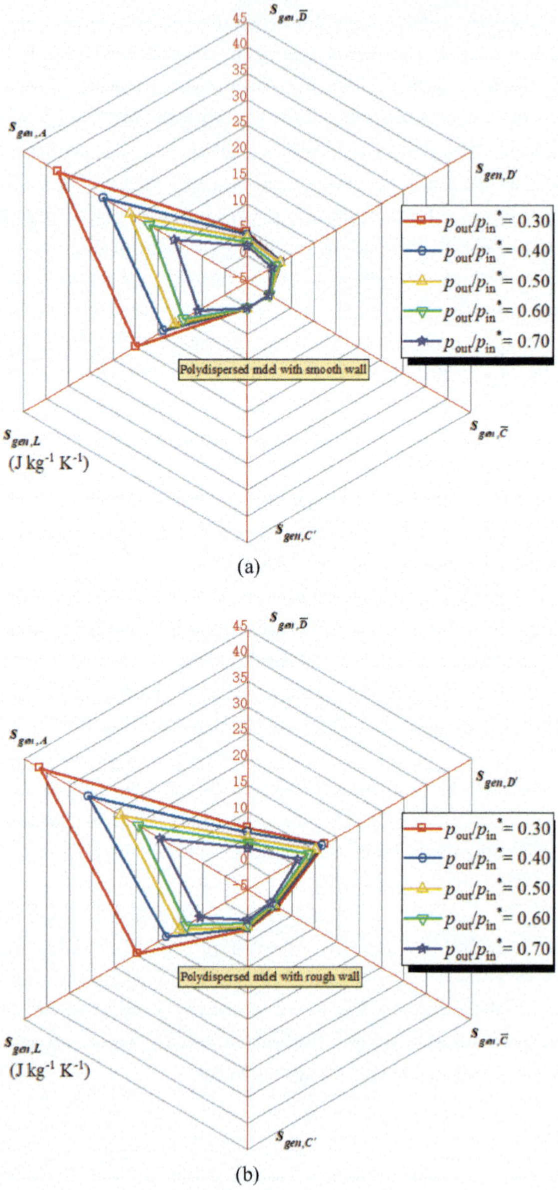

Fig. 5.16 Radar diagrams illustrating multi-component entropy generation in polydisperse model simulations of wet steam flow, comparing smooth and rough walls across various back pressure conditions: **a** with smooth wall; **b** with rough wall

Table 5.6 Entropy generations (J kg^{-1} K^{-1}) at the outlet based on the polydisperse, monodispersed, and dry models with different back pressure ratio p_{out}/p^*_{in}

p_{out}/p^*_{in}	Models	$s_{gen,\overline{D}}$	$s_{gen,D'}$	$s_{gen,\overline{C}}$	$s_{gen,C'}$	$s_{gen,L}$	$s_{gen,A}$	s_{gen}	e_D (kJ/kg)
0.3	QMOM, rough	6.930	12.382	1.800	2.605	19.643	41.517	84.877	25.293
	QMOM	4.619	2.699	0.165	0.063	19.775	37.317	64.638	19.262
	Mono	4.663	2.605	0.170	0.092	22.723	29.641	59.895	17.849
	Dry flow	6.525	3.131	0.384	0.909	–	5.480	16.428	4.896
0.4	QMOM, rough	5.947	11.782	1.355	2.454	13.153	30.755	65.446	19.503
	QMOM	4.139	2.692	0.127	0.050	13.737	27.186	47.932	14.284
	Mono	4.527	2.820	0.130	0.063	15.508	21.857	44.906	13.382
	Dry flow	4.486	4.535	0.265	0.791	–	4.921	14.998	4.469
0.5	QMOM, rough	4.872	10.194	0.976	2.080	10.236	24.704	53.062	15.812
	QMOM	3.456	2.569	0.083	0.036	11.106	21.151	38.402	11.444
	Mono	3.706	2.443	0.088	0.023	11.940	17.261	35.459	10.567
	Dry flow	4.073	5.484	0.191	0.796	–	3.157	13.701	4.083
0.6	QMOM, rough	3.913	8.596	0.654	1.474	8.712	19.619	42.969	12.805
	QMOM	2.677	1.436	0.059	0.006	9.299	16.938	30.415	9.064
	Mono	3.050	1.455	0.075	0.009	9.303	14.247	28.138	8.385
	Dry flow	3.494	5.662	0.144	0.627	–	1.815	11.742	3.499
0.7	QMOM, rough	2.944	6.257	0.328	0.819	5.570	14.570	30.489	9.086
	QMOM	1.934	0.705	0.038	0.006	5.886	11.326	19.894	5.928
	Mono	2.139	0.744	0.039	0.007	5.908	9.725	18.562	5.532
	Dry flow	2.662	2.709	0.080	0.232	–	0.355	6.038	1.799

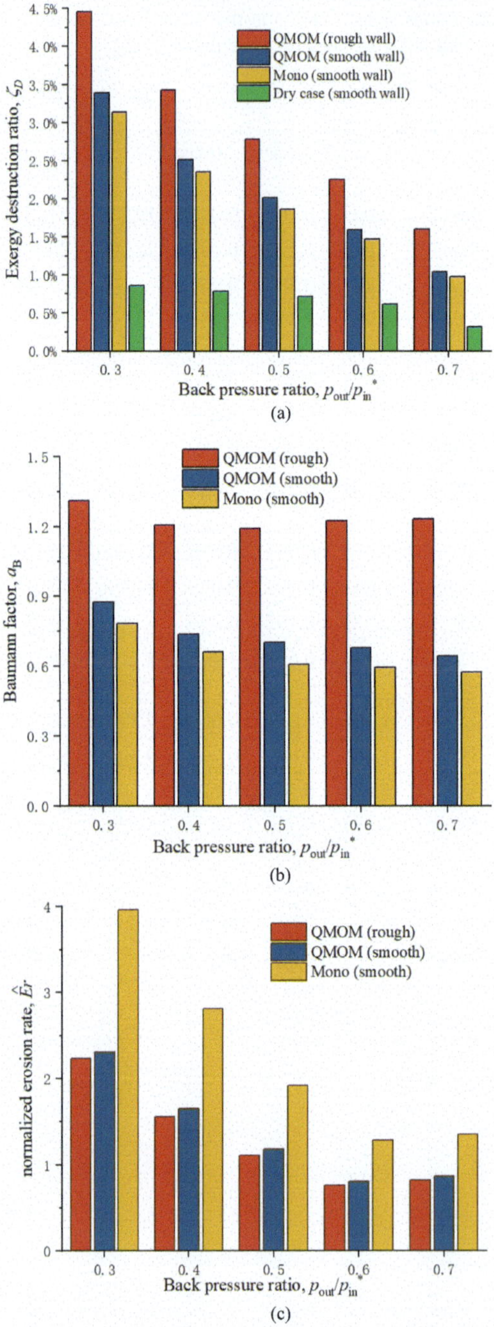

Fig. 5.17 Mass-weighted mean values of exergy destruction efficiency, Baumann factor, and normalized erosion rate for both wet and dry conditions across various back pressure and wall roughness settings. **a** exergy destruction ratio ζ_D; **b** Baumann factor a_B; **c** normalized erosion rate $\hat{E}r$

References

1. Abadi SNR, Ahmadpour A, Abadi SMNR, Meyer JP (2017) CFD-based shape optimization of steam turbine blade cascade in transonic two phase flows. Appl Therm Eng 112:1575–1589
2. Afzalifar A, Turunen-Saaresti T, Grönman A (2018) Comparison of moment-based methods for representing droplet size distributions in supersonic nucleating flows of steam. J Fluids Eng 140(2):021301
3. Aliabadi MAF, Lakzian E, Jahangiri A, Khazaei I (2020) Numerical investigation of effects polydispersed droplets on the erosion rate and condensation loss in the wet steam flow in the turbine blade cascade. Appl Therm Eng 164:114478
4. Bakhtar F, Mahpeykar MR, Abbas KK (1995) An investigation of nucleating flows of steam in a cascade of turbine blading-theoretical treatment. Trans ASME J Fluids Eng. 117(1):138–144
5. De Souza LGM, Janiga G, John V, Thévenin D (2010) Reconstruction of a distribution from a finite number of moments with an adaptive spline-based algorithm. Chem Eng Sci 65(9):2741–2750
6. Diemer RB, Olson JH (2002) A moment methodology for coagulation and breakage problems: part 2-moment models and distribution reconstruction. Chem Eng Sci 57(12):2211–2228
7. Ding H, Wang C, Wang G (2017) Transient conjugate heat transfer in critical flow nozzles. Int J Heat Mass Transf 104:930–942
8. Dykas S, Wróblewski W (2011) Single- and two-fluid models for steam condensing flow modeling. Int J Multiph Flow 37(9):1245–1253
9. Gerber AG, Kermani MJ (2004) A pressure based Eulerian–Eulerian multi-phase model for non-equilibrium condensation in transonic steam flow. Int J Heat Mass Transf 47(10–11):2217–2231
10. Gordon RG (1968) Error bounds in equilibrium statistical mechanics. J Math Phys 9(5):655–663
11. Hughes FR, Starzmann J, White AJ, Young JB (2016) A comparison of modeling techniques for polydispersed droplet spectra in steam turbines. J Eng Gas Turb Power 138(4):042603
12. Hulburt HM, Katz S (1964) Some problems in particle technology: a statistical mechanical formulation. Chem Eng Sci 19(8):555–574
13. Jasim NM (2023) Probabilistic derivation of droplet velocity using quadrature method of moments. Open Eng 13(1):20220407
14. John V, Angelov I, Öncül AA, Thévenin D (2007) Techniques for the reconstruction of a distribution from a finite number of its moments. Chem Eng Sci 62(11):2890–2904
15. Lee BE, Riu KJ, Shin SH, Kwon SB (2003) Development of a water droplet erosion model for large steam turbine blades. KSME Int J 17:114–121
16. Lehnigk R, Bainbridge W, Liao Y, Lucas D, Niemi T, Peltola J, Schlegel F (2022) An open-source population balance modeling framework for the simulation of polydisperse multiphase flows. AIChE J 68(3):e17539
17. McGraw R (1997) Description of aerosol dynamics by the quadrature method of moments. Aerosol Sci Technol 27(2):255–265
18. Mirhoseini MS, Boroomand M (2017) Multi-objective optimization of hot steam injection variables to control wetness parameters of steam flow within nozzles. Energy 141:1027–1037
19. Moore MJ (1973) Predicting the fog-drop size in wet-steam turbines. Wet Stream 37: 101–109
20. Starzmann J, Casey MM, Mayer JF, Sieverding F (2014) Wetness loss prediction for a low pressure steam turbine using computational fluid dynamics. Proc Inst Mech Eng Part A J Power Energy 228(2):216–231
21. White AJ (2003) A comparison of modelling methods for polydispersed wet steam flow. Int J Numer Meth Eng 57(6):819–834
22. White AJ, Hounslow MJ (2000) Modelling droplet size distributions in polydispersed wet-steam flows. Int J Heat Mass Transf 43(11):1873–1884
23. White AJ, Young JB (1993) Time-marching method for the prediction of two-dimensional, unsteady flows of condensing steam. J Propul Power 9(4):579–587

24. Wongsarivej P, Tanthapanichakoon W (2015) A model for a non-uniform spray evaporator taking into account the effect of non-isothermal polydisperse droplets. Int J Heat Mass Transf 90:1170–1177
25. Xiao Q, Tsai HM, Papamoschou D (2007) Numerical investigation of supersonic nozzle flow separation. AIAA J 45(3):532–541
26. Xu P (1998) Truncated SVD methods for discrete linear ill-posed problems. Geophys J Int 135(2):505–514

Chapter 6
Three-Field Two-Fluid Model with Homogeneous and Heterogeneous Condensation

6.1 Introduction

The two primary processes in supersonic separators are condensation and swirling flow separation [29]. These processes are facilitated by specific components, namely the supersonic nozzle, which induces condensation, and the vortex generator, which creates the swirling flow. Supersonic separators exhibit complex characteristics involving three fields (gas, liquid droplets, and liquid film) and two fluids (gas and liquid) [15]. These characteristics encompass intricate phenomena such as phase changes (condensation and evaporation), interphase slip, liquid film dynamics, swirling flows, and shockwaves [5]. Although many researchers have performed extensive flow field analyses and structural optimizations of supersonic separators, numerous challenges still require comprehensive investigation. This is especially true for understanding the behavior of liquid droplets and liquid films within the supersonic separators. The interactions and dynamics of these components are critical to improving the efficiency and effectiveness of the separation processes, necessitating further detailed studies to address the unresolved issues.

There are three primary approaches to studying the functioning and efficiency of supersonic separators: theoretical analysis, experimental research, and computational simulation. Conducting experiments using natural gas can be both prohibitively expensive and potentially hazardous. Therefore, air is commonly used as a substitute feed carrier gas in low-pressure experimental setups instead of natural gas [27, 28]. This substitution generally has a negligible impact on the reliability and validity of the experimental results. For example, the experimental findings by Ma et al. [18] demonstrated that the injection of foreign droplets induced heterogeneous condensation, which effectively enhanced the separation efficiency. Additionally, Cao and Yang [2] conducted an experimental study on a novel supersonic separator featuring an ellipsoidal center body to evaluate its dehydration performance. Their results indicated that the maximum dew point depression could be as high as 34.9 °C with a pressure recovery coefficient of 20.6%. Furthermore, Wang et al. [27] carried

H. Ding et al., *Supersonic Separators for Sustainable Gas Purification*, Green Energy and Technology, https://doi.org/10.1007/978-3-031-85213-8_6

out experimental research on an innovative supersonic separator incorporating a reflux channel. The results showed that the cylindrical drainage structure could significantly reduce the interaction between shock waves and the boundary layer, thereby improving the overall separation performance. In another study, de Oliveira Arinelli et al. [3, 4] conducted process simulations using Aspen HYSYS to investigate offshore natural gas processing with high CO_2 content based on supersonic separators. Simultaneously, computational fluid dynamics (CFD) simulations are employed to model the distribution characteristics of the flow field by establishing appropriate numerical models. Consequently, numerous scholars have developed a variety of numerical models specifically tailored for supersonic separators, aiming to simulate and optimize their performance with greater accuracy and efficiency.

The first commonly used model for analyzing supersonic separators is a two-fluid model based on the Dispersed Particle Method (DPM), which is also referred to as the Eulerian–Lagrangian approach. This method is capable of tracking the trajectories of individual droplets within the flow. Wen et al. [30] developed a detailed numerical model by combining the RNG k–ε turbulence model with the Dispersed Particle Method (DPM) to accurately calculate both the flow field and the trajectories of droplets within a supersonic separator. Their comprehensive results indicated that when the separation section length is ten times the diameter of the throat, the separation efficiency can achieve a remarkable 95%. Furthermore, Yang and Wen [34] conducted an in-depth study on the behavior of particles within a supersonic separator characterized by strong swirling flow. Their findings demonstrated that the majority of the particles were effectively separated by centrifugal forces, either to the wall or directly into the moisture outlet, achieving a separation efficiency of 80% for droplets with a size of 1.5 μm. In another significant study, Liu et al. [16] applied the compressible Navier–Stokes equations coupled with the Dispersed Particle Method (DPM) to predict the separation efficiency. Their results showed that increasing the droplet size within a certain range significantly improved the separation efficiency. Jiang et al. [12] conducted an extensive analysis of the effect of droplet size and inlet velocity on separation efficiency. They discovered that droplets of approximately 1 μm in size were effectively separated under the given conditions. Moreover, Shooshtari and Shahsavand [23] developed a numerical model that accounted for the actual distribution of droplet sizes to simulate a more realistic trajectory of condensation droplets. However, it is important to note that the existing DPM-based two-fluid models have significant limitations. These models do not consider the processes of condensation and evaporation of droplets within the supersonic separator, thereby limiting their overall accuracy and applicability. This gap highlights the need for further refinement and development of models to include these critical processes for more accurate predictions and optimizations.

Another widely utilized model for analyzing supersonic separators is the Eulerian two-fluid model, often referred to as the Eulerian–Eulerian approach. This sophisticated method incorporates the crucial processes of condensation and evaporation, providing a comprehensive framework for studying these phenomena. Dykas and Wróblewski [8] employed the Eulerian–Eulerian approach to simulate homogeneous condensing flow in both low- and high-pressure supersonic nozzles. Their study meticulously examined the performance of the numerical algorithm in relation to various gas and liquid properties, offering valuable insights into the dynamics of such systems. Patel et al. [22] combined the Eulerian–Eulerian model with a modified turbulence model to investigate the irreversible heat and mass transfer losses occurring during the condensation process within nozzles and turbine cascades. This combination allowed for a detailed exploration of the complex interactions between the different phases. Additionally, White and Hounslow [33] and Ding et al. [6] conducted numerical simulations to determine the distribution of polydisperse droplets in condensing flows, further enhancing the understanding of these intricate processes. Edathol et al. [10] used Eulerian–Eulerian models to predict the behavior of homogeneous condensing flow within a supersonic nozzle. Their findings underscored the need for more experimental validations under varying geometric and boundary conditions to ensure the effectiveness of the non-equilibrium condensation model. Abadi et al. [1] developed a Eulerian–Eulerian model to analyze wet steam flow in high-pressure nozzles. The results indicated that a decreased superheat level initiates the nucleation process earlier, providing key insights into the operational dynamics of such systems. Wen et al. [31, 32] advanced the field by developing a Eulerian two-fluid model to study the complex flow field structure where non-equilibrium condensation and shock coexist in a supersonic separator. They conducted an in-depth analysis of the impact of swirl strength on the condensation process, contributing significantly to the understanding of these interactions. Liu et al. [17] focused on the influence of the pressure recovery process on droplet condensation and evaporation, revealing that the interaction between shock waves and boundary layers reduces liquefaction efficiency. Under actual operational conditions, saturated natural gas typically contains foreign liquid droplets before entering the dehydration equipment, leading to heterogeneous condensation within the supersonic separator. To address this, Niknam et al. [19] conducted heterogeneous condensation experiments to assess the dehydration efficiency of a supersonic separator. They established a two-phase heat and mass transfer model in a 2D asymmetric domain to predict dehydration efficiency, providing a more accurate representation of real-world conditions. Shooshtari and Shahsavand [24] hypothesized specific rates of salt particle injection into the supersonic separator. This led to the development of a Eulerian multi-fluid model that considers both homogeneous and heterogeneous condensation, demonstrating that the particle injection rate significantly impacts separation efficiency. Despite these advancements, it is noteworthy that the aforementioned Eulerian two-fluid models generally do not account for interphase slip and droplet drag. Ding et al. [5] developed a homogeneous/heterogeneous condensation model that includes interphase slip and droplet drag, validated for high computational accuracy and used to elucidate droplet behavior characteristics. Nevertheless, these numerical models only represent two fields (gas and droplets), thereby neglecting the formation of liquid films. Recently, no researcher has developed a comprehensive three-field (gas, droplets, and liquid film) two-fluid model that includes the liquid film within the context of a supersonic separator. This highlights a significant gap in the current research landscape and underscores the need for further development to achieve a more holistic understanding of these complex systems.

As previously noted, the current computational models for supersonic separators lack consideration for the formation of a liquid film. Furthermore, the existing three-field two-fluid model neglects the condensation effect, rendering it unsuited for direct application in supersonic separators. To explore the intricate mechanisms of heat and mass transfer among the gas, liquid droplets, and liquid film within these separators, this chapter introduces an innovative three-field two-fluid model based on the Eulerian–Eulerian approach. This novel model incorporates both homogeneous and heterogeneous condensation phenomena along with interphase slip, aspects not previously addressed in the literature. Subsequently, the study investigates the coupled heat and mass transfer processes and predicts the separation efficiency. The proposed model offers a robust framework for analyzing three-field heat and mass transfer dynamics, encompassing condensation effects and strong swirl characteristics, thereby optimizing the performance of supersonic separators.

6.2 Three-Field Two-Fluid Model

The Euler–Euler three-field two-fluid model for a supersonic separator encompasses the governing equations for the gas phase, a model for homogeneous and heterogeneous condensation, an Eulerian wall film model, and the coupling of heat and mass transfer between the gas, liquid droplets, and liquid film phases. Figure 6.1 illustrates the schematic representation of the three-field two-fluid system within the supersonic separator. This model is developed based on several key assumptions: (i) the gas studied is moist gas, as a substitute for moist natural gas in the research; (ii) the droplets are assumed to be spherical, and the coalescence and collision between droplets are taken into account; (iii) both homogeneous and heterogeneous condensation is considered, and it is assumed that the latent heat released by condensation is entirely absorbed by the gas phase; (iv) both homogeneous and heterogeneous condensation processes, as well as interphase slip, are taken into account; (v) due to the strong swirling flow present in the system, the drag force's impact on the two-phase momentum equations is considered; (vi) slip between gas and liquid phases, phase change between gas phase and liquid film, and deposition and entrainment between liquid droplets and liquid film are considered.

6.2.1 Gas Phase Model

The conservation equations for mass, momentum, and energy of the gas phase, which is a mixture of air and vapor, are governed by

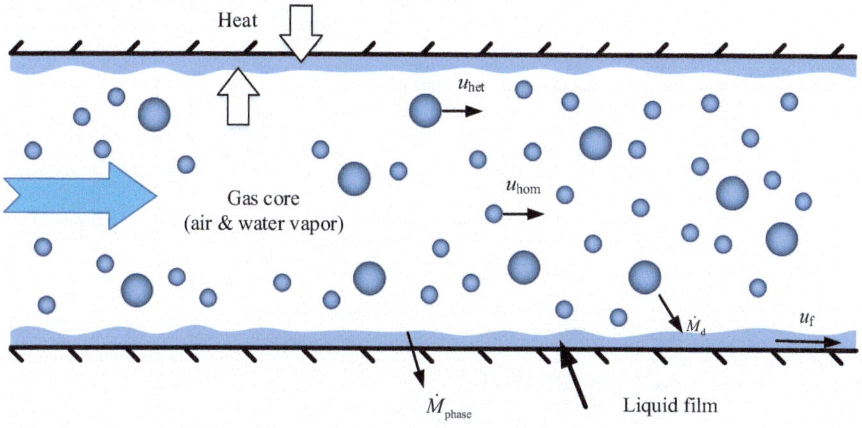

Fig. 6.1 Three-field two-fluid supersonic separator modeling diagram

$$\frac{\partial}{\partial t}(\alpha_g \rho_g) + \nabla \cdot (\alpha_g \rho_g u_g) = -(\dot{m}_{hom} + \dot{m}_{het}) \tag{6.1}$$

$$\frac{\partial}{\partial t}(\alpha_g \rho_g u_g) + \nabla \cdot (\alpha_g \rho_g u_g u_g) = -\alpha_g \nabla p + \nabla \cdot (\alpha_g \tau_{eff}) + \alpha_g \rho_g \vec{g} - (\dot{m}_{hom} + \dot{m}_{het}) u_g - \vec{F}_D \tag{6.2}$$

$$\frac{\partial}{\partial t}(\alpha_g \rho_g E) + \nabla \cdot [\alpha_g (\rho_g E + p) u_g] = \nabla \cdot \left[\alpha_g \lambda_{eff} \nabla T_g - \alpha_g \sum_i h_i \vec{J}_i + \alpha_g (\tau_{eff} \cdot u_g) \right] + (\dot{m}_{hom} + \dot{m}_{het}) h_{lg} \tag{6.3}$$

where the subscript 'g', 'hom', 'het', and 'i' represent gas phase, homogeneous nucleus, heterogeneous nucleus, and species, respectively. α_g is the volume fraction of gas phase, $\alpha_g = 1 - \alpha_{het} - \alpha_{hom}$, u and $\vec{g}$ is the velocity and gravity vectors. $\vec{J_i}$ is the diffusion flux of species i. The ρ, p, T, E, h_i, and h_{lg} denote the density, pressure, temperature, total energy, enthalpy of species i, and latent heat, respectively. The drag force is $\vec{F}_D = \vec{F}_{D,hom} + \vec{F}_{D,het}$. The $\dot{m}_{hom}$ and $\dot{m}_{het}$ (kg m^{-3} s^{-1}) are droplet mass changing rates for homogeneous and heterogenous condensation, respectively. λ_{eff} is the effective thermal conductivity, $\lambda_{eff} = \lambda + \lambda_{tu}$, where λ_{tu} represents turbulent thermal conductivity. The effective stress tensor is described by

$$\tau_{eff} = \mu_{eff} \left[\left(\nabla u_g + \nabla u_g^T \right) - \frac{2}{3} \nabla \cdot u_g I_{te} \right] \tag{6.4}$$

where I_{te} is the unit tensor, μ_{eff} is the effective molecular viscosity, $\mu_{eff} = \mu + \mu_{tu}$. The turbulent viscosity μ_{tu} and turbulent thermal conductivity λ_{tu} is calculated by the turbulence model. For simulating swirling flow, the Reynolds stress model is applied to calculate the turbulence flow.

The species transport equation of the gas phase is

$$\frac{\partial}{\partial t}(\alpha_g \rho_g Y_i) + \nabla \cdot (\alpha_g \rho_g u_g Y_i) = -\nabla \cdot \left(\alpha_g \vec{J}_i\right) + S_{\mathrm{ma,g}} \tag{6.5}$$

where Y_i represents the mass fraction of species i. $S_{\mathrm{ma,g}}$ is the mass source term where the value is $-(\dot{m}_{\mathrm{hom}} + \dot{m}_{\mathrm{het}})$ for the species of vapor.

6.2.2 *Dispersed Droplet Model*

The volumetric concentration of dispersed homogeneous and heterogeneous droplet numbers, N_{hom} and N_{het} (m^{-3}) are expressed by

$$\frac{\partial N_{\mathrm{hom}}}{\partial t} + \nabla \cdot (n_{\mathrm{hom}} u_{\mathrm{hom}}) = J - \dot{N}_{\mathrm{col}} \tag{6.6}$$

$$\frac{\partial N_{\mathrm{het}}}{\partial t} + \nabla \cdot (n_{\mathrm{het}} u_{\mathrm{het}}) = 0 \tag{6.7}$$

where J (m^{-3} s^{-1}) is homogeneous nucleation rate. $\dot{N}_{\mathrm{col}}$ is the source term due to collision.

The volume fractions of homogeneous and heterogeneous droplets, α_{hom} and α_{het} are expressed as

$$\frac{\partial}{\partial t}(\alpha_{\mathrm{hom}} \rho_l) + \nabla \cdot (\alpha_{\mathrm{hom}} \rho_l u_{\mathrm{hom}}) = \dot{m}_{\mathrm{hom}} - \dot{m}_{\mathrm{col}} \tag{6.8}$$

$$\frac{\partial}{\partial t}(\alpha_{\mathrm{het}} \rho_l) + \nabla \cdot (\alpha_{\mathrm{het}} \rho_l u_{\mathrm{het}}) = \dot{m}_{\mathrm{het}} + \dot{m}_{\mathrm{col}} \tag{6.9}$$

where ρ_l is the droplet density. $\dot{m}_{\mathrm{col}}$ is the mass source term due to collision and coalescence. The effective density of homogeneous and heterogeneous droplets is $\rho_{\mathrm{hom}} = \alpha_{\mathrm{hom}} \rho_l$ and $\rho_{\mathrm{het}} = \alpha_{\mathrm{het}} \rho_l$.

The momentum conservation equations of homogeneous and heterogeneous droplets are expressed by

$$\frac{\partial}{\partial t}(\alpha_{\mathrm{hom}} \rho_l u_{\mathrm{hom}}) + \nabla \cdot (\alpha_{\mathrm{hom}} \rho_l u_{\mathrm{hom}} u_{\mathrm{hom}}) = -\alpha_{\mathrm{hom}} \nabla p + \alpha_{\mathrm{hom}} \rho_l \vec{g} + (\dot{m}_{\mathrm{hom}} - \dot{m}_{\mathrm{col}}) u_{\mathrm{hom}} + \vec{F}_{\mathrm{D,hom}} \tag{6.10}$$

$$\frac{\partial}{\partial t}(\alpha_{\text{het}}\rho_{\text{l}}u_{\text{het}}) + \nabla\cdot(\alpha_{\text{het}}\rho_{\text{l}}u_{\text{het}}u_{\text{het}}) = -\alpha_{\text{het}}\nabla p + \alpha_{\text{het}}\rho_{\text{l}}\vec{g} + (\dot{m}_{\text{het}} + \dot{m}_{\text{col}})u_{\text{het}} + \vec{F}_{\text{D,het}} \tag{6.11}$$

where $\vec{F}_{\text{D,hom}}$ and $\vec{F}_{\text{D,het}}$ are drag force between gas and water droplet and $\vec{F}_{\text{D,hom}} + \vec{F}_{\text{D,het}} = \vec{F}_{\text{D}}$.

6.2.3 Eulerian Wall Film Model

The mass, momentum, and energy conservation equation of the two-dimensional film in three-dimensional domain are as follows

$$\frac{\partial}{\partial t}(\rho_{\text{l}}\delta) + \nabla_{\text{sur}}\cdot(\rho_{\text{l}}\delta u_{\text{f}}) = (\dot{m}_{\text{d}} - \dot{m}_{\text{f}}) \tag{6.12}$$

$$\frac{\partial}{\partial t}(\rho_{\text{l}}\delta u_{\text{f}}) + \nabla_{\text{sur}}\cdot(\rho_{\text{l}}\delta u_{\text{f}}u_{\text{f}}) = -\delta\cdot\nabla_{\text{sur}}p_{\text{l}} + \rho_{\text{l}}\delta\vec{g}_{\tau} + \frac{3}{2}\tau_{\text{gf}} - \frac{3\mu u_{\text{f}}}{\delta} + \vec{F}_{\text{D}} + (\dot{m}_{\text{d}} - \dot{m}_{\text{f}})u_{\text{f}} \tag{6.13}$$

$$\frac{\partial}{\partial t}(\rho_{\text{l}}\delta h_{\text{f}}) + \nabla_{\text{sur}}\cdot(\rho_{\text{l}}\delta h_{\text{f}}u_{\text{f}}) = \frac{\lambda_{\text{l}}}{\delta}(T_{\text{sur}} + T_{\text{w}} - 2T_{\text{m}}) + (\dot{m}_{\text{d}} - \dot{m}_{\text{f}})h_{\text{lg}} \tag{6.14}$$

where ∇_{sur} is the surface gradient. δ is the film thickness. u_{f} is the film velocity vector. $\dot{m}_{\text{d}}$ and $\dot{m}_{\text{f}}$ represent deposition rate, the film phase change rate (kg m^{-2} s^{-1}), respectively. $\vec{g}_{\tau}$ is the gravity component parallel to the film. The third and fourth terms on the right-hand side of Eq. (6.13) represent viscous shear stress on gas-film and film-wall interfaces. h_{f} is the enthalpy of the liquid film, p_{l} is the pressure between the gas and liquid phases, λ_{l} is the thermal conductivity, and μ is the dynamic viscosity. τ_{gf} is the viscous shear stress at the gas–liquid film interface. T_{sur}, T_{w}, and T_{m} are the film surface, wall, and film half depth temperature, respectively.

6.2.4 Interphase Exchange Between Gas and Droplet

The homogeneous/heterogeneous droplet mass changing rates (condensation and evaporation rate) between the gas phase and dispersed droplets are calculated by

$$\dot{m}_{\text{hom}} = J\rho_{\text{l}}\frac{4\pi r_{\text{c}}^3}{3} + n_{\text{hom}}\rho_{\text{l}}4\pi r_{\text{hom}}^3\frac{\text{d}r_{\text{hom}}}{\text{d}t} \tag{6.15}$$

$$\dot{m}_{\mathrm{het}} = n_{\mathrm{het}} \rho_{\mathrm{l}} 4\pi r_{\mathrm{het}}^3 \frac{\mathrm{d}r_{\mathrm{het}}}{\mathrm{d}t} \tag{6.16}$$

The homogeneous nucleation rate J ($\mathrm{m}^{-3}\ \mathrm{s}^{-1}$) is calculated by Kantrowitz [14]

$$J = \frac{\upsilon \rho_{\mathrm{g}}^2}{S_{\mathrm{v}}} \sqrt{\frac{2\sigma}{\pi m_{\mathrm{m}}^5}} \exp\left(-\frac{16\pi}{3} \frac{\upsilon^2 \sigma^3}{(\kappa_{\mathrm{B}} T_{\mathrm{g}})^3 (\ln(S_{\mathrm{v}}))^2}\right) \tag{6.17}$$

where, υ and m_{m} represent volume and mass of single water molecule. k_{B} represents Boltzmann's constant, σ ($\mathrm{N\ m}^{-1}$) is liquid surface tension. S_{v} is the supersaturation of vapor. The critical droplet size of homogeneous nucleation is calculated by

$$r_{\mathrm{c}} = \frac{2\sigma}{\rho_{\mathrm{l}} R_{\mathrm{g}} T_{\mathrm{g}} \ln(S_{\mathrm{v}})} \tag{6.18}$$

where R_{g} represent specific gas constant. The mean radii r_{hom} and r_{het} are expressed as

$$r_{\mathrm{hom}} = \left(\frac{3\alpha_{\mathrm{hom}}}{4\pi N_{\mathrm{hom}}}\right)^{\frac{1}{3}}, \quad r_{\mathrm{het}} = \left(\frac{3\alpha_{\mathrm{het}}}{4\pi N_{\mathrm{het}}}\right)^{\frac{1}{3}} \tag{6.19}$$

The growth rate of droplet in condensation and evaporation processes is calculated by

$$\frac{\mathrm{d}r}{\mathrm{d}t} = \frac{1}{\rho_{\mathrm{l}} h_{\mathrm{lg}}} (T_{\mathrm{r}} - T_{\mathrm{v}}) \sum_{i=1}^{2} \alpha_{\mathrm{k}} \tag{6.20}$$

where the subscript 'v' represents vapor. α_{k} is the heat transfer coefficient between the water droplet and species i [5]. The droplet surface temperature T_{r} is calculated by

$$T_{\mathrm{r}} = T_{\mathrm{d}} - \Delta T \frac{r_{\mathrm{c}}}{r} \tag{6.21}$$

where ΔT is subcooling, T_{d} is the dew point temperature at p_{v}.Generally, the size and slip velocity of heterogeneous droplets (foreign droplet) are significantly greater than those of homogeneous droplets, since the diameter of homogeneous droplets is generally below 0.1 μm [13, 23]. Thus, the heterogeneous droplets as collectors collide with the surrounding smaller size homogeneous droplets. The mean collision frequency f is expressed as [20]

$$f = N_{\mathrm{hom}} \pi (r_{\mathrm{het}} + r_{\mathrm{hom}})^2 \left| u_{\mathrm{het}} - u_{\mathrm{g}} \right| \tag{6.22}$$

Therefore, the source term of Eq. (6.6) is

$$\dot{N}_{\text{col}} = N_{\text{het}} f \tag{6.23}$$

The homogeneous droplets are collected by heterogeneous droplets, whose mass is transferred to the heterogeneous droplets, thereby the heterogeneous droplet radius increases gradually. The mass source term due to coalescence is calculated by

$$\dot{m}_{\text{col}} = N_{\text{het}} f \rho_{\text{l}} \frac{4\pi r_{\text{hom}}^3}{3} \tag{6.24}$$

The drag force between the continuous and dispersed droplets in Eq. (6.2) is defined as [21]

$$\vec{F}_{\text{D,hom}} = \frac{9\mu_{\text{g}}}{2r_{\text{hom}}^2} C_{\text{D}} \frac{\text{Re}_{\text{p,hom}}}{24} \alpha_{\text{hom}} \left(u_{\text{g}} - u_{\text{hom}}\right) \tag{6.25}$$

$$\vec{F}_{\text{D,het}} = \frac{9\mu_{\text{g}}}{2r_{\text{het}}^2} C_{\text{D}} \frac{\text{Re}_{\text{p, het}}}{24} \alpha_{\text{het}} \left(u_{\text{g}} - u_{\text{het}}\right) \tag{6.26}$$

where the relative Reynolds number $\text{Re}_{\text{p,hom}}$ and $\text{Re}_{\text{p,het}}$ are expressed as

$$\text{Re}_{\text{p,hom}} = \frac{2\rho_{\text{g}} r_{\text{hom}} \left|u_{\text{hom}} - u_{\text{g}}\right|}{\mu_{\text{g}}} \tag{6.27}$$

$$\text{Re}_{\text{p,het}} = \frac{2\rho_{\text{g}} r_{\text{het}} \left|u_{\text{het}} - u_{\text{g}}\right|}{\mu_{\text{g}}} \tag{6.28}$$

The drag coefficient CD is expressed as [25]

$$C_{\text{D}} = \frac{24}{\text{Re}_{\text{p}}}, \ \text{Re}_{\text{p}} \leq 1 \tag{6.29}$$

$$C_{\text{D}} = \frac{24}{\text{Re}_{\text{p}}} \left(1 + 0.15\text{Re}_{\text{p}}^{0.687}\right), \ 1 < \text{Re}_{\text{p}} \leq 1000 \tag{6.29a}$$

$$C_{\text{D}} = 0.44, \ \text{Re}_{\text{p}} > 1000 \tag{6.29b}$$

or lower Reynolds numbers, the molecular viscous force becomes predominant. Nonetheless, when the particle diameter is sufficiently small to be comparable to the mean free path length of gas molecules (Knudsen number Kn > 1), the drag coefficient needs to be adjusted using the Cunningham slip correction factor, $C_{\text{D}} = \frac{24}{\text{Re}_{\text{p}} C_{\text{c}}}$. This correction factor is determined by the equation provided in Dykas and Wróblewski [9]

$$C_c = 1 + 2\mathrm{Kn}_g\left(1.257 + 0.4e^{\frac{-1.1}{(2\mathrm{Kn}_g)}}\right) \tag{6.30}$$

6.2.5 Coupling of Gas Core with Liquid Film

The droplets that deposit on the wall will progressively form a liquid film. Consequently, their mass and momentum are transferred from the dispersed phase and incorporated into the liquid film equations as source terms at the boundary of the wall film. The rate of droplet deposition, which represents the mass source term of the liquid film (kg m^{-2} s^{-1}), is described in Yuan et al. [35]

$$\dot{m}_d = \dot{m}_{d,hom} + \dot{m}_{d,het} = (\alpha_{hom}\rho_l u_{hom} + \alpha_{het}\rho_l u_{het}) \cdot \vec{n} \tag{6.31}$$

where $\vec{n}$ is film area normal.

The momentum source term of liquid film is

$$S_{mo,f} = S_{mo,f,hom} + S_{mo,f,het} = \dot{m}_{d,hom} u_{t,hom} + \dot{m}_{d,het} u_{t,het} \tag{6.32}$$

where film tangential velocity vector is $u_t = u - (u \cdot \vec{n})\vec{n}$.

$$\dot{m}_f = \frac{1}{\delta} \frac{\frac{\rho_g M_v}{D}}{\frac{\rho_g M_v}{D} + C_{phase}} C_{phase}(Y_s - Y_v) \tag{6.33}$$

where D represents cell-center-to-wall distance. M_v is the mass diffusivity of vapor. C_{phase} is the phase change constant. Y_s represent saturation mass fraction of vapor. Y_v is the mass fraction of vapor at the cell-center of near-wall cell.

6.3 Experiment and Validation

This section initially provides a comprehensive introduction to the typical structure and operating principles of the supersonic separator. Following this, an experimental platform for supersonic condensation and separation is constructed. Based on this platform, both experimental validation and Computational Fluid Dynamics (CFD) validation are conducted. These validations collectively demonstrate and confirm the accuracy and reliability of the developed model.

6.3.1 Typical Structure of Supersonic Separator

Supersonic separators combine low-temperature condensation and cyclone separation functions. They mainly consist of a swirl generator, a Laval nozzle, a gas–liquid separation section, and a diffuser. The typical structure is shown in Fig. 6.2. The gas–liquid mixture to be separated enters the supersonic separator at a certain temperature and pressure through the inlet, passing through the swirl generator where it is induced into a swirling flow by the stator blades with a certain deflection angle. The gas–liquid mixture then flows through the Laval nozzle, where it is accelerated, depressurized, and cooled in the converging section of the nozzle, creating a low-temperature condensation environment in the flow field. When the mixture reaches the throat of the nozzle, it accelerates to supersonic speed, with the temperature dropping by 50–80 K compared to the inlet temperature. During continuous cooling, the saturation of condensable components in the gas phase gradually increases, reaching supersaturation and starting to condense, forming droplets that enter the liquid phase. Some condensable components near the wall also condense into a liquid film. Subsequently, in the diverging section of the Laval nozzle, due to the swirling action, large mass droplets with significant inertia are centrifuged to the wall, depositing and forming a liquid film. The liquid film flows along the pipe wall and is eventually discharged from the liquid outlet, while the separated gas enters the diffuser, where its pressure and temperature are restored before exiting through the dry gas outlet.

Throughout the process, the gas phase flows axially along the center of the pipeline as a continuous phase, while part of the liquid phase flows axially along the inner wall of the pipeline as a thin liquid film. Another part of the liquid phase exists as entrained droplets within the gas core. This flow pattern is known as annular flow. Ideally, the liquid film flows along the pipe wall, and only dry gas enters the diffuser. However, in reality, the annular flow of the gas–liquid two phases (gas and liquid) and the three fields (gas, droplets, and liquid film) in the supersonic separator involves complex heat and mass transfer processes. These include non-equilibrium condensation between the gas phase and droplets, phase changes between the gas phase and liquid film, and deposition and entrainment between droplets and the liquid film. Not all condensable components will condense, and not all droplets will be centrifugally separated. Some small mass droplets will escape into the diffuser and re-evaporate into the gas phase. These factors all affect the actual separation efficiency, necessitating further research and optimization.

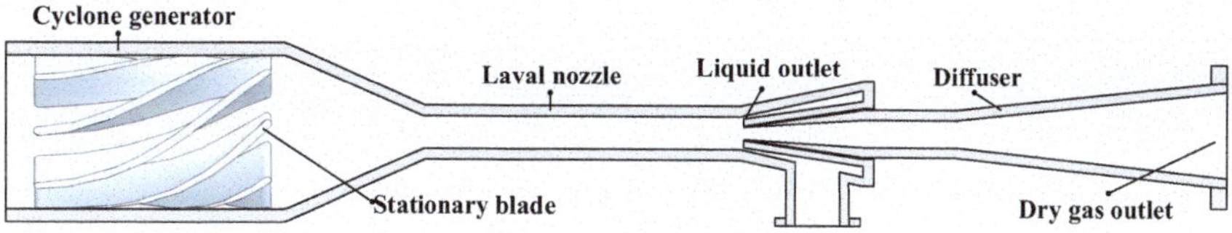

Fig. 6.2 Typical structure and flow state of supersonic separator

6.3.2 Experiment Platform

This subsection starts with an in-depth introduction to the operational workflow and performance indicators of the gas–liquid two-phase flow control system for supersonic condensation and separation behavior. Following this, it describes the design and fabrication of the Laval nozzle and the supersonic separator. Finally, it provides a detailed overview of the devices and relevant parameters used in the measurement system. The entire experimental setup lays a comprehensive foundation for subsequent theoretical research endeavors.

Control System of Gas–Liquid Two-Phase Flow

The experimental platform has been constructed, and a Laval nozzle and supersonic separator have been designed and fabricated for experimental validation. Figure 6.3 depicts the flow chart of the experimental platform. The experimental medium consists of air and water, and the entire system is divided into three parts: the gas–liquid two-phase flow control system, the experimental objects, and the measurement system. The gas–liquid two-phase flow control system is capable of delivering gas–liquid mixtures with adjustable temperature, humidity, and pressure to the experimental objects. The experimental objects include the Laval nozzle and supersonic separator. Experimental research on the Laval nozzle can provide initial insights into vapor condensation characteristics, while experimental research on the supersonic separator can comprehensively evaluate condensation and separation performances, and provide data support for subsequent theoretical research. The measurement system is used to measure the parameters of the gas phase, droplets, and liquid film.

The gas phase system configuration employs the SRC-30SA screw air compressor for experiments. This compressor converts mechanical energy into pneumatic pressure to adjust the intake pressure. It has a rated discharge volume of 210 m^3/h and a rated discharge pressure of 0.8 MPa. Due to impurities in the gas coming from the outside environment, two sets of oil separators are used to remove oil droplets, and two sets of moisture separators remove water. The purified gas is stored in a

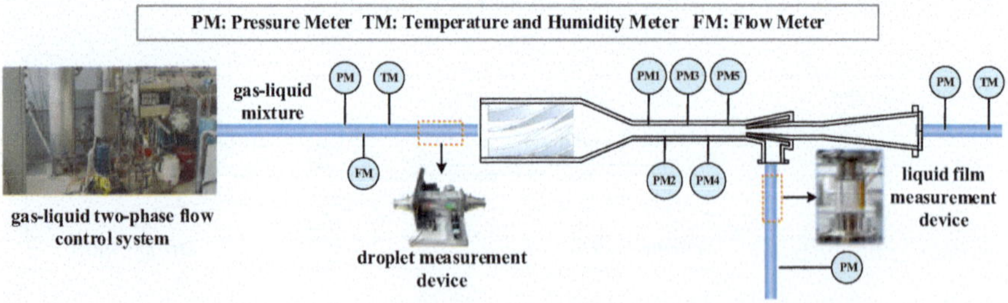

Fig. 6.3 Schematic diagram of the experimental platform for supersonic separation

gas storage tank. Intake pressure is controlled by a pressure regulating valve, and an A-level precision filter downstream ensures the purity of the experimental gas. Subsequently, gas temperature and humidity are adjusted: a gas heater with a rated power and voltage of 25 kW and 380 V, respectively, controls gas temperature within the range of 0–100 °C. A humidifier is used to control gas humidity: liquid is atomized into fine mist droplets (diameter between 10 and 100 μm) by high-pressure micro-fog nozzles and sprayed into the evaporator. The multi-stage evaporator structure rapidly evaporates the mist droplets, thereby adjusting the gas humidity within a range of 0–100%.

The device for adjusting the inlet liquid content is connected to a water storage tank. The water discharges from the storage tank, passes through a filter for purification, and enters a metering pump. The metering pump has a water pressure adjustment range of 0–10 MPa and a flow rate adjustment range of 0–17 L/h, which can be controlled in conjunction with a frequency converter to adjust the liquid intake. A calibration column on the pipeline monitors whether the liquid intake meets the standard, with a calibration accuracy of $\pm$ 2%. A pulsation dampener is used to weaken pressure fluctuations in the pipeline, maintaining pressure fluctuations within $\pm$ 5%. The water supply is divided into two paths: one supplies high-pressure mist nozzles in the evaporator tank to generate moist air, meeting experimental pressure, temperature, and humidity conditions, and directly feeds into the homogeneous Laval nozzle or supersonic separator for homogeneous condensation experiments. The other path of the water supply is transported via high-pressure hoses to the rear pipeline of the evaporator tank, where high-pressure mist nozzles generate heterogeneous mist droplets, mixing with moist air to form gas–liquid two-phase flow for heterogeneous condensation and separation experiments. The physical apparatus of the gas–liquid two-phase flow control system is shown in Fig. 6.4.

Design of Separator

The structure requirements for the Laval nozzle and supersonic separator are as follows: (1) The Laval nozzle interior must not produce shock waves; (2) The swirl generator must generate sufficient tangential velocity; (3) The liquid drainage structure should minimally affect the flow and facilitate smooth separation of the liquid film. The geometry information on nozzle S2 has been introduced in Chap. 4.

Figure 6.5 shows the structure of the designed supersonic separator. First, the swirl generator in the supersonic separator was parametrically modeled. The swirl generator has 8 stationary blades, each with a height of 13.5 mm and a thickness of 3 mm. The leading and trailing edge radii of the blades are both 1.5 mm, and the axial chord length is 80 mm. A second-order Bézier curve was chosen for parameterizing the mid-curve:

$$f(t) = (1 - t)^2 P_0 + 2t(1 - t)P_1 + t^2 P_2, \quad t \in [0, 1] \tag{6.34}$$

Fig. 6.4 Physical diagram of partial experimental apparatus

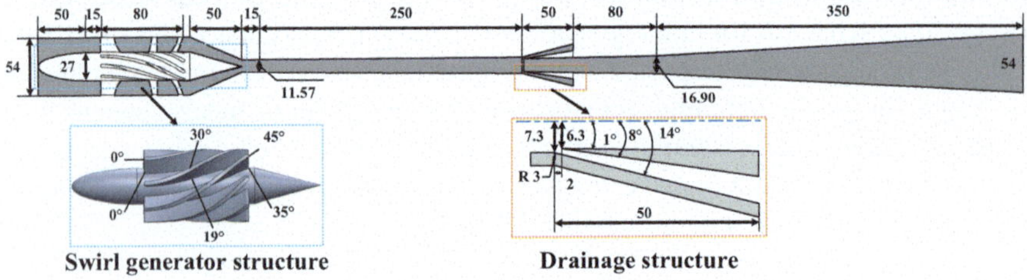

Fig. 6.5 Dimensional diagram of the supersonic separator structure

where P_0, P_1, and P_2 respectively denote the starting point, control point, and endpoint of the mid-curve. The parameter t is calculated as the axial distance from P_1 to P_0 divided by the axial distance from P_1 to P_2. The mid-curve of the stationary blades is divided into inner and outer mid-curves, both with an inlet angle set to 0°. The installation angles are 19° and 30° respectively, and the outlet angles are 35° and 45°. The stationary blades are mounted on a diffuser cone to allow smooth fluid entry into the stationary blade area. The diffuser cone has a total length of 205 mm, and a maximum diameter of 27 mm, with a front elliptical area of 50 mm and a straight pipe section of 15 mm, followed by a rear straight pipe section of 10 mm and a tapered core area of 50 mm. The overall structure of the swirl generator is depicted in Fig. 6.5.

The presence of the diffuser cone gives the inlet section of the Laval nozzle a annular shape, with an inner diameter of 27 mm and an outer diameter of 54 mm. The contraction section of the Laval nozzle is 50 mm long. The throat cross-sectional

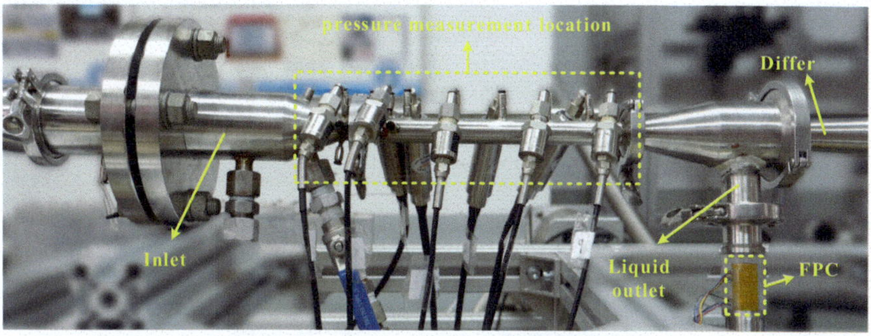

Fig. 6.6 The experimental photograph of the supersonic separator

area of the Laval nozzle is 105.07 mm^2, with a throat diameter of 11.57 mm. Behind the throat, there is a straight section of 15 mm length for securing and connecting clamps. The expansion ratio (AR), defined as the ratio of the nozzle exit area to the throat area, is 1.6, and the length of the nozzle expansion section is 250 mm. The inlet and outlet diameters of the expansion diffuser are 12.6 mm and 54 mm respectively, located 2 mm behind the exit of the Laval nozzle. The expansion diffuser consists of two stages with expansion angles of 1° and 2°, respectively. The outer wall of the moisture outlet connects to the nozzle exit with an expansion angle of 14°, while the inner wall connects to the entrance of the expansion diffuser with an expansion angle of 8°. The experimental photograph of the designed supersonic separator is shown in Fig. 6.6.

Measurement System

For measuring gas phase temperature, humidity, pressure, and flow rate, a signal measurement and acquisition system is utilized, as depicted in Fig. 6.7. This system employs pressure gauges, temperature and humidity sensors, and a vortex flow meter to measure specific positions within the experimental setup. Monitoring, display, and adjustment of operating parameters are managed by the MCGS SCADA system. The temperature and humidity sensor used is the Vaisala HMP110, capable of measuring temperatures from − 40 to 80 °C (± 5%) and humidity from 0 to 100% (± 5%). Additionally, pressure taps are installed along the walls of the Laval nozzle and the supersonic separator nozzle to measure surface pressure data under different conditions, providing insights into condensation within the nozzles. Pressure signals are acquired using high-frequency response micro-pressure sensors with a measurement range of 0–8 atm, frequency response from 0 to 20 kHz, and accuracy of 2%. These pressure gauges are connected to signal conditioning boards, converting current signals into voltage signals. Data acquisition is carried out using an NI USB-6353 data acquisition card, transmitting data to LabVIEW software for real-time display, storage, and post-processing.

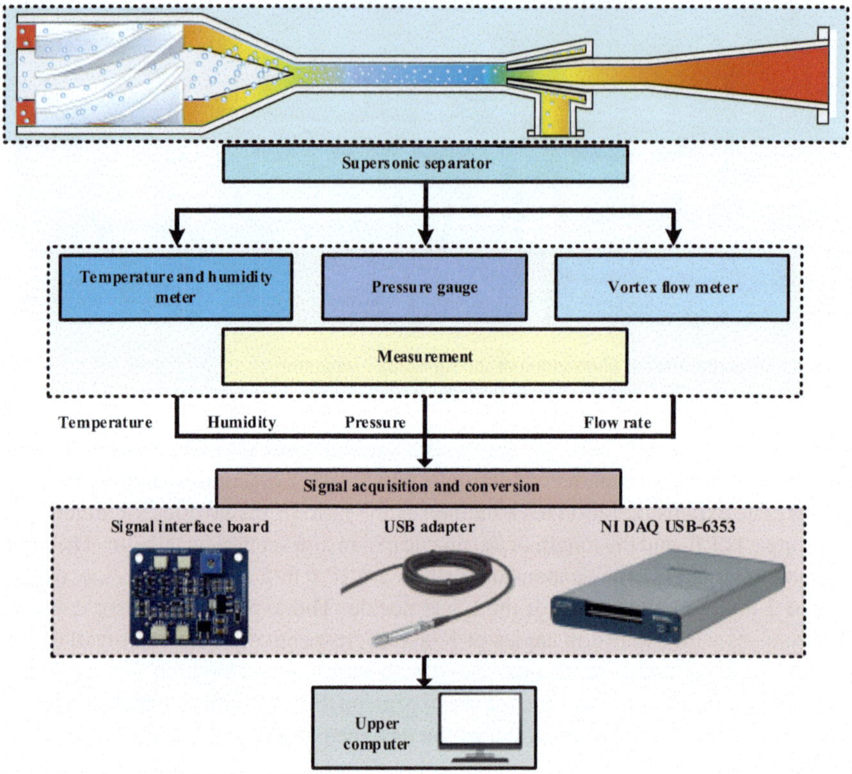

Fig. 6.7 Signal measurement and acquisition system

The droplet measurement device, as shown in Fig. 6.8a, employs the multi-wavelength extinction method to measure droplet sizes ranging from 0.01 to 10 μm. Due to scattering effects from the droplets, the original light source in the measurement area experiences attenuation. By measuring the original and transmitted light intensities, the droplet size distribution can be inversely calculated. The thickness of the liquid film at the liquid phase outlet is measured using a flexible printed circuit (FPC) based liquid film sensor, as depicted in Fig. 6.8b. Its basic structure includes a pair of transmitter–receiver electrodes. When these electrodes are covered with a conductive liquid of a certain thickness and subjected to a specific electrical potential, a weak electric current is generated between the electrodes, which correlates with the thickness of the conductive liquid.

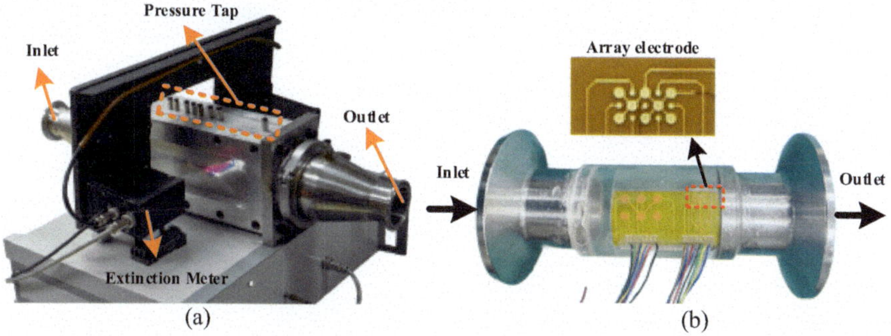

Fig. 6.8 Physical diagram of the droplet liquid film measurement device: **a** extinction method for droplet measurement; **b** FPC liquid film measurement

6.3.3 Experimental Validation

Based on the flow and heat transfer characteristics within the supersonic separator, the experimental validation is structured into three main components: validation of supersonic condensation flow in Laval nozzle which has been reported in Chap. 4, validation of intense swirling flow in the cyclone separator, and confirmation of liquid film flow and phase transitions within the rectangular channel. This approach sequentially assesses the model's accuracy in predicting condensing flow behavior, swirl separation efficiency, and liquid film dynamics.

Dispersed Phase Behaviors in Cyclone Separator

The geometry and grid of the cyclone separator are presented in Fig. 6.9. A structured hexahedral grid has been generated, achieving grid independence with 218,277 cells. The particles released for the simulation represent typical cement raw materials with a characteristic diameter of 29.9 μm. Both the air and particles enter the cyclone separator at the same velocity and are separated due to centrifugal forces. For more information on additional boundary conditions, please refer to the article by Wang et al. [25].

The tangential velocity distribution depicted in Fig. 6.10a for the cyclone separator reveals notable findings. Specifically, the comparison between simulated and experimental data shows a maximum discrepancy of approximately $\pm$ 5% in tangential velocity. As illustrated in Fig. 6.10b, the assessment of collection efficiency across various inlet velocities underscores optimal agreement between simulation and experimental results at inlet velocities ranging from 15 to 30 m s^{-1}. These outcomes robustly affirm the fidelity of the developed model in accurately simulating particle dynamics within intense swirling flows.

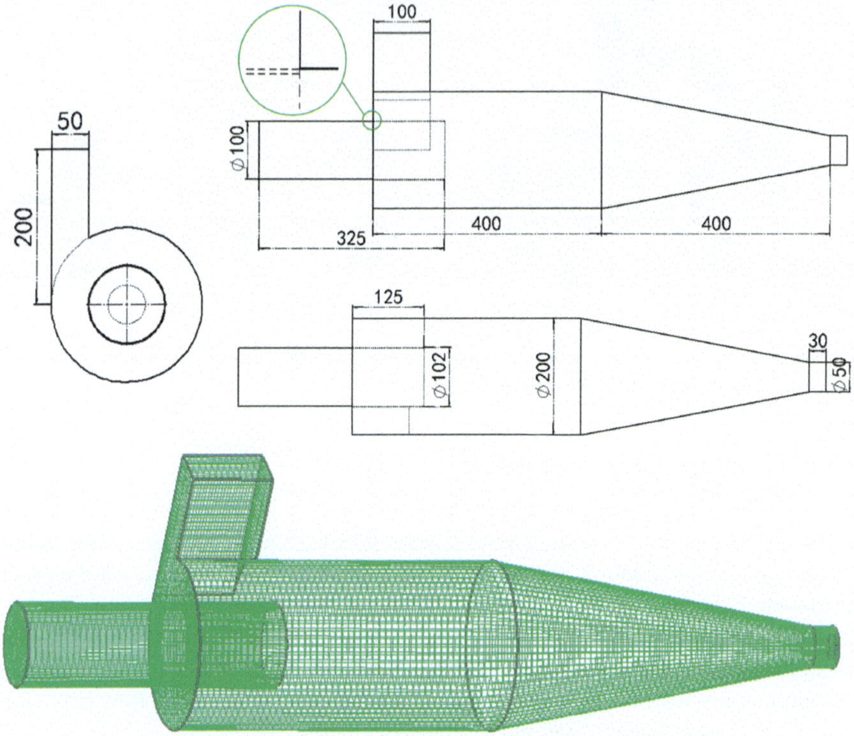

Fig. 6.9 Geometry and grid diagram of cyclone separator (mm)

Liquid Film Flow and Film Evaporation

The computational domain and grid structure of the rectangular channel are illustrated in Fig. 6.11. A structured hexahedral grid has been meticulously generated, particularly refined near the channel walls to ensure accurate boundary layer calculations ($y^+ < 1$). Additionally, the grid is further refined within the initial liquid film height along the bottom wall to enhance precision in modeling liquid film flow and evaporation dynamics. Following a rigorous grid independence test, the total number of grid cells is established at 960,687. In this simulation setup, a downward liquid film flow initiates from the channel's top, while upward airflow commences from the channel's bottom, creating a countercurrent interaction with the liquid film. The inlet air temperature is set at 45 °C, and the outlet pressure is maintained at 1 atm. Initial conditions include predefined film mass flux and incoming film temperature settings. The bottom wall's heating temperature is carefully adjusted to facilitate the evaporation of the liquid film. For detailed boundary conditions, please refer to the works of Du et al. [7].

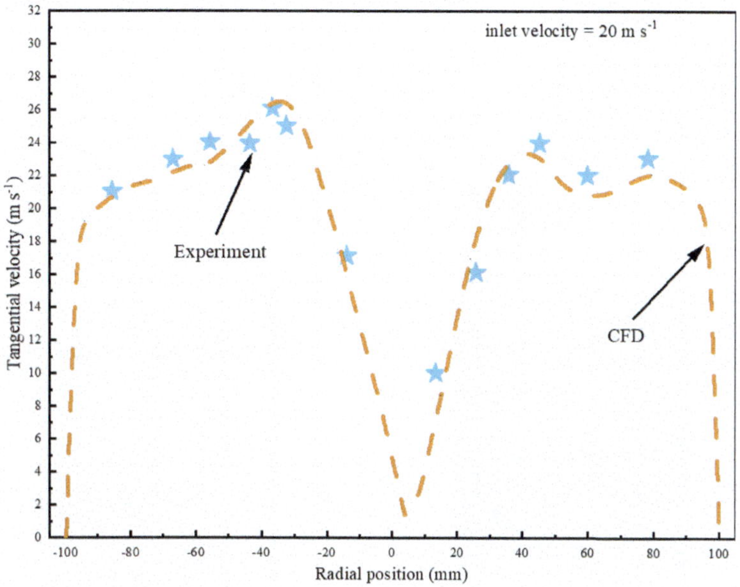

(a)

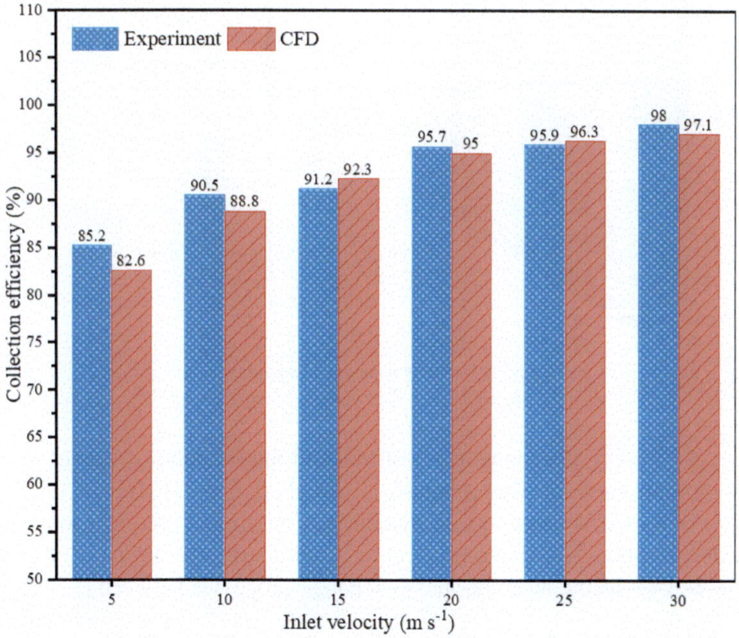

(b)

Fig. 6.10 Comparison between CFD and experiment in cyclone separator: **a** comparison of tangential velocity; **b** comparison of collection efficiency

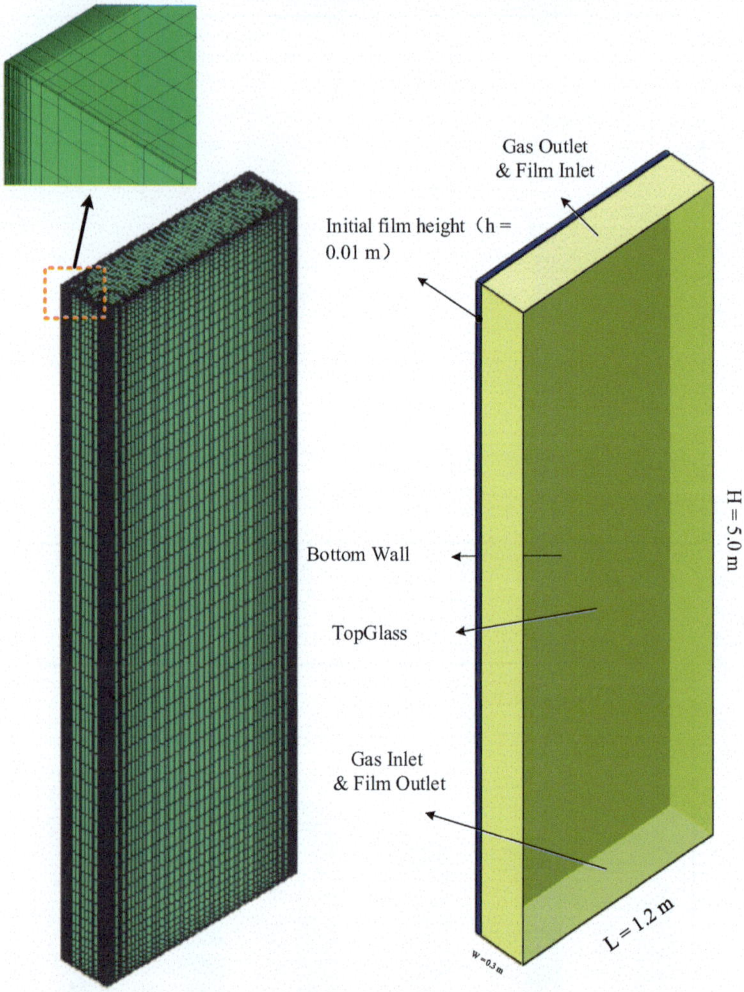

Fig. 6.11 Computational domain and grid of rectangular channel

Figure 6.12 presents the average heat flux and the evaporation rate of the liquid film on the bottom wall. The comparison reveals that the discrepancies between simulation and experimental results are confined within ± 10%. This indicates that the developed model is capable of accurately simulating the behavior of the liquid film flow as well as the heat and mass exchange processes between the liquid film and the moist air.

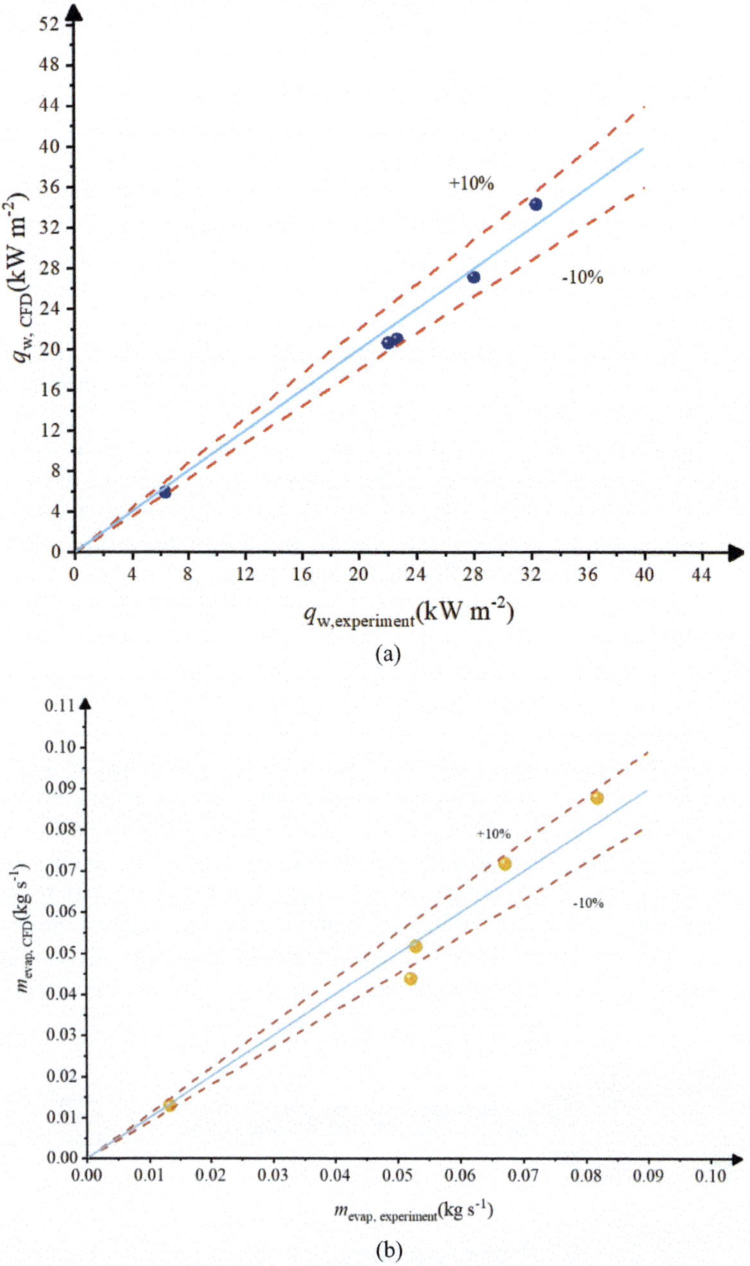

Fig. 6.12 Comparison between CFD and experiment in the bottom wall of rectangular channel: **a** comparison of the average surface heat flux; **b** comparison of the average surface liquid film evaporation rate

6.4 Flow Field and Droplet Behaviors

Figure 6.13 provides a visual representation of the supersonic separator's geometry and grid configuration. The three-dimensional computational domain is defined as the annular passage situated between the inner central body and the exterior shell, featuring a throat cross-sectional area of 105.07 mm^2. The grid employed is of the structured hexahedral variety, which has been meticulously refined in the regions near the walls to satisfy the requirements for boundary layer calculations. Following a grid independence test, it was determined that the total number of grid cells utilized is 549,900.

The standard boundary conditions include an inlet pressure P_{in} of 250 kPa, an inlet temperature T_{in} of 30 °C, and a relative humidity Φ_{in} of 100%, corresponding to a partial vapor pressure $P_{\text{v,in}}$ of 4246 Pa. The outlet pressure P_{out} is set to 100 kPa. The parameters $\rho_{\text{het,in}}$ and $n_{\text{het,in}}$ are specifically designated to establish the boundary conditions for heterogeneous condensation, representing the mass and number concentrations of foreign droplets introduced into the supersonic separator. There are two models of supersonic separators, A and B, with Separator B having an inlet diameter 10 mm smaller. Detailed settings for each of the four cases (Case 1 through Case 4) are provided in Table 6.1. As illustrated in Table 6.1, Cases 1–4 has the identical radii for the foreign droplets, $r_{\text{het,in}}$, but have different mass concentrations of these droplets. Prior research indicates that the optimal radius for foreign droplets in separator A is approximately 1.0 μm, while for separator B, the optimal radius is around 0.3 μm.

The velocity streamline of supersonic separator A is depicted in Fig. 6.14. Following the expansion and cooling processes within the supersonic nozzle, the moist gas continues to travel at supersonic speed through the nozzle's divergent section. The peak gas phase velocity and subcooling degree can attain values of 514.4 m/s and 45.2 °C, respectively. At the location of liquid separation, a sharp decline in velocity is observed in Fig. 6.14, signifying the presence of a shock wave. Nonetheless, the velocity at this liquid separation point does not fully diminish to subsonic speed, instead, it further accelerates to supersonic speed within the diffuser.

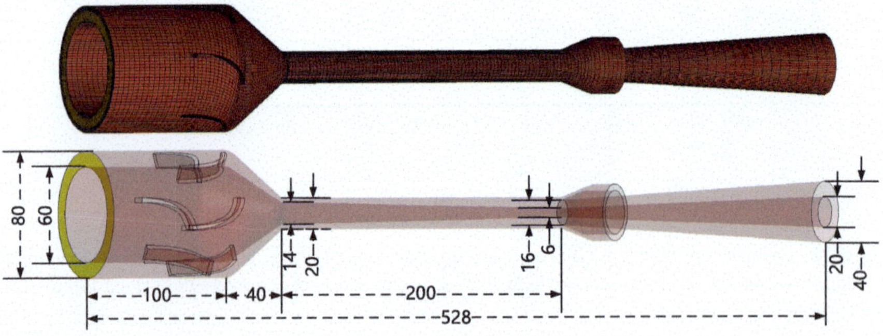

Fig. 6.13 Geometry and grid diagram of supersonic separator A

Table 6.1 Cases of different foreign droplets in supersonic separators

Cases	Separators	$r_{het,in}$ (μm)	$\rho_{het,in}$ (kg m^{-3})	$n_{het,in}$ (m^{-3})	$Q_{het,in}$ (ml min^{-1})
Case 1	Separator A	1.0	0.001	2.39×10^{11}	1.73
Case 2	Separator A	1.0	0.01	2.39×10^{12}	17.09
Case 3	Separator A	1.0	0.05	1.19×10^{13}	84.14
Case 4	Separator A	1.0	0.1	2.39×10^{13}	165.97

The shock wave phenomenon at the liquid separation point impacts the heat and mass exchange processes between the gas and liquid phases, leading to the evaporation of condensed droplets, which should be minimized to enhance separation efficiency.

The tangential velocities of supersonic separators with varying inlet diameters are illustrated in Fig. 6.15, correlating closely with separation efficiency. Supersonic separator A exhibits a high tangential velocity distribution near the liquid separation point, while for separator B, it is closer to the throat. The reduced height of the swirling blades decreases swirl intensity, as detailed in Table 6.2. Supersonic separator B achieves a maximum swirl strength of 0.42, approximately double that of supersonic separator A.

The dehydration efficiency of the supersonic separator can be assessed by analyzing the mole fraction and partial pressure of vapor at the dry gas outlet. As

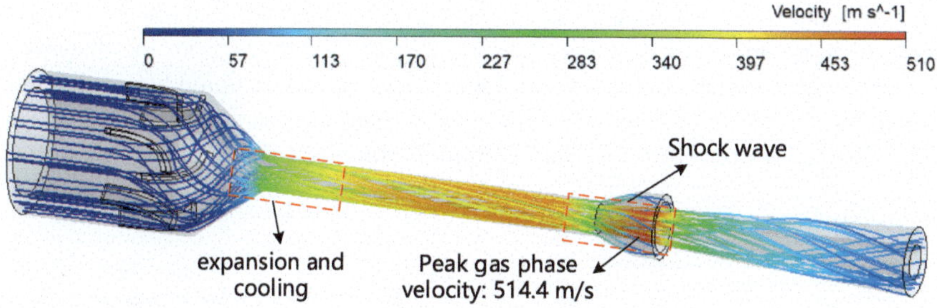

Fig. 6.14 Velocity streamline of the gas phase in supersonic separator A (Case 2)

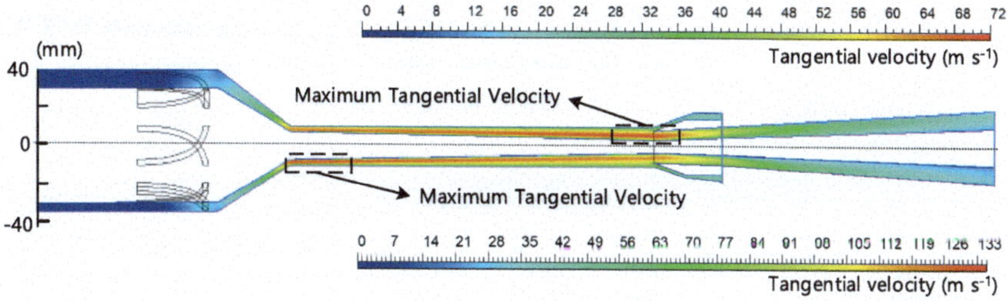

Fig. 6.15 The contours of tangential velocity u_t on axial profile (Case 2): in supersonic separators A and B

Table 6.2 Maximum tangential velocity and centrifugal acceleration of different supersonic separators

Separators	Inlet diameter (mm)	$u_{t,max}$ (m s^{-1})	$a_{c,max}$ (m s^{-2})	S_{max} (–)
A	80	71.74	1.04×10^6	0.23
B	70	134.43	3.8×10^6	0.42

illustrated in Fig. 6.16, following supersonic separation, the majority of saturated vapor is expelled through the wet gas outlet, thereby achieving high separation efficiency. Figure 6.16 clearly shows that the condensation of vapor leads to a rapid decline in partial pressure near the nozzle throat. As the gas flow approaches the liquid separation zone, a significant number of droplets in the near-wall region evaporate due to the effects of the hot gas flow and shock waves, which is detrimental to the efficiency of water separation.

The distribution of droplets within the supersonic separator is illustrated in Fig. 6.17. As observed from Fig. 6.17a, droplet condensation initiates near the nozzle throat, with the highest rate of droplet mass change reaching 412.0 kg m^{-3} s^{-1}. Evaporation of droplets primarily takes place in the near-wall region and at the wet gas outlet, where the peak evaporation rate is – 283.1 kg m^{-3} s^{-1}.

Figure 6.17b, c demonstrate that droplets progressively move toward the wall and develop into a liquid film due to the action of centrifugal force. Unlike the numerical model that does not account for the presence of a liquid film (refer to Ding et al. [5]), once the dispersed droplets are centrifuged to the wall, they do not remain in the form of dispersed droplets. Instead, they are separated from the dispersed droplets and accumulate to form a liquid film on the wall. This liquid film undergoes heat and mass exchange processes with the gas phase. The model that excludes the liquid film tends to overestimate the separation efficiency. Therefore, as illustrated

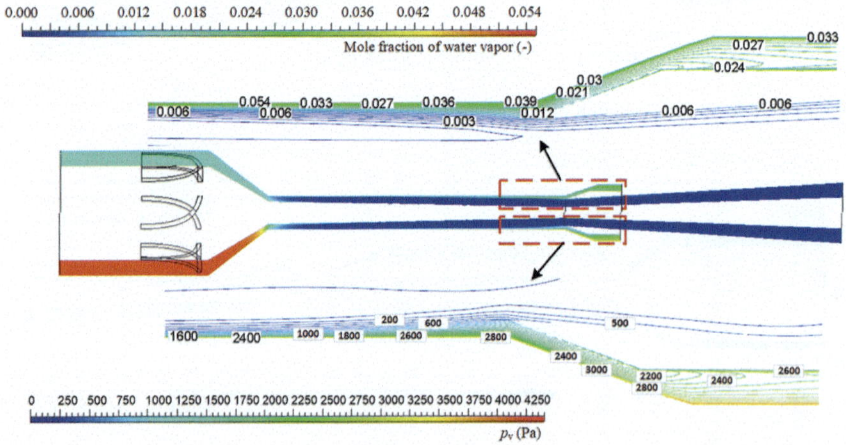

Fig. 6.16 The distributions of mole fraction of vapor χ_v and partial pressure of vapor (Case 2)

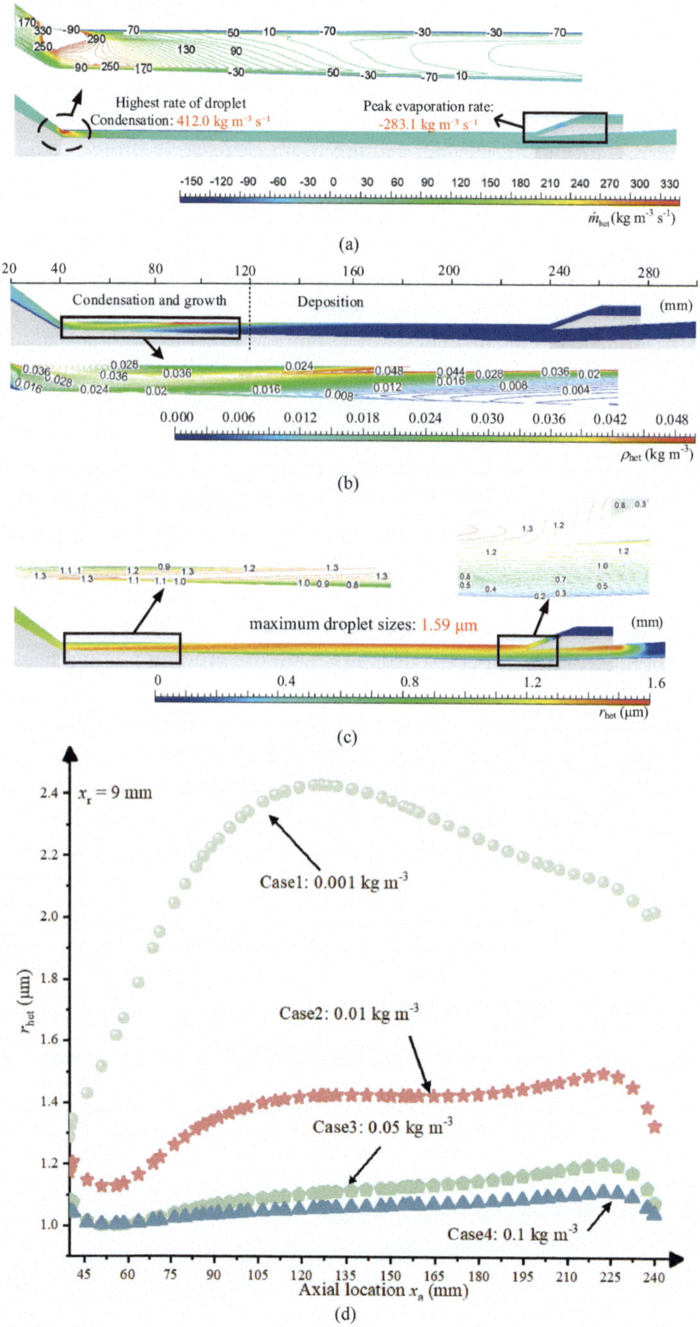

Fig. 6.17 The parameter distribution of droplet phase inside supersonic separator: **a** the contour of droplet mass changing rate $\dot{m}_{het}$ on axial profile (Case 2); **b** the contour of effective density ρ_{het} on axial profile (Case 2); **c** the contour of droplet radius r_{het} on axial profile (Case 2); **d** axial distributions of droplet radius r_{het} at $x_r = 9$ mm with $\rho_{het,in}$ variation

in Fig. 6.17b, droplets commence condensation and growth at the nozzle throat. The effective density and radius of the droplets increase, but the effective density starts to decrease after the axial position $x_a = 120$ mm, indicating that the majority of dispersed droplets are depositing to form a liquid film. Under the influence of centrifugal force, droplets coalesce, leading to an increase in droplet size, with the maximum size reaching 1.6 μm.

Figure 6.17c, d show the droplet size distribution along the axial direction under varying $\rho_{het,in}$ values. It is observed that when the $\rho_{het,in}$ value is smaller, the droplets tend to be larger in size. The maximum droplet sizes for Cases 1–4 are 2.50 μm, 1.59 μm, 1.51 μm, and 1.48 μm, respectively.

Figure 6.18 illustrates the droplet surface temperature, the gas phase temperature, and the latent heat, showcasing the heat transfer between dispersed droplets and the gas phase. In the divergent section of the nozzle, the gas phase expands, leading to a gradual decrease in temperature, which in turn causes the wet steam to condense and form dispersed droplets. The latent heat released during condensation is transferred to the gas phase, resulting in a slight increase in its temperature. Compared to the droplet surface temperature, the gas phase temperature is lower in the low-temperature zone and higher in the high-temperature zone. During the condensation process, heat is transferred from the dispersed droplets to the gas phase, with the maximum latent heat release reaching 5.3×10^8 J m^{-3}. Conversely, during evaporation, heat is transferred from the gas phase to the dispersed droplets, with the minimum latent heat being -3.4×10^8 J m^{-3}.

The interphase slip velocity characterizes the momentum coupling between the droplet and the gas phase. Defined as the difference between the droplet velocity and the gas velocity, the slip velocity's axial and centrifugal (radial) components are shown in Fig. 6.19. It is evident that the slip velocity progressively increases along the flow direction before the liquid separation position, indicating a growing drag

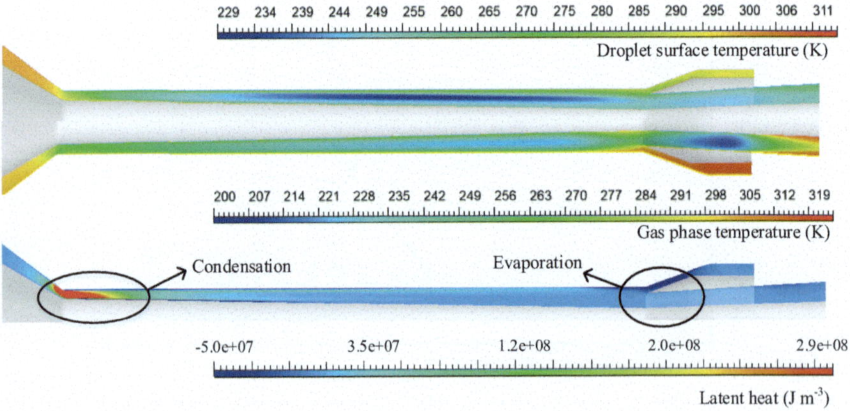

Fig. 6.18 The heat transfer of supersonic separator A include droplet surface temperature, gas phase temperature and latent heat

force between the gas phase and the dispersed droplets. The slip velocity for $\rho_{het,in} = 0.01$ kg m^{-3} is higher than that for $\rho_{het,in} = 0.1$ kg m^{-3}. As shown in Fig. 6.17d, this occurs because larger droplet sizes ($\rho_{het,in} = 0.01$ kg m^{-3}) result in greater droplet inertia. Under the influence of centrifugal force, the centrifugal slip velocity rises gradually, with the centrifugal velocity of droplets exceeding that of the gas phase by up to 24.9 m s^{-1}, enabling effective gas–liquid separation.

In Fig. 6.20, the streamlines and effective density of heterogeneous droplets for Case 2 with $r_{het,in} = 1.0$ μm are illustrated using both single-fluid and multi-fluid models. In the single-fluid model, a no-slip condition for fluid velocity is assumed. Consequently, the liquid phase streamlines form a circular pattern within the cyclone, as depicted in Fig. 6.20a, and the effective density of heterogeneous droplets remains almost uniform. Due to the no-slip assumption in the single-fluid model, the heat transfer coefficient is lower compared to that in the multi-fluid model. The droplet number density for heterogeneous droplets ($n_{het,in} = 2.39 \times 10^{12}$ m^{-3} in this scenario) is typically insufficient to inhibit homogeneous nucleation in the single-fluid model, resulting in 8% heterogeneous condensation and 90% homogeneous condensation. Conversely, in the multi-fluid slip model, the streamlines in the cyclone section exhibit centrifugal motion, as shown in Fig. 6.20b. The heat transfer coefficient around the heterogeneous droplets significantly increases due to interphase slip velocity. In this context, homogeneous condensation nearly vanishes. The effective density of heterogeneous droplets near the shell wall surges sharply due to centrifugal force, enhancing separation performance.

6.5 Liquid Film Behaviors

Figures 6.21, 6.22 and 6.23 presents the distribution of the liquid film in supersonic separator A. In Fig. 6.21, condensed droplets are visibly driven by centrifugal force to adhere to the wall surface, forming a liquid film that subsequently exits via the wet gas outlet. Notably, the liquid film appears thicker at the corner of the wet gas outlet, as illustrated in Fig. 6.23, where a circular flow pattern of the liquid film is evident. Across Cases 1–4, the swirling liquid film thickness at the wet gas outlet measures 21 μm, 47 μm, 74 μm, and 89 μm, respectively, with increasing $\rho_{het,in}$ resulting in thicker liquid films. At $x_a = 130$ mm, the maximum velocity of the liquid film reaches 1.9 m s^{-1}. The film phase change rate reflects the mass exchange between the liquid film and the gas phase: a negative rate indicates condensation, while a positive rate denotes evaporation. As depicted in Fig. 6.21, between x_a = 120–220 mm, vapor phase condensation contributes to an increase in the liquid film mass.

Figure 6.22 illustrates the film temperature and heat flux, which depict the heat transfer dynamics of the liquid film. This includes the coupled heat exchange between dispersed droplets and the gas phase, alongside wall heat conduction. Near the axial position $x_a = 120$ mm, dispersed droplets gradually coalesce to form a liquid film. Here, the temperature of the liquid film decreases to a minimum of 304.1 K due to droplet deposition, coinciding with a peak heat flux of 0.74 MW m^{-2}. Following this,

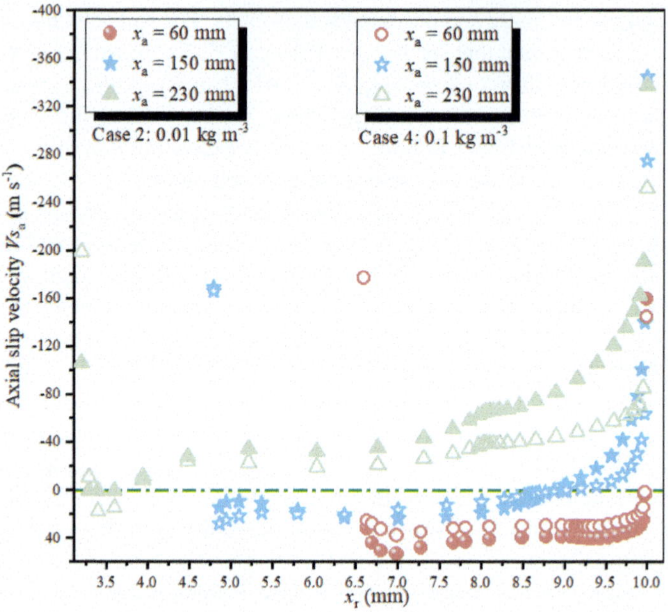

(a)

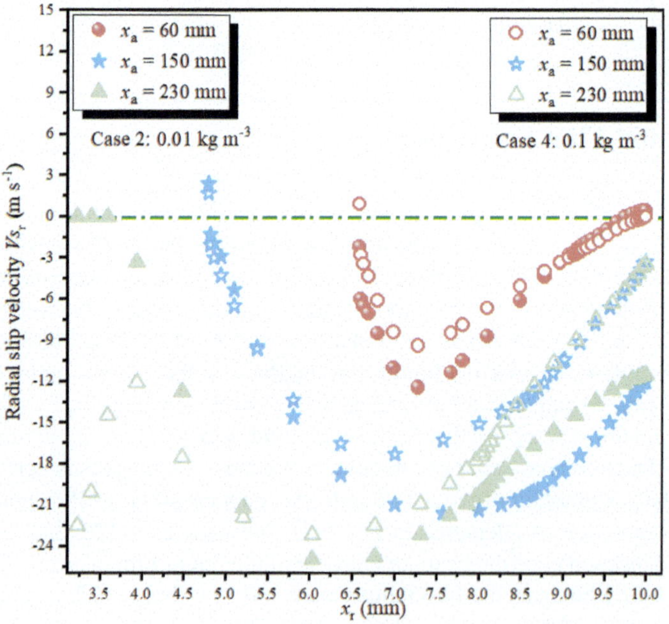

(b)

Fig. 6.19 Distribution of axial and centrifugal slip velocity components with different $\rho_{\mathrm{het,in}}$: **a** axial slip velocity for Case 2 and Case 4; **b** centrifugal slip velocity for Case 2 and Case 4

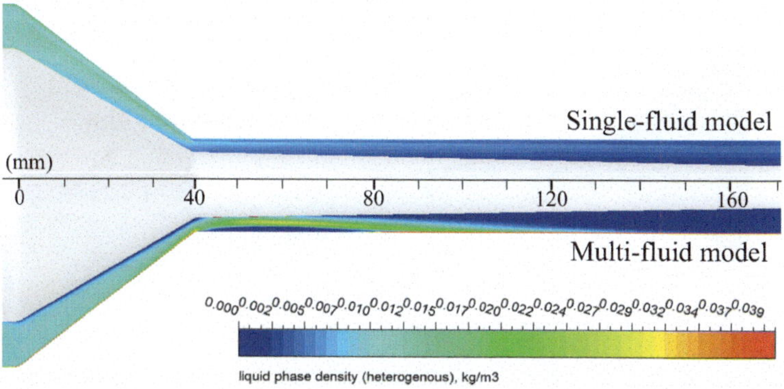

(a)

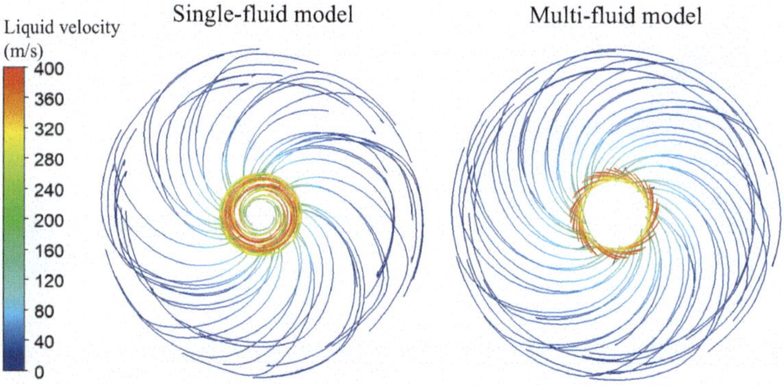

(b)

Fig. 6.20 Comparisons of streamline and effective density of heterogeneous droplets between single-fluid (no-slip) model (8% heterogeneous + 90% homogeneous condensations at x_a = 240 mm) and multi-fluid (slip) model (91% heterogeneous + 0% homogeneous condensation at x_a = 240 mm) for Case 2: **a** comparisons of effective density of heterogeneous droplets; **b** comparisons of streamline of heterogeneous droplets

heat is transferred from the wall to the cooler liquid film, causing its temperature to increase. Beyond the shock wave, the liquid film undergoes evaporation, leading to a decrease in its temperature.

The interfacial wave velocity plays a pivotal role in the mass transfer dynamics of the liquid film. The frequency and velocity of these waves are determined using the cross-correlation principle, which involves the implementation of a cross-correlation algorithm. This principal hinges on analyzing the similarity between signals upstream and downstream, thereby converting the measurement of wave velocity into a time interval measurement. Initially, time-dependent thickness measurements δ_1 and δ_2 of the liquid film are collected at positions x_1 and x_2, separated by a distance x_l, with a sampling frequency set at 40 kHz. Subsequently, fluctuations y_1 and y_2 in the liquid film thickness are computed, followed by their conversion into the

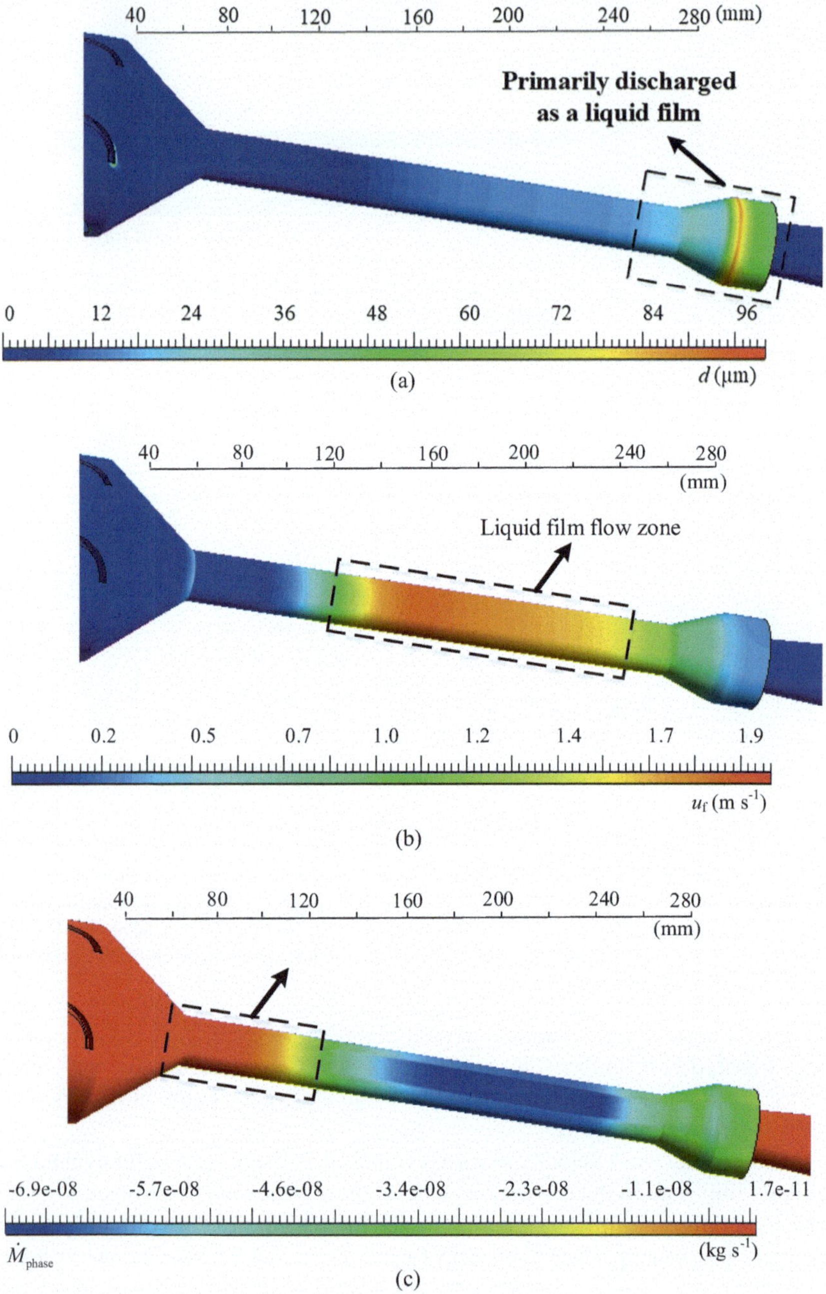

Fig. 6.21 Distribution of liquid film parameters in supersonic separator A (Case 2): **a** d; **b** u_{f}; **c** $\dot{M}_{\mathrm{phase}}$

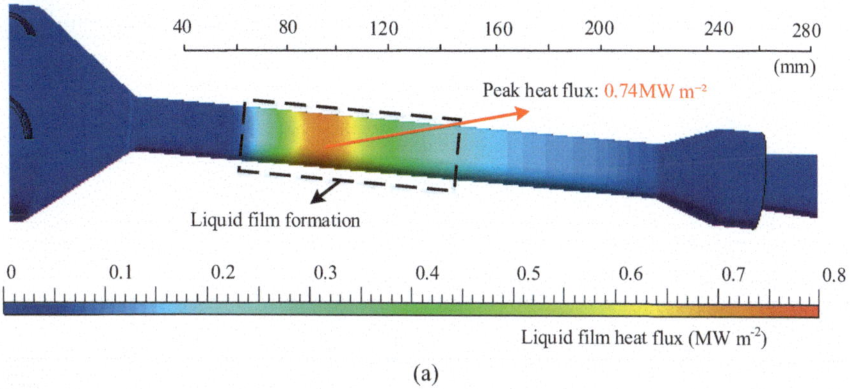

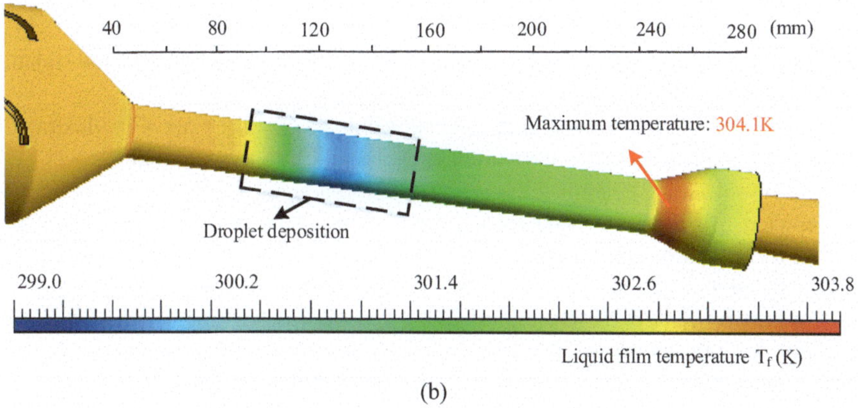

Fig. 6.22 Liquid film distribution and heat transfer in supersonic separator A (Case 2): **a** liquid film heat flux; **b** liquid film temperature

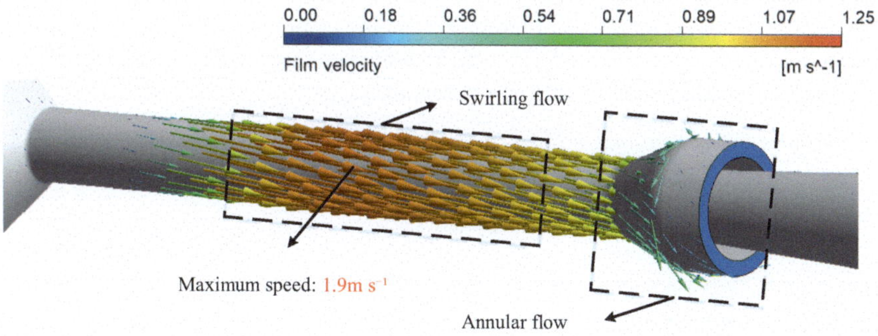

Fig. 6.23 Liquid film distribution of supersonic separator A

time–frequency domain. The cross-correlation function $R_{12}(\tau)$ used to quantify the correlation between signals y_1 and y_2, is expressed as [26]

$$R_{12}(\tau) = \lim_{T\to\infty} \frac{1}{T} \int_0^T y_1(\tau) y_2(t - \tau)\mathrm{d}t \tag{6.35}$$

where y_1 represents the fluctuation of film thickness at position x_1, and y_2 represents the fluctuation of film thickness at position x_2.

The interfacial wave velocity u_w is calculated by

$$u_w = \frac{x_l}{\tau_0} \tag{6.36}$$

where the time τ_0 corresponding to the peak of $R_{12}(\tau)$ is the transit time of the signal, $x_l = x_2 - x_1$ is transit distance.

The process of calculating the interfacial wave velocity for Case 4 is illustrated in Fig. 6.24. Time-dependent measurements of liquid film thickness at the wet gas outlet positions $x_1 = 270$ mm and $x_2 = 275$ mm (separated by a distance $x_l = 5$ mm) are depicted in Fig. 6.24a. The fluctuations in the interfacial wave are shown in Fig. 6.24b, c, where it is evident that y_1 lags behind y_2. The dominant frequency of the interfacial wave, $f_w = 11.07$ Hz (for Case 4), is obtained through time–frequency domain transformation. Figure 6.24d shows that the transit time $\tau_0 = 3.35$ ms is determined using the cross-correlation function, which reaches a maximum cross-correlation coefficient of 0.93. Therefore, the interfacial wave velocity is calculated to be 1.49 m s^{-1}.

Table 6.3 illustrates the dominant frequency and interfacial wave velocity corresponding to different $\rho_{het,in}$ values. As $\rho_{het,in}$ increases, the thickness of the liquid film also tends to increase. Higher $\rho_{het,in}$ values result in elevated frequencies and velocities of the interfacial wave, thereby enhancing instability within the supersonic separator. Moreover, the correlation coefficient decreases with increasing $\rho_{het,in}$, indicating a rise in the complexity of interfacial waves.

6.6 Dehydration Performance

Dew point depression (ΔT_d) and water removal rate (η_v), known as key indicators of dehydration performance, are widely used in evaluation. ΔT_d quantifies the temperature difference between the ambient air and the dew point, reflecting operational efficiency. Meanwhile, η_v measures the effectiveness of separating water from the gas stream, indicating overall separation efficiency. The ΔT_d is expressed as

$$\Delta T_d = T_{d,in} - T_{d,dry} \tag{6.37}$$

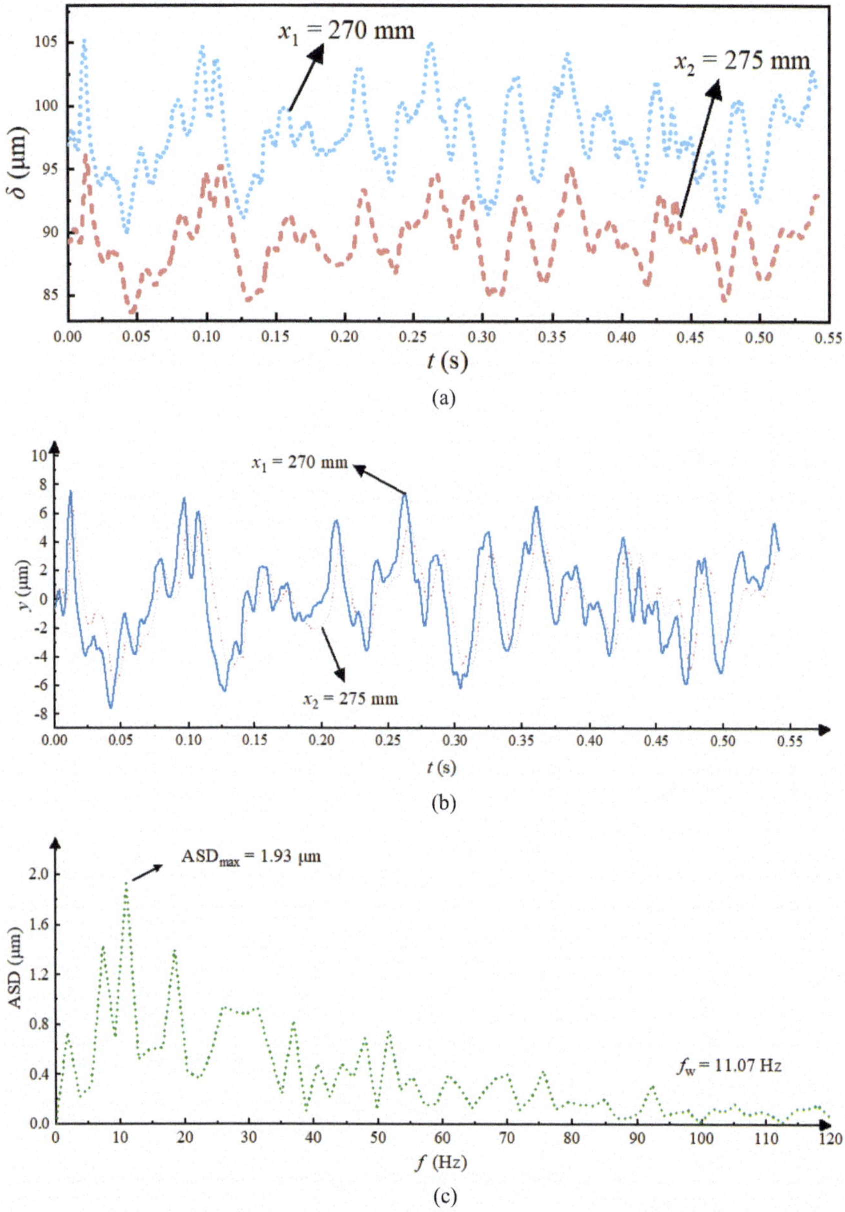

Fig. 6.24 The calculation process of the interfacial wave velocity (Case 4): **a** raw data of liquid film thickness obtained by data collection; **b** time domain diagram of the fluctuation of liquid film thickness; **c** frequency domain diagram of the fluctuation of liquid film thickness; **d** calculation of transit time using cross-correlation function

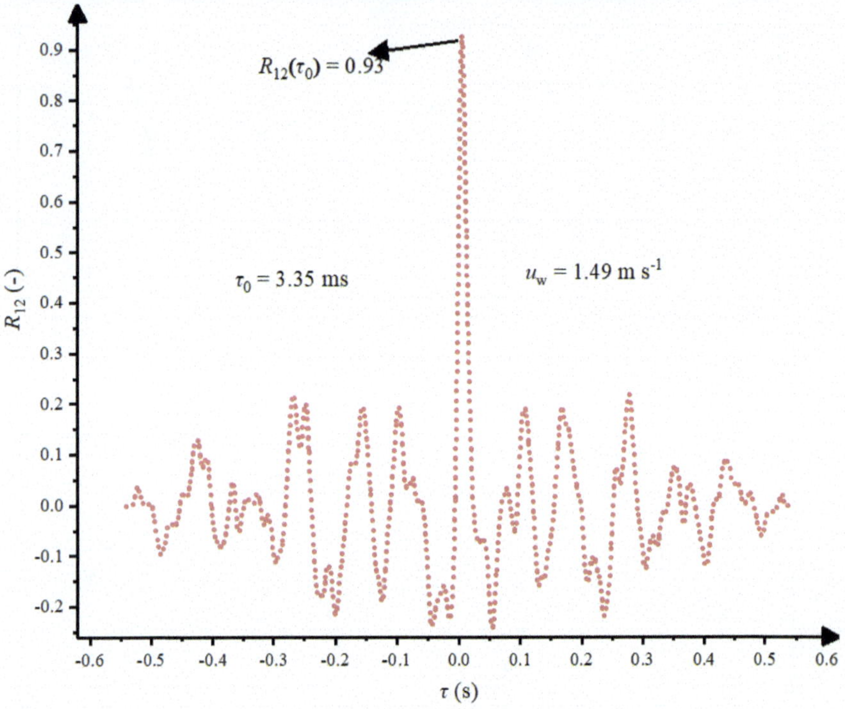

(d)

Fig. 6.24 (continued)

Table 6.3 Frequency and cross-correlation velocity of interfacial waves of liquid film at liquid outlet with different $\rho_{het,in}$

Cases	δ_{wet} (μm)	f_w (Hz)	τ_0 (ms)	$R_{12}(\tau_0)$ (–)	u_w (m s^{-1})
Case 1	21	5.24	15.40	0.98	0.32
Case 2	47	7.97	9.60	0.97	0.52
Case 3	74	10.00	5.75	0.95	0.87
Case 4	89	11.07	3.35	0.93	1.49

where T_d (°C) is the dew point temperature of gas phase. The separation efficiency η_v is expressed as

Table 6.4 Dehydration performance of supersonic separator A

Cases	$p_{v,dry}$ (Pa)	$T_{d,dry}$ (°C)	$\Delta T_{d,dry}$ (°C)	η_v (%)
Case 1	754.63	2.93	27.07	55.58
Case 2	252.97	− 10.32	40.32	85.11
Case 3	369.09	− 6.01	36.01	78.27
Case 4	754.15	2.91	27.09	55.60

$$\eta_v = \frac{\chi_{v,in} - \chi_{v,dry}}{\chi_{v,in}} \times 100\% \quad (6.38)$$

where, χ_v is the mole fraction of vapor.

Table 6.4 presents the dehydration performance details for supersonic separator A. The effective vapor partial pressure at the dry gas outlet ($p_{v,dry}$) encompasses both the effective partial pressure of vapor and liquid droplets, with its minimum value noted as 252.97 Pa. The dew point at the dry gas outlet ($T_{d,dry}$) can drop as low as − 10.32 °C. In comparison, supersonic separator B (Case 6) shows $p_{v,dry}$ at 200.14 Pa and $T_{d,dry}$ at − 12.93 °C.

Figure 6.25 presents a histogram depicting the dehydration performance for various $\rho_{het,in}$ values. It is evident that the optimal effective density of foreign droplets, $\rho_{het,in}$, is 0.01 kg m^{-3}. This indicates that to achieve higher dehydration efficiency, $\rho_{het,in}$ should be moderate. When $\rho_{het,in}$ is too low, the swirl separation effect is insufficient, allowing more liquid droplets to remain in the gas stream and enter the dry gas outlet, thus reducing separation efficiency. Conversely, an excessively high $\rho_{het,in}$ can lead to adverse effects such as thickening the liquid film, which hinders separation and increases instability within the supersonic separator. The model that disregards the liquid film tends to overestimate separation efficiency. For instance, when $\rho_{het,in}$ is set at 0.01 kg m^{-3}, the maximum separation efficiency and dew point depression for separator A are recorded at 85.11% (compared to 86.71% without considering the liquid film and 40.32 °C, respectively). Similarly, the separation efficiency for supersonic separator B reaches 88.22% (93.91% without considering the liquid film, with a corresponding dew point depression of 42.93 °C).

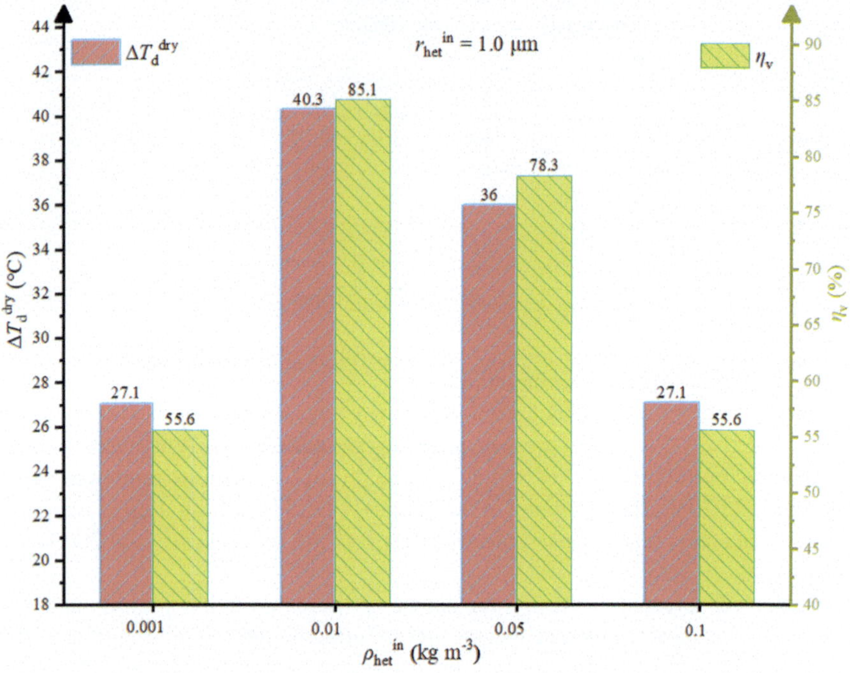

Fig. 6.25 Comparison of separation efficiency of different $\rho_{het,in}$

References

1. Abadi SNR, Kouhikamali R, Atashkari K (2015) Two-fluid model for simulation of supersonic flow of wet steam within high-pressure nozzles. Int J Therm Sci 96:173–182
2. Cao X, Yang W (2015) The dehydration performance evaluation of a new supersonic swirling separator. J Nat Gas Sci Eng 27:1667–1676
3. de Oliveira Arinelli L, de Medeiros JL, de Melo DC, Teixeira AM, Brigagão GV, Passarelli FM et al (2019) Carbon capture and high-capacity supercritical fluid processing with supersonic separator: natural gas with ultra-high CO_2 content. J Nat Gas Sci Eng 66:265–283
4. de Oliveira Arinelli L, Teixeira AM, de Medeiros JL, Ofélia de Queiroz FA (2019) Supersonic separator for cleaner offshore processing of natural gas with high carbon dioxide content: environmental and economic assessments. J Clean Prod 233:510–521
5. Ding H, Sun C, Wang C, Wen C, Tian Y (2020) Prediction of dehydration performance of supersonic separator based on a multi-fluid model with heterogeneous condensation. Appl Therm Eng 171:115074
6. Ding H, Tian Y, Wen C, Wang C, Sun C (2021) Polydispersed droplet spectrum and exergy analysis in wet steam flows using method of moments. Appl Therm Eng 182:116148
7. Du K, Hu P, Hu Z (2020) Numerical investigation of water film evaporation with the countercurrent air in the asymmetric heating rectangular channel for passive containment cooling system. Sci Technol Nucl Install 2020(1):5924325
8. Dykas S, Wróblewski W (2012) Numerical modelling of steam condensing flow in low and high-pressure nozzles. Int J Heat Mass Transf 55(21–22):6191–6199
9. Dykas S, Wróblewski W (2013) Two-fluid model for prediction of wet steam transonic flow. Int J Heat Mass Transf 60:88–94

10. Edathol J, Brezgin D, Aronson K, Kim HD (2020) Prediction of non-equilibrium homogeneous condensation in supersonic nozzle flows using Eulerian–Eulerian models. Int J Heat Mass Transf 152:119451
11. Han J, Feng J, Hou T, Chen W, Peng X (2021) Numerical and experimental study on gas-water separators for a PEMFC system. Int J Green Energy 18(5):490–502
12. Jiang W, Bian J, Wu A, Gao S, Yin P, Hou D (2018) Investigation of supersonic separation mechanism of CO_2 in natural gas applying the discrete particle method. Chem Eng Process Process Intensif 123:272–279
13. Jurski K, Géhin E (2003) Heterogeneous condensation process in an air water vapour expansion through a nozzle—experimental aspect. Int J Multiph Flow 29(7):1137–1152
14. Kantrowitz A (1951) Nucleation in very rapid vapor expansions. J Chem Phys 19(9):1097–1100
15. Li H, Anglart H (2016) Prediction of dryout and post-dryout heat transfer using a two-phase CFD model. Int J Heat Mass Transf 99:839–850
16. Liu X, Liu Z, Li Y (2014) Investigation on separation efficiency in supersonic separator with gas-droplet flow based on DPM approach. Sep Sci Technol 49(17):2603–2612
17. Liu Y, Cao X, Yang J, Li Y, Bian J (2021) Energy separation and condensation effects in pressure energy recovery process of natural gas supersonic dehydration. Energy Convers Manage 245:114557
18. Ma Q, Hu D, He G, Hu S, Liu W, Xu Q, Wang Y (2009) Performance of inner-core supersonic gas separation device with droplet enlargement method. Chin J Chem Eng 17(6):925–933
19. Niknam PH, Mortaheb HR, Mokhtarani B (2018) Dehydration of low-pressure gas using supersonic separation: experimental investigation and CFD analysis. J Nat Gas Sci Eng 52:202–214
20. O'Rourke PJ (1981) Collective drop effects on vaporizing liquid spray. Ph.D. thesis, Princeton University
21. Ounis H, Ahmadi G, McLaughlin JB (1991) Brownian diffusion of submicrometer particles in the viscous sublayer. J Colloid Interface Sci 143(1):266–277
22. Patel Y, Patel G, Turunen-Saaresti T (2015) Influence of turbulence modelling on non-equilibrium condensing flows in nozzle and turbine cascade. Int J Heat Mass Transf 88:165–180
23. Shooshtari SR, Shahsavand A (2018) Numerical investigation of water droplets trajectories during natural gas dehydration inside supersonic separator. J Nat Gas Sci Eng 54:131–142
24. Shooshtari SR, Shahsavand A (2018) Optimal operation of refrigeration oriented supersonic separators for natural gas dehydration via heterogeneous condensation. Appl Therm Eng 139:76–86
25. Wang B, Xu DL, Chu KW, Yu AB (2006) Numerical study of gas–solid flow in a cyclone separator. Appl Math Model 30(11):1326–1342
26. Wang C, Zhao N, Feng Y, Sun H, Fang L (2018) Interfacial wave velocity of vertical gas-liquid annular flow at different system pressures. Exp Therm Fluid Sci 92:20–32
27. Wang Y, Yu Y, Hu D (2020) Experimental investigation and numerical analysis of separation performance for supersonic separator with novel drainage structure and reflux channel. Appl Therm Eng 176:115111
28. Wen C, Cao X, Yang Y, Zhang J (2011) Swirling effects on the performance of supersonic separators for natural gas separation. Chem Eng Technol 34(9):1575–1580
29. Wen C, Cao X, Yang Y, Li W (2012) Numerical simulation of natural gas flows in diffusers for supersonic separators. Energy 37(1):195–200
30. Wen C, Cao X, Yang Y, Zhang J (2012) Evaluation of natural gas dehydration in supersonic swirling separators applying the discrete particle method. Adv Powder Technol 23(2):228–233
31. Wen C, Ding H, Yang Y (2020) Optimisation study of a supersonic separator considering nonequilibrium condensation behaviour. Energy Convers Manage 222:113210
32. Wen C, Karvounis N, Walther JH, Ding H, Yang Y (2020) Non-equilibrium condensation of water vapour in supersonic flows with shock waves. Int J Heat Mass Transf 149:119109
33. White AJ, Hounslow MJ (2000) Modelling droplet size distributions in polydispersed wet-steam flows. Int J Heat Mass Transf 43(11):1873–1884

34. Yang Y, Wen C (2017) CFD modeling of particle behavior in supersonic flows with strong swirls for gas separation. Sep Purif Technol 174:22–28
35. Yuan S, Fan Y, Chen B, Li J, Gao L, Zhang S (2020) Forming and stripping of the wall film and the influence on gas–liquid separation. Asia-Pac J Chem Eng 15(3):e2447

Chapter 7
Euler–Lagrange–Euler Approach for Homogeneous and Heterogeneous Condensing Droplets and Films in Supersonic Separator

7.1 Introduction

To date, four feasible numerical methods have been demonstrated in the literature: the Euler–Euler method, the Euler–Lagrange method, the Euler–Euler–Euler method, and the Euler–Lagrange–Euler method [5]. The Euler–Euler approach treats both the gas phase and the droplets as continuous phases, solving them using Euler's equations. This model effectively calculates evaporation and condensation processes [22]. Dykas and Wróblewski [10] employed the Euler–Euler method to simulate homogeneous condensation flow in both low-pressure and high-pressure supersonic nozzles, examining the gas–liquid performance of the nozzle. Building on this, they revised the drag model and incorporated interphase slip into the Euler–Euler model [11]. Liu et al. [18] developed a Euler–Euler mathematical model for methane-steam condensation, studying the energy separation and condensation characteristics of wet natural gas. Their findings revealed that the interaction of the shock wave with the boundary layer reduces liquefaction efficiency. Wen et al. [29] investigated the flow field structure when homogeneous condensation and shock coexist in supersonic flow, analyzing the impact of swirl intensity on the condensation process. Shooshtari and Shahsavand [23] created a Euler–Euler model that includes both homogeneous and heterogeneous condensation, discovering the effect of heterogeneous salt particles on the condensation and separation within the supersonic separator. Ding et al. [6] further explored the behavior of homogeneous and heterogeneous droplets in a supersonic separator by considering interphase slip and droplet drag. Their results indicate that a moderate size of heterogeneous droplets ensures optimal separation performance.

In contrast to the Euler–Euler model, the Euler–Lagrange model uses a Lagrange approach to simulate droplet motion. This model allows for easy tracking of individual droplet trajectories and collisions between multiple droplets [14]. Wen et al. [28] applied the Euler–Lagrange model to numerically calculate the natural gas flow field and particle separation characteristics in a supersonic cyclone, predicting

H. Ding et al., *Supersonic Separators for Sustainable Gas Purification*, Green Energy and Technology, https://doi.org/10.1007/978-3-031-85213-8_7

the trajectories of particles of different sizes. Liu et al. [17] combined the Euler–Lagrange model with a three-dimensional fluid numerical model of the supersonic separator, exploring the impact of various structural parameters and operating conditions on separation efficiency. Bai et al. [2] utilized the Euler–Lagrange model to simulate particle-laden supersonic branching flow, uncovering key features of this flow. Yang and Wen [30] analyzed particle motion in strongly swirling supersonic flow, finding that gas flow accelerates to supersonic speeds, creating low-pressure and low-temperature conditions ideal for gas removal. Wang et al. [26] used the Euler–Lagrange model to predict droplet trajectories and the separation efficiency of the supersonic separator, showing that increasing droplet size enhances separation efficiency, with particles sized 2–4 μm being particularly sensitive to separation.

It is evident that the Euler–Euler model and the Euler–Lagrange model are two widely used techniques currently. However, despite numerous studies confirming the viability of these models, they share a limitation: only the motion of the gas and droplets is calculated [5, 8]. The formation and development of the liquid film on the wall are either simplified or not considered, making these models inaccurate for simulating supersonic annular flow. To address this issue, efforts to couple these two models with the liquid film model have commenced [15, 19]. In Chap. 6, the Euler–Euler–Euler model is described in detail, revealing that: the Euler–Euler–Euler model excels in simulating supersonic condensation annular flow, as it comprehensively accounts for the changes in gas, droplets, and liquid film. However, challenges arise in calculating discrete quantities such as single droplet trajectories, deposition of individual droplets, droplet collisions and coalescence, as well as droplet and film breakup. These limitations hinder further quantitative studies [16, 33]. In contrast, the Euler–Lagrange–Euler method, which is based on the Euler–Lagrange framework, is adept at tracking local particles [1]. Yuan et al. [32] investigated the re-entrainment mechanisms of wall films, solving the gas core with the Euler–Lagrange method and studying the formation and stripping of the liquid film using the Euler wall-film model. They observed that the average thickness and separation efficiency of the film decrease as the product of the continuous phase's viscosity and density increases. Han et al. [12] developed a Euler–Lagrange–Euler simulation model to examine gas-phase flow, droplet trajectories, and liquid film distributions in various separators within the PEMFC system, ultimately identifying a separator structure more suitable for this system. Deng et al. [5] effectively simulated the gas–liquid separation process in an axial flow cyclone using the Euler–Lagrange method alongside the film model, demonstrating that over time, the liquid film would cover the entire separation segment space.

As noted previously, despite the advantages of the Euler–Euler model, Euler–Lagrange model, and Euler–Euler–Euler model, the Euler–Lagrange–Euler model is the best fit for simulating swirling annular flow. To thoroughly investigate the properties of gas, droplets, and liquid film in supersonic swirling condensation annular flow, as well as their heat-mass interactions, Ding et al. [9] developed an innovative modified Euler–Lagrange–Euler model. This model was introduced into the numerical study of supersonic separators, and for the first time, the homogeneous condensation properties and their impact on flow were calculated. The model's capabilities were validated through experiments. Subsequently, the model was applied to study condensation flow in a supersonic nozzle and swirling condensation annular flow in a supersonic separator. The evolution and interaction of gas, droplets, and liquid film were analyzed, with a focus on the interaction of homogeneous and heterogeneous condensation. Finally, the enhancement of separation capacity by heterogeneous droplets in the presence of homogeneous droplets within the supersonic separator was examined.

7.2 Euler–Lagrange–Euler Model

Swirling condensation annular flow in a supersonic separator was modeled using a coupled approach combining the Eulerian–Lagrangian method and the Eulerian liquid film model. The traditional Euler–Lagrange–Euler model does not account for homogeneous droplets and has limitations in handling complex mass and heat transfer characteristics. In this chapter, the conventional Euler–Lagrange–Euler model was modified to include homogeneous condensation within the mathematical framework. This enhancement incorporates a more accurate nucleation and growth model, addressing the traditional model's deficiencies in mass and heat transfer. The modified Euler–Lagrange–Euler model primarily includes control and tracking equations for gas, droplets, and liquid film, along with the necessary coupling processes among these three fields. As shown in Fig. 7.1, the model is based on the following assumptions: (1) An ideal gas is assumed, with vapor as the condensable gas. (2) Both homogeneous and heterogeneous condensation of droplets are considered. (3) Droplets are assumed to be spherical and can evaporate back into vapor. (4) Collision and coalescence between droplets, as well as droplet breakup, are taken into account. (5) Droplet deposition, film separation, and film stripping are included. (6) The interaction between gas and film primarily involves evaporation and condensation [7].

7.2.1 Modeling the Gas, Droplets, and Liquid Film

The Reynolds-averaged Navier–Stokes (RANS) equations and the species transport equation are employed to predict the gas phase, incorporating continuity, momentum, and energy equations. The governing equations for the gas phase are calculated by

$$\frac{\partial \rho_{\mathrm{g}}}{\partial t} + \nabla\left(\rho_{\mathrm{g}}\vec{u}_{\mathrm{g}}\right) = -\left(O_{\mathrm{pg}} + O_{\mathrm{fg}}\right) \tag{7.1}$$

$$\frac{\partial}{\partial t}\left(\rho_{\mathrm{g}}\vec{u}_{\mathrm{g}}\right) + \nabla\left(\rho_{\mathrm{g}}\vec{u}_{\mathrm{g}}\vec{u}_{\mathrm{g}}\right) = \nabla\left[\mu_{\mathrm{g}}\left(\nabla\vec{u}_{\mathrm{g}} + \vec{u}_{\mathrm{g}}^{\mathrm{T}} - \frac{2}{3}\nabla\vec{u}_{\mathrm{g}}\mathrm{Kr}\right)\right]$$

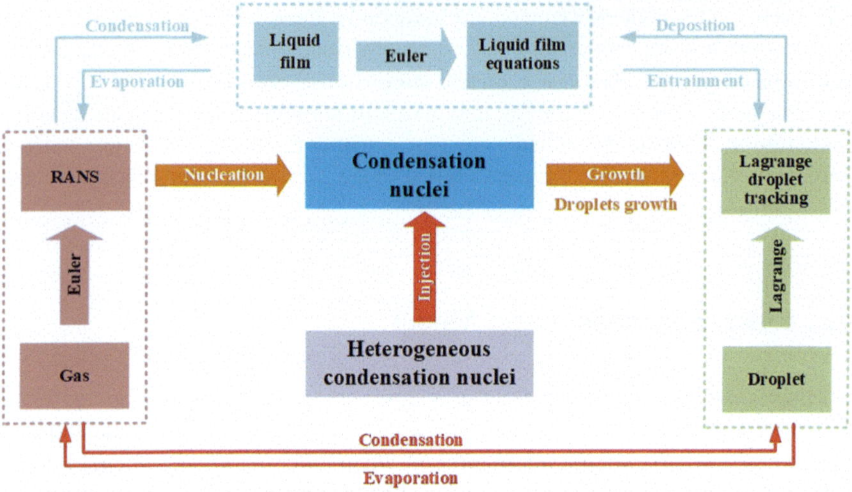

Fig. 7.1 Schematic diagram of the modified Euler–Lagrange–Euler model for modeling supersonic swirling condensation annular flow in the supersonic separator

$$- \nabla p_g + \rho_g \vec{g} + U_{pg} - \left(O_{pg} + O_{fg}\right)\vec{v}_f \quad (7.2)$$

$$\frac{\partial}{\partial t}(\rho_g E) + \nabla\left[\vec{v}_g\left(\rho_g E + p_g\right)\right] = \nabla\left(-\lambda_{eff}\nabla T_g + \tau_e \vec{u}_g\right) + Q_{pg} + \left(O_{pg} + O_{fg}\right)h_v \quad (7.3)$$

$$\frac{\partial\left(\rho_g Y_i\right)}{\partial t} + \nabla\left(\rho_g \vec{u}_g Y_i\right) = -\nabla\vec{J}_i + O_i \quad (7.4)$$

$$\vec{J}_i = -\left(\rho_g D_{i,m} + \frac{\mu_g}{Sc}\right)\nabla Y_i - D_{T,i}\frac{\nabla T_g}{T_g} \quad (7.5)$$

where ρ, $\vec{u}$, T and p are the density, velocity, temperature and pressure, respectively. The subscripts g, p, f and i represent gas, particles, film and species index. μ, Kr, E, λ_{eff}, τ_e, Y, Sc and $\vec{J}$ are the dynamic viscosity, Kronecker delta number, total energy, effective thermal conductivity, stress tensor, mass fraction, Turbulent Schmidt number and diffusion flux, respectively. m represents mass. O, U, and Q represent transfer source terms of mass, momentum and energy. Their subscripts determine their transfer direction. For example, O_{pg} represents the mass transferred from droplets to the gas phase. and their transfer direction also depends on the subscripts. $D_{i,m}$ is the mass diffusion coefficient for species i in the mixture, $D_{T,i}$ is the thermal diffusion coefficient.

The droplet trajectory equation is

$$m_p \frac{d\vec{u}_p}{dt} = m_p\left(\vec{u}_g - \vec{u}_p\right)\frac{18\mu}{\rho_p d_p^2 C_c} + m_p\vec{g} \tag{7.6}$$

where d is the diameter. And $\vec{g}$ is the acceleration of gravity. C_c is the Cunningham correction to Stokes' drag law [21], which is described as:

$$C_c = 1 + \frac{2\lambda_{mg}}{d_p}\left(1.257 + 0.4e^{-\frac{1.1d_p}{2\lambda_{mg}}}\right) \tag{7.7}$$

where λ_{mg} is the mean free path of a gas molecule. Collision and coalescence between homogeneous and heterogeneous droplets can occur when their paths intersect, affecting droplet properties as measured by the O'Rourke algorithm [20]. This algorithm assumes that two droplets can collide only if they enter the same computational cell. When droplets meet the conditions for collision, their likely outcomes include coalescence, bouncing, or breaking up. When homogeneous droplets collide with heterogeneous ones, the significant size difference makes it highly probable that the homogeneous droplets will be absorbed by the heterogeneous droplets. This likelihood is determined by the final calculations of the O'Rourke algorithm. The algorithm considers factors such as relative velocity, droplet size, and impact parameter to predict the result of each collision event, thus providing a detailed understanding of the interactions between different types of droplets within the system.

To simulate the liquid film behavior, the Eulerian wall film model is employed. The governing equation for the liquid film is

$$\frac{\partial(\rho_f\delta)}{\partial t} + \nabla(\rho_f\vec{u}_f\delta) = \delta\left(O_{pf} + O_{gf}\right) \tag{7.8}$$

$$\begin{aligned}\frac{\partial}{\partial t}(\rho_f\delta u_f) + \nabla(\rho_f\delta\vec{u}_f\vec{u}_f) = &-\delta\nabla p_{g,mix} + \rho_f\delta\vec{g} + \frac{3}{2}\tau_{gf}\\ &- \frac{3\mu_f\vec{u}_f}{\delta} + \delta\left(O_{pf} + O_{gf}\right)\vec{u}_f\end{aligned} \tag{7.9}$$

$$\begin{aligned}\frac{\partial}{\partial t}(\rho_f\delta h_f) + \nabla(\rho_f\delta h_f\vec{u}_f) = &\frac{\lambda_e}{\delta}(T_{sur} + T_{wall} - 2T_{mid})\\ &+ \delta\left(O_{pf} + O_{gf}\right)h_f\end{aligned} \tag{7.10}$$

where δ, $p_{g,mix}$, τ_{gf}, h_f, and h_v are the film thickness, gas mixture pressure, viscous shear stress, enthalpy of film and latent heat of vapor, respectively. T_{sur}, T_{wall}, T_{mid} are the temperature of the film surface, wall, and film half-depth, respectively.

7.2.2 Modeling the Droplet Nucleation and Growth

The transfers of mass and momentum from the gas phase to the discrete droplet phase can be determined by analyzing the change in mass and momentum of a droplet as it moves through each control volume:

$$O_{\mathrm{gp}} = \frac{\Delta m_{\mathrm{p}}}{m_{\mathrm{p},0}} \frac{q_{\mathrm{p},0}}{V_{\mathrm{cell}}} \tag{7.11}$$

$$U_{\mathrm{gp}} = \sum O_{\mathrm{gp}} \left(\vec{u}_{\mathrm{g}} - \vec{u}_{\mathrm{p}}\right) \frac{18\mu}{\rho_{\mathrm{p}} d_{\mathrm{p}}^2 C_{\mathrm{c}}} \tag{7.12}$$

where V_{cell} is the cell volume. $m_{\mathrm{p},0}$ and $q_{\mathrm{p},0}$ are the initial mass (kg) and initial mass flow rate (kg s^{-1}) of droplets at the cell entry.

For homogeneous droplets, their mass increase depends on:

$$\Delta m_{\mathrm{p}} = m_{\mathrm{nucleation}} + m_{\mathrm{growth}} \tag{7.13}$$

For heterogeneous droplets, their mass increase depends on:

$$\Delta m_{\mathrm{p}} = m_{\mathrm{growth}} \tag{7.14}$$

where $m_{\mathrm{nucleation}}$ is the increase in droplet mass due to homogeneous nucleation, and m_{growth} is the increase in droplet mass due to homogeneous or heterogeneous growth.

A single droplet nucleates and causes a change in mass of

$$m_{\mathrm{nucleation}} = \frac{4}{3} J \rho_{\mathrm{p}} \pi r_{\mathrm{c}}^3 V_{\mathrm{cell}} \Delta t \tag{7.15}$$

where Δt represents the time step.

7.2.3 Modeling the Droplet-Film and Gas-Film Mass-Heat Interaction

The calculation equation for O_{gf} is as follows [25]:

$$O_{\mathrm{gf}} = \frac{(\rho_{\mathrm{m}} M_{\mathrm{v}}/\delta)}{\frac{\rho_{\mathrm{g,mix}} M_{\mathrm{v}}}{\delta} + C_{\mathrm{phase}}} C_{\mathrm{phase}} \left(Y_{\mathrm{v}} - Y_{\mathrm{v,s}}\right) \tag{7.16}$$

where $\rho_{\mathrm{g,mix}}$, M_{v} and C_{phase} are the density of the gas mixture, mass diffusivity of the vapor and phase change constant. Y_{v} and $Y_{\mathrm{v,s}}$ represent the vapor mass fraction and the saturation vapor mass fraction. The subscript s represents saturation.

When $Y_v > Y_{v,s}$, C_{phase} can be calculated as

$$C_{phase} = C_{con} \tag{7.17}$$

When $Y_v < Y_{v,s}$, C_{phase} can be calculated as

$$C_{phase} = C_{evap}\delta \times 10^4 \tag{7.18}$$

where C_{con} and C_{evap} are condensation and evaporation constant.

$Y_{v,s}$ is computed as:

$$Y_{v,s} = \frac{p_{v,s}}{p_g} \frac{M_{w,v}}{M_{w,g,mix}} \tag{7.19}$$

where $M_{w,v}$ and $M_{w,g,mix}$ represent the molecular weight of the vapor and the gas mixture, respectively. $p_{v,s}$ is solely dependent on temperature and is calculated as follows [24]:

$$\begin{aligned} \log_{10} p_{v,s} = & -2.1794 + 0.02953 \times T - 9.1837 \times 10^{-5} \times T^2 \\ & + 1.4454 \times 10^{-7} \times T^3 \end{aligned} \tag{7.20}$$

O_{pf} is computed as:

$$O_{pf} = O_{de} - O_{se} - O_{st} \tag{7.21}$$

where O_{de} denotes the mass exchange from droplet deposition. O_{se} refers to the mass source from liquid film separation, occurring near the separation port of the supersonic separator when the following two conditions are satisfied:

$$\varphi_{se} > \varphi_c, \ \mathrm{We}_f > \mathrm{We}_c \tag{7.22}$$

where φ_{se} and We_f are the separation angle of the liquid outlet and the Weber number based on the film.

When there is a significant relative velocity between the gas core and the liquid film on the wall, Kelvin–Helmholtz waves develop and amplify on the film surface, ultimately leading to the stripping of droplets from the surface. This process is referred to as film stripping. O_{st} is computed as [32]:

$$O_{st} = \frac{C \times 9 \times \sqrt[3]{16} \times \pi \omega_{wf} \rho_f}{V_{cell}} \left(\frac{\mu_f \sqrt{\sigma/\rho_f}}{\beta \rho_g \vec{u}_g^2} \right)^{2/3} \tag{7.23}$$

where C and ω_{wf} are the mass coefficient and the wave frequency. ω_{wf} is defined by

$$\omega_{\mathrm{wf}} = 0.384\sqrt{\frac{\rho_{\mathrm{g}}}{\rho_{\mathrm{f}}}\frac{\rho_{\mathrm{g}}\vec{u}_{\mathrm{g}}^{3}}{\sigma}} \tag{7.24}$$

It is also important to clarify that in this chapter, the superscripts and subscripts in are used to denote the inlet, whereas the superscripts and subscripts out indicate the outlet.

7.2.4 Solution Strategy

The boundary conditions for the simulation are defined using pressure at the inlet and outlet. At the inlet, discrete droplets are introduced, with their initial velocity set to match the velocity of the surrounding gas phase. Additionally, the initial diameter and mass flow rate of these foreign droplets can be specified, allowing for precise control over their characteristics as they enter the system.

Figure 7.2 shows the solution strategy for the proposed modified Euler–Lagrange–Euler model. The overall domain calculation progresses through the alternating computations of gas, droplets, and liquid film until convergence is reached. In this approach, the gas core calculation provides source term information for the liquid film, and then the liquid film calculation supplies source term data back to the gas core flow. For discrete droplet computations, data about heterogeneous and homogeneous condensation nuclei are stored in external files, allowing droplet nuclei to be injected into the flow field at each time step by referencing these files. Heterogeneous condensation nuclei remain constant, while homogeneous condensation nuclei are updated at each time step based on whether the vapor supersaturation $S > 1$ is met in each computational cell. Initially, the homogeneous condensation nuclei file is empty. After several time-step iterations, homogeneous condensation nuclei begin to form and are recorded, with their mass determined by Eq. (7.15). The RSM turbulence model is used due to its high accuracy in swirling flow [13]. The governing equations, as well as the turbulent kinetic energy and dissipation rate equations, are solved using the second-order upwind scheme. Pressure inlet and outlet conditions are employed as boundary conditions. When a droplet hits the wall, it is either collected by the liquid film or breaks apart and bounces off. Droplets reaching the gas outlet are considered to have escaped. The accuracy of water mass balance, accounting for phase changes and mass exchange, is maintained at better than 1%.

7.3 Validation

To ensure the accuracy of the simulation model, it is essential to validate it through experimental testing. This section provides a brief overview of the experimental process used for this validation and presents a comparison between the experimental results and the results obtained from the simulation. By doing so, we verify the

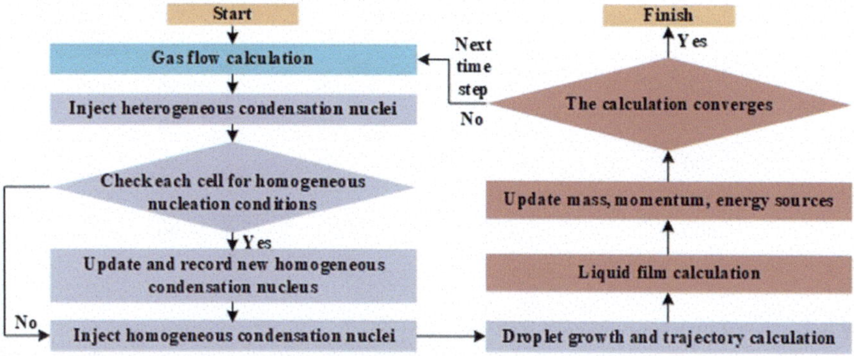

Fig. 7.2 Solution strategy of the modified Euler–Lagrange–Euler model

accuracy of the model that was established in Sect. 7.2. The comparison demonstrates that the Euler–Lagrange–Euler model developed in Sect. 7.2 exhibits a high level of accuracy. Consequently, this model proves to be highly effective for analyzing and understanding the complex mass and heat transfer phenomena occurring within the supersonic separator.

7.3.1 Introduction to the Experimental Apparatus

To validate the proposed model, experiments were conducted using a specific setup, shown in Fig. 6.3. The gas control subsystem is capable of generating gas with adjustable pressure, temperature, and humidity. Simultaneously, the droplet control subsystem creates heterogeneous droplets with controllable particle size and concentration. Once mixed, the gas and droplets flow into either the supersonic nozzle or the supersonic separator to start the experiment. Sensors for measuring gas, droplets, and liquid film are strategically placed along the pipeline, and all data are displayed in real-time on the host computer. The specific components, principles, and schematic diagrams of the apparatus are detailed in Chap. 6 and will not be repeated here.

Based on these experimental setups, a series of experiments for the supersonic nozzle and supersonic separator were meticulously conducted under the specific conditions listed in Tables 7.1 and 7.2. Among them, p_{in}, p_{out}, T_{in}, T_{out} RH_{in}, d_{p}^{in} and q_{p}^{in} represent the inlet pressure, outlet pressure, inlet temperature, outlet temperature, inlet humidity, inlet heterogeneous droplet diameter and inlet heterogeneous droplet mass flow rate. These experiments were performed with the inlet pressure p_{in} and inlet temperature T_{in} representing stagnation parameters, while maintaining a constant outlet static pressure of 100 kPa throughout the testing process.

Table 7.1 Conditions for supersonic nozzle experiments

Case	p_{in} (kPa)	p_{out} (kPa)	T_{in} (K)	RH_{in} (%)
a	300	100	303	80
b	500	100	303	80
c	300	100	323	80
d	300	100	323	54
e	300	100	323	90

Table 7.2 Conditions for supersonic separator experiments

Case	p_{in} (kPa)	p_{out} (kPa)	T_{in} (K)	T_{out} (K)	RH_{in} (%)	d_{p}^{in} (μm)	q_{p}^{in} (kg/s)
1	300	100	300	300	100	2.2	0.0001
2	300	100	300	300	100	2.2	0.0010
3	300	100	300	300	100	2.2	0.0020

7.3.2 *Validation of Supersonic Condensation Flow*

To validate the proposed model for supersonic condensation flow, Ding et al. [9] conducted experiments. Conduct experiments using the supersonic nozzle model shown in Chap. 4, and then create a model, divide the grid, and perform CFD simulations. Figure 7.3 shows that the pressure ratio curves of the five sets of results correspond well, indicating that an increase in p_{in}, RH_{in}, or T_{in} advances the onset position of homogeneous condensation, under constant conditions. This demonstrates that the proposed model effectively predicts the homogeneous nucleation and growth process in supersonic condensation flow.

7.3.3 *Validation of Supersonic Condensation Annular Flow*

To validate the proposed model for supersonic swirling condensation annular flow, an extensive series of experiments were conducted. Figure 7.4 provides a detailed illustration of the dimensions and mesh configuration of the supersonic separator utilized in the study. The separator was meticulously modeled using a structured hexahedral mesh, which was specifically refined near the walls and throat to accurately capture the intricate flow details in these critical areas.

The process of determining the optimal mesh involved conducting Grid Convergence Index (GCI) tests, as referenced in Baker et al. [3]. These tests revealed that the optimal number of grid cells required to ensure sufficient resolution and accurate simulation results was 557,560. In the provided figure, the z-axis is used to represent the axial direction of the flow, allowing for a clear visualization of the

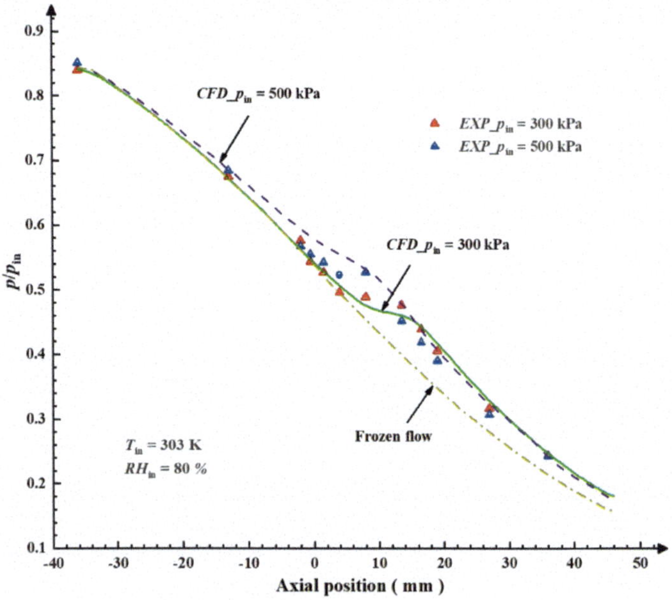

(a)

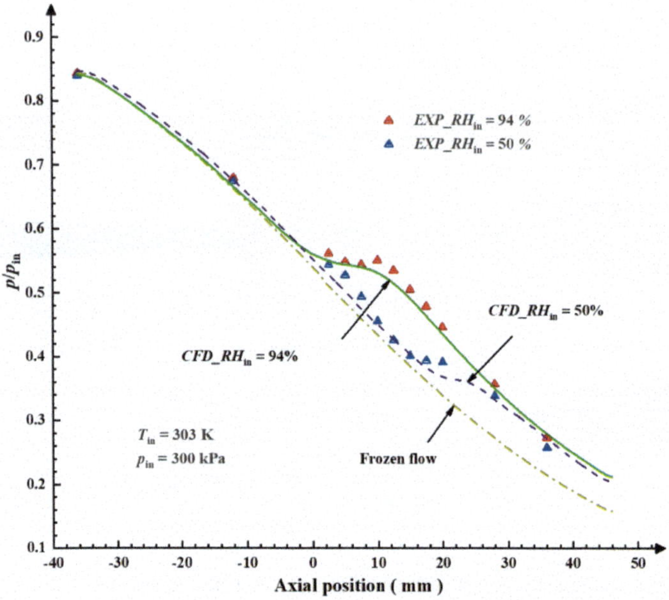

(b)

Fig. 7.3 Experimental and simulation results of axial pressure distribution under different experimental conditions: **a** keep $T_{in} = 303$ K, $RH_{in} = 80\%$, change p_{in}; **b** keep $T_{in} = 303$ K, $p_{in} = 300$ kPa, change RH_{in}; **c** keep $RH_{in} = 80\%$, $p_{in} = 300$ kPa, change T_{in}

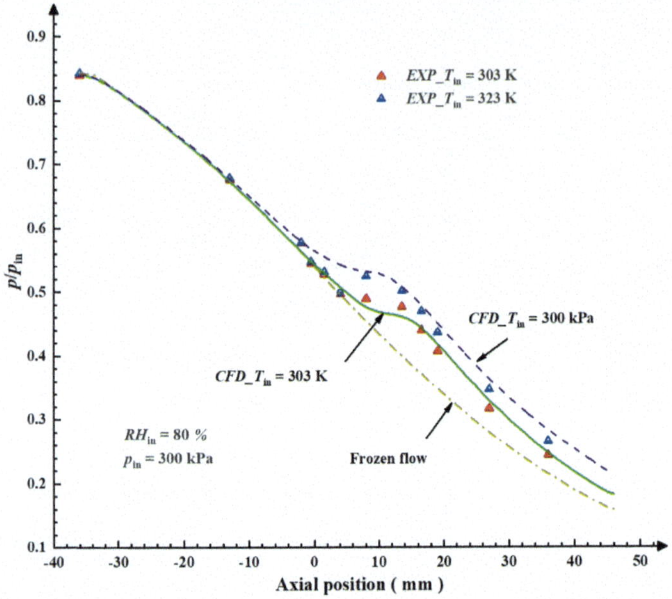

(c)

Fig. 7.3 (continued)

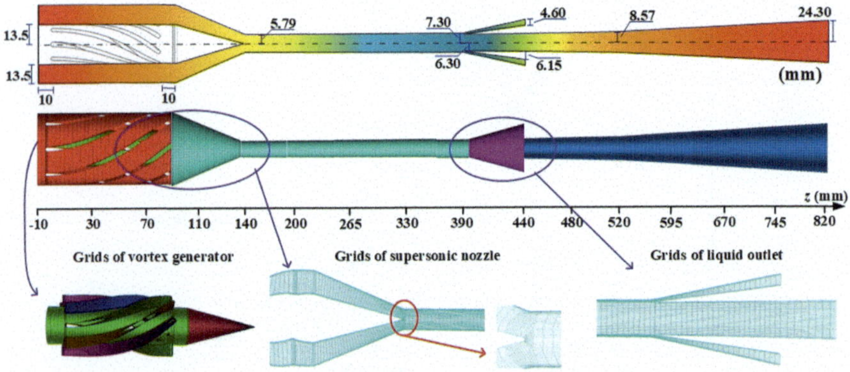

Fig. 7.4 Size and mesh of the supersonic separator

flow dynamics. This thorough approach to mesh refinement and validation highlights the reliability and precision of the experimental model employed in this study, reinforcing its credibility and the validity of the results obtained.

Figure 7.5 presents the gas pressure distribution as obtained from both experimental data and the CFD model for case 1. The experimental results closely match the CFD data, and the consistency is evaluated using the R-Square (R^2) metric.

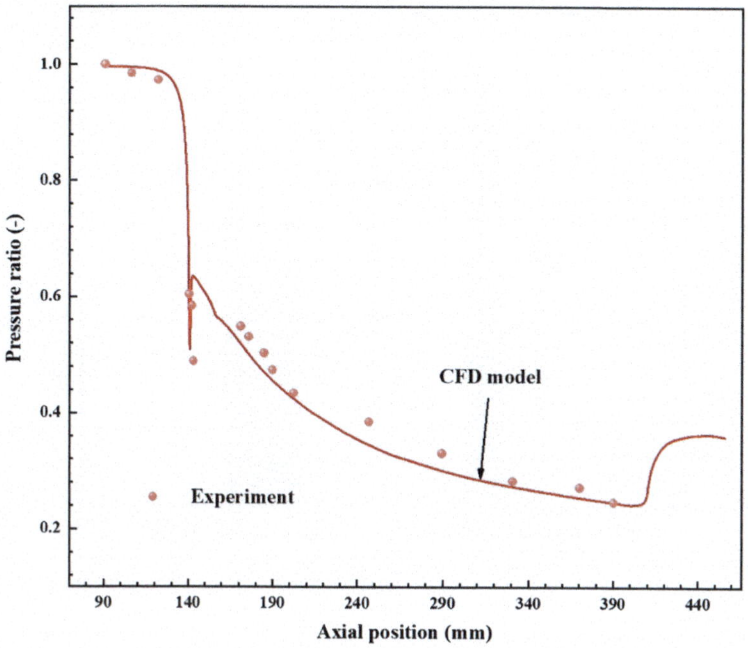

Fig. 7.5 Gas pressure distribution along the wall surface of the supersonic nozzle obtained by experiment and CFD model for case 1

The R^2 value of the pressure ratio is 0.963, demonstrating that the established CFD model effectively captures the gas behavior and gas-droplet phase changes within the supersonic separator.

Figure 7.6 displays the computed results for vapor removal rate (η_v) and dew point depression (ΔT_d). The parameter η_v measures the alteration in vapor mole fraction within the gas phase, where a higher η_v denotes superior separation performance. The formula for η_v is expressed as:

$$\eta_v = \frac{x_{v,in} - x_{v,out}}{x_{v,in}} \times 100\% \tag{7.25}$$

where $x_{v,in}$ is the inlet vapor mole fraction, and $x_{v,out}$ is the vapor mole fraction at the gas outlet. ΔT_d is calculated as the difference between the inlet and the gas outlet dew point, which is expressed as

$$\Delta T_d = T_{d,in} - T_{d,out} \tag{7.26}$$

In case 1, the CFD model yielded η_v and ΔT_d values of 73.07% and 20.04 K, respectively, compared to the experimental values of 77.27% and 22.95 K. The prediction errors for both are within a reasonable range, indicating the model's reliability.

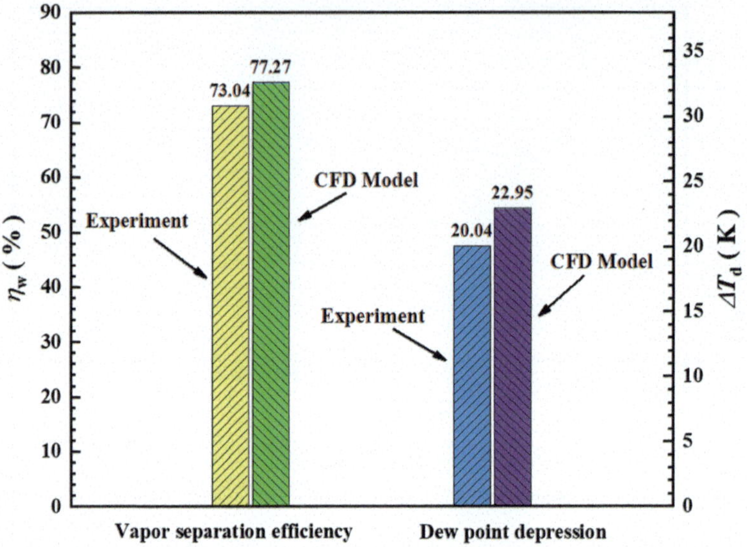

Fig. 7.6 Water separation efficiency and dew point depression obtained by experiment and CFD model for case 1

Figure 7.7 presents the CFD model's prediction results for liquid film behavior. Specifically, Fig. 7.7a–c illustrates the variation in film thickness at the liquid outlet for cases 1–3 over a period of 120 s during the experiment. The processed experimental results are juxtaposed with the simulation data, as illustrated in Fig. 7.7d.

For cases 1–3, the experimental measurements of film thickness were recorded as 18.80 μm, 52.71 μm, and 89.24 μm, respectively, while the corresponding simulated film thicknesses were determined to be 12.74 μm, 50.29 μm, and 88.15 μm. The close agreement between these experimental and simulated results confirms that the Euler–Lagrange–Euler model accurately predicts liquid film behavior. This validation underscores the robustness of the model and demonstrates its capability to simulate complex physical phenomena with a high degree of accuracy. Consequently, the Euler–Lagrange–Euler model is established as a highly reliable tool for studying supersonic swirling condensation annular flow, providing valuable and precise insights into the dynamics and efficiency of the separation process under various operational conditions.

7.4 Coupling of Homogeneous and Heterogeneous Droplets

The model validation results indicate that the established modified Euler–Lagrange–Euler model produces reasonable data. In this section, this modified model was utilized to simulate and analyze the flow characteristics of supersonic condensation

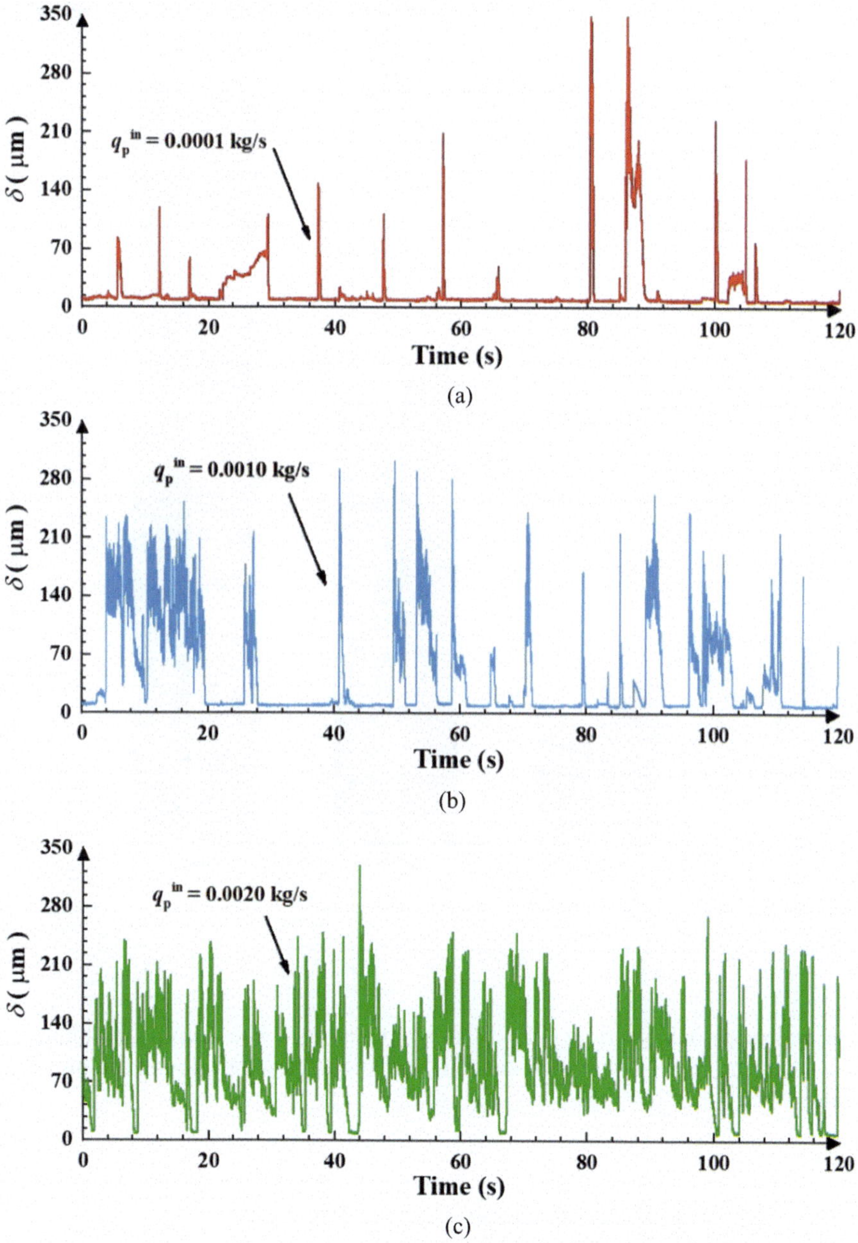

Fig. 7.7 Film thickness of liquid outlet obtained by experiment and CFD model: **a** case 1; **b** case 2; **c** case 3; **d** comparison results between experimental values and simulation values

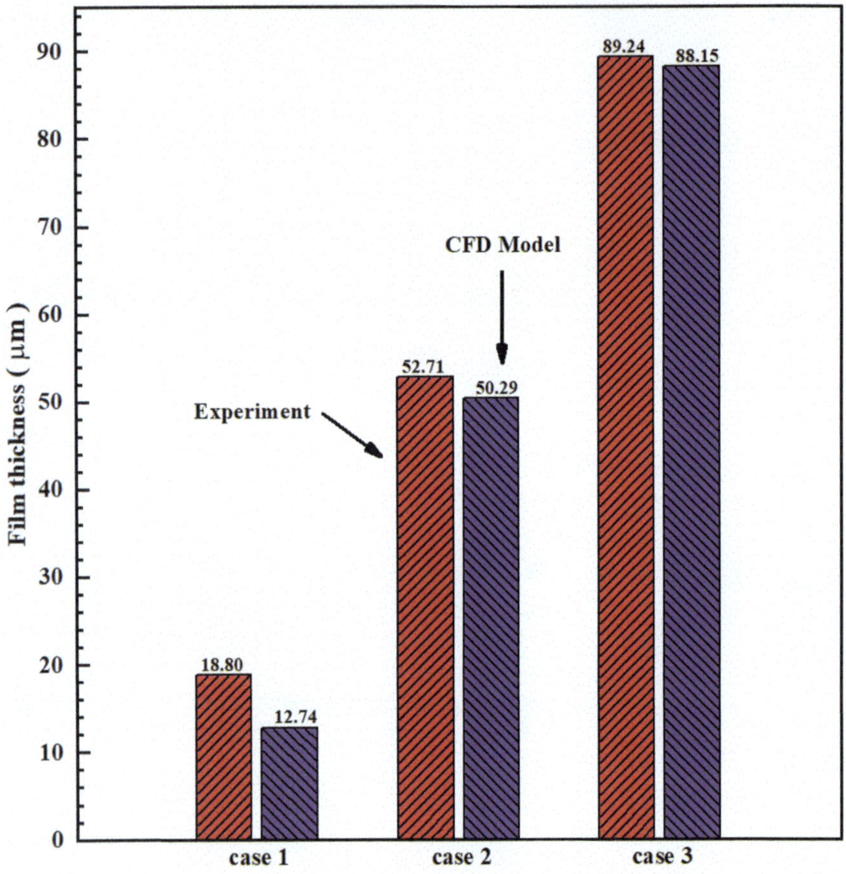

(d)

Fig. 7.7 (continued)

flow and supersonic swirling condensation annular flow. The coupling relationship between homogeneous and heterogeneous droplets inside a supersonic separator was studied.

7.4.1 Effect of Heterogeneous Droplet Diameter on Condensation

Figure 7.8 demonstrates the influence of foreign droplets on the homogeneous condensation within the flow field. The presence of heterogeneous droplets growth delays the onset of the homogeneous condensation. Following this, the impact of the initial diameter of heterogeneous droplets ($d_{\mathrm{p}}^{\mathrm{in}}$) on the condensation flow is analyzed.

Figure 7.9 shows how variations in $d_{\mathrm{p}}^{\mathrm{in}}$ affect gas-phase parameters in the supersonic flow. It is found that smaller heterogeneous droplet diameters lead to a higher-pressure ratio and a greater reduction rate of vapor mass fraction y_{v} in the nozzle's convergent section.

The presence of heterogeneous droplets allows vapor molecules to condense on their surfaces as soon as the vapor deviates from the saturation line. As a result, the pressure increase is more gradual compared to the abrupt rise seen with only homogeneous condensation. This behavior highlights the significant role those heterogeneous droplets play in moderating the condensation process, leading to a more controlled and steady increase in pressure, which is beneficial for the stability and efficiency of the supersonic flow system.

Figure 7.10 is the distribution of axial subcooling ΔT and Sauter diameter d_{32} for different heterogeneous droplet diameters. ΔT is calculated by $\Delta T = T_{\mathrm{v,s}} - T_{\mathrm{g}}$, where $T_{\mathrm{v,s}}$ is the saturation temperature of vapor. d_{32} is calculated as follows [31]:

$$d_{32} = \frac{\sum n_i d_{\mathrm{p}}^3}{\sum n_i d_{\mathrm{p}}^2} \tag{7.27}$$

In Fig. 7.10, the normalized $d_{32,\mathrm{het}}$ is calculated by subtracting $d_{\mathrm{p}}^{\mathrm{in}}$ from the current $d_{32,\mathrm{het}}$. As $d_{\mathrm{p}}^{\mathrm{in}}$ decreases, homogeneous condensation flow diminishes, while heterogeneous condensation flow becomes more significant. When $d_{\mathrm{p}}^{\mathrm{in}} = 1.0\,\mu m$, vapor molecules condense directly on the provided condensation cores, achieving thermodynamic equilibrium and completely suppressing homogeneous condensation.

In this case, the location of maximum subcooling is near the nozzle outlet. For $d_{\mathrm{p}}^{\mathrm{in}} = 2.0\,\mu m$, as the heterogeneous droplets grow large enough, they lose their ability to act as condensation cores. The flow has not yet reached thermodynamic equilibrium, leading to both homogeneous and heterogeneous condensation occurring together in the flow field. The maximum subcooling position lies between the maximum subcooling positions of the homogeneous case and the $d_{\mathrm{p}}^{\mathrm{in}} = 1.0\,\mu m$ case.

7.4.2 *Effect of Heterogeneous Droplet Mass Flow Rate on Condensation*

The concentration of inlet heterogeneous droplets also significantly impacts the supersonic condensation flow. Figure 7.11 illustrates the distribution of $y_{\mathrm{v,in}} - y_{\mathrm{v}}$ and p/p_{in} in the supersonic nozzle. An increase in the inlet droplet mass flow rate ($q_{\mathrm{p}}^{\mathrm{in}}$) corresponds to an increase in p/p_{in} and a higher rate of y_{v} reduction.

Figure 7.12 shows the distribution of ΔT and d_{32} for different heterogeneous droplet mass flow rates. As $q_{\mathrm{p}}^{\mathrm{in}}$ increases, more condensation cores are available for vapor molecules to condense and grow on, thereby strengthening the inhibitory effect on homogeneous condensation. Consequently, as $q_{\mathrm{p}}^{\mathrm{in}}$ rises, the maximum subcooling

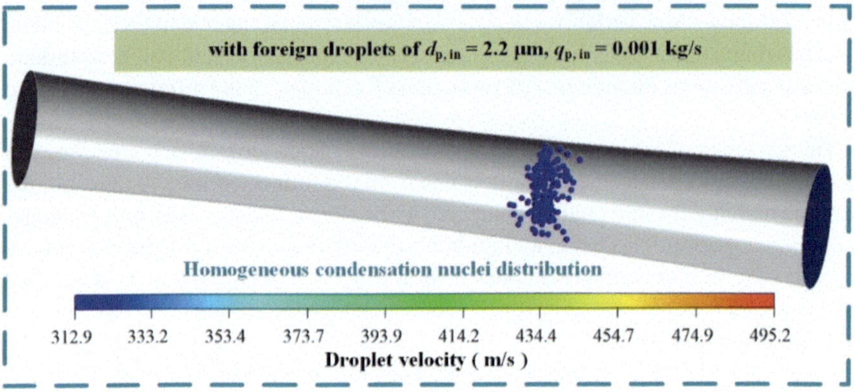

(a)

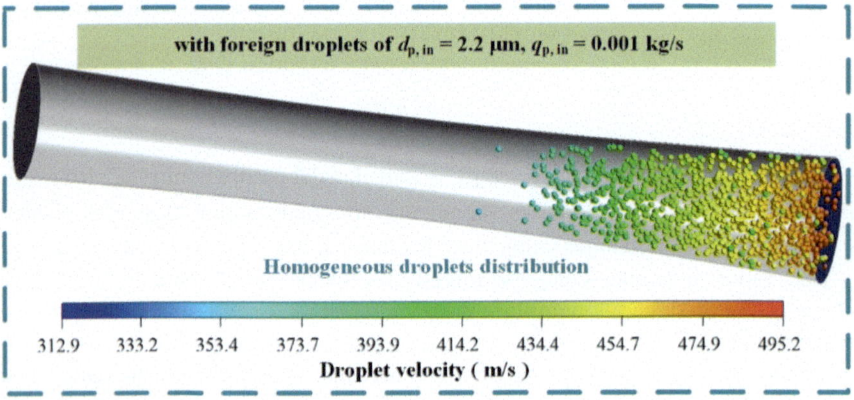

(b)

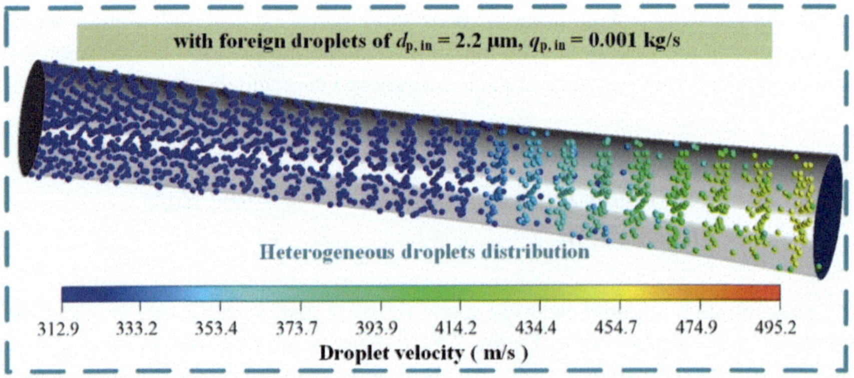

(c)

Fig. 7.8 Distributions of homogeneous cores, homogeneous droplets and heterogeneous droplets: **a** homogeneous condensation nuclei distribution; **b** homogeneous droplets distribution; **c** heterogeneous droplets distribution; **d** homogeneous droplets nuclei distribution without foreign droplets; **e** homogeneous condensation distribution without foreign droplets

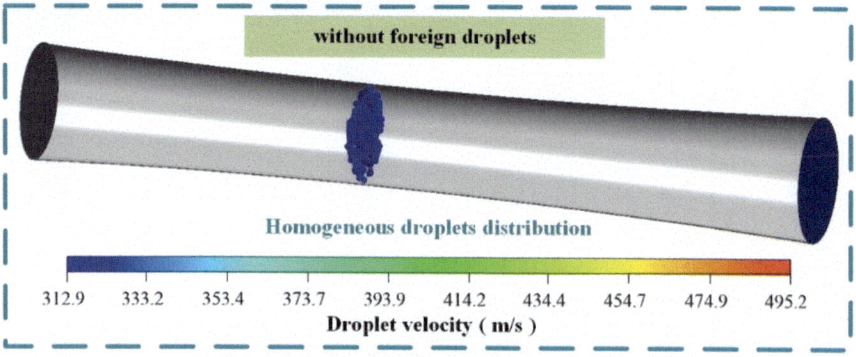

(d)

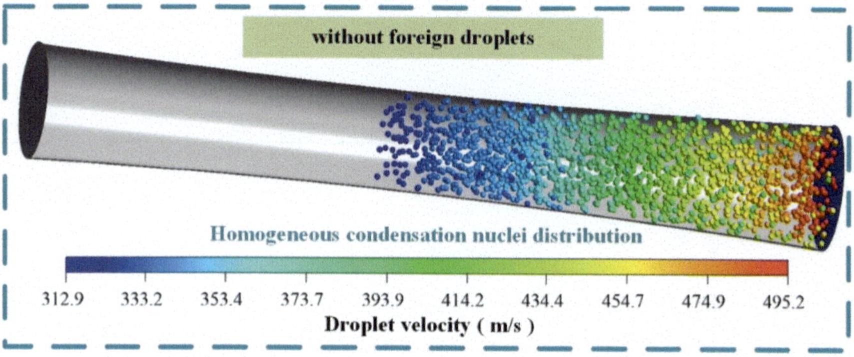

(e)

Fig. 7.8 (continued)

position and the onset position of homogeneous condensation shift closer to the outlet. For $q_{\mathrm{p}}^{\mathrm{in}} = 0.001$ kg/s, 0.0015 kg/s, and 0.002 kg/s, the homogeneous condensation onset positions are 2.5 mm, 12.5 mm, and 20.1 mm, respectively.

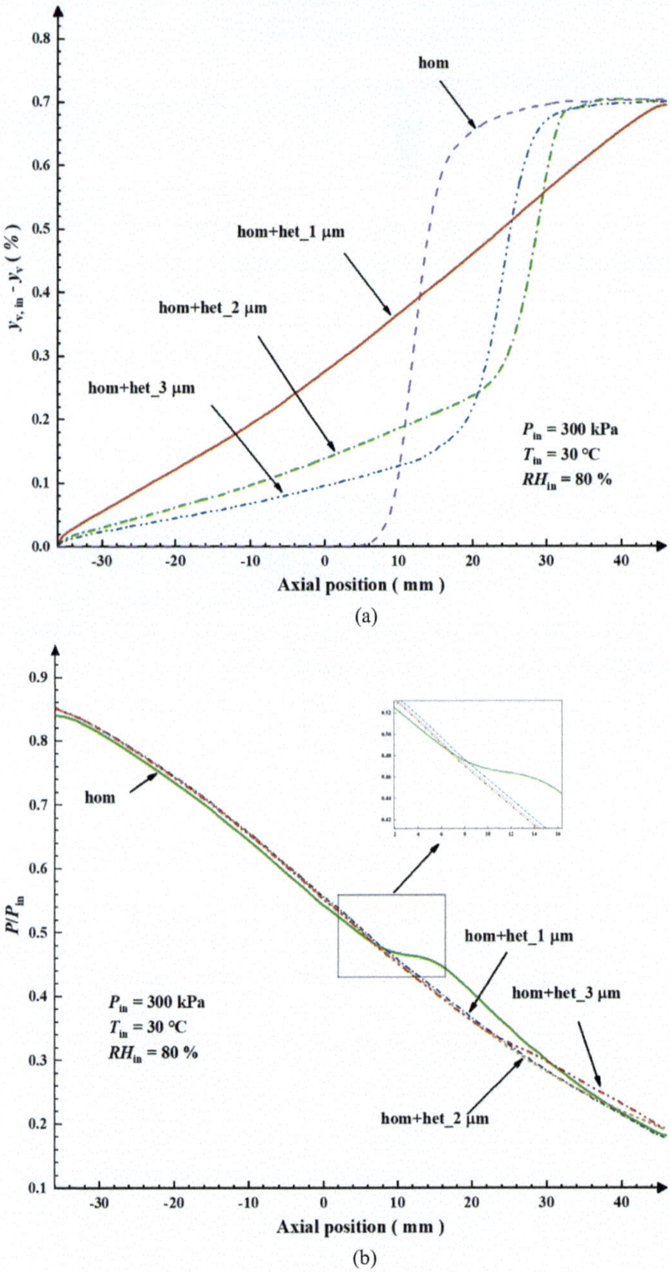

Fig. 7.9 Influence of heterogeneous droplet diameter on vapor mass fraction and pressure ratio: **a** axial relative mass fraction of vapor; **b** axial pressure ratio along the path

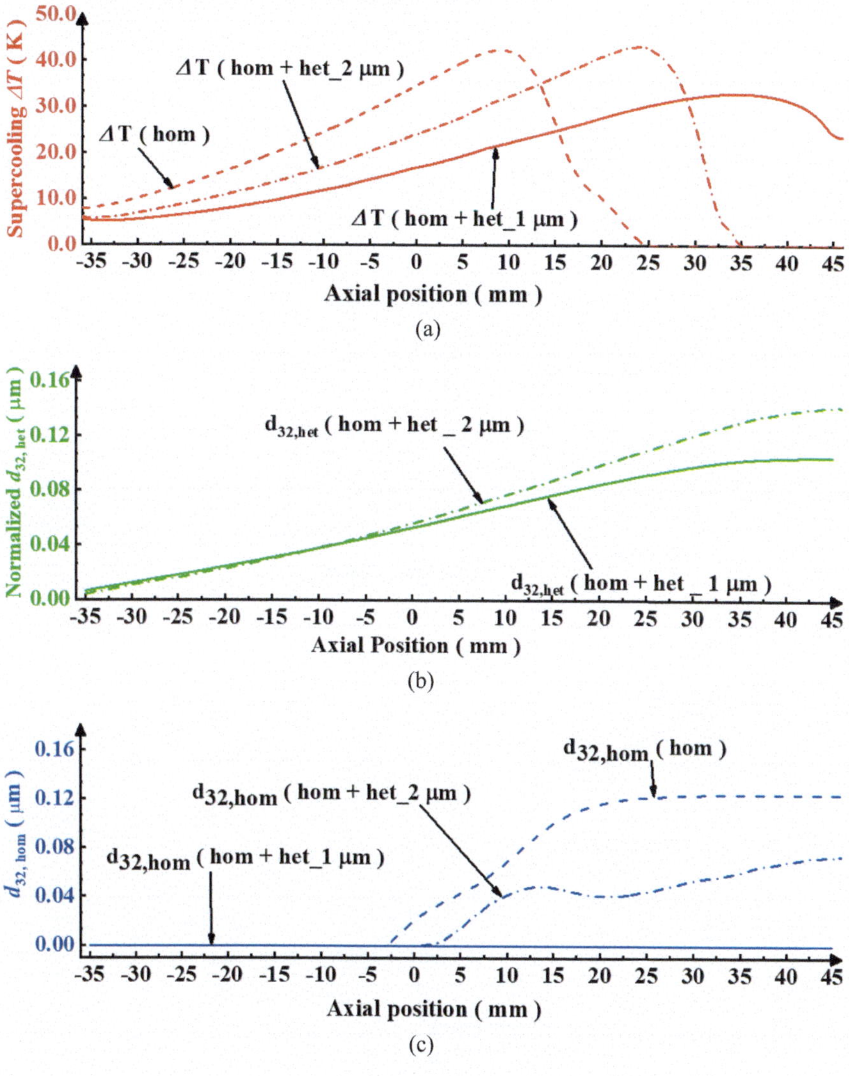

Fig. 7.10 Influence of heterogeneous droplet diameter on subcooling and droplet Sauter diameter: **a** subcooling ΔT; **b** normalized $d_{32,\text{het}}$; **c** $d_{32,\text{hom}}$

This shift indicates that higher concentrations of heterogeneous droplets not only provide more condensation cores but also delay the homogeneous condensation process, pushing it further downstream. This effect is crucial for optimizing the design and operation of supersonic separators, ensuring that the condensation process is controlled and efficient.

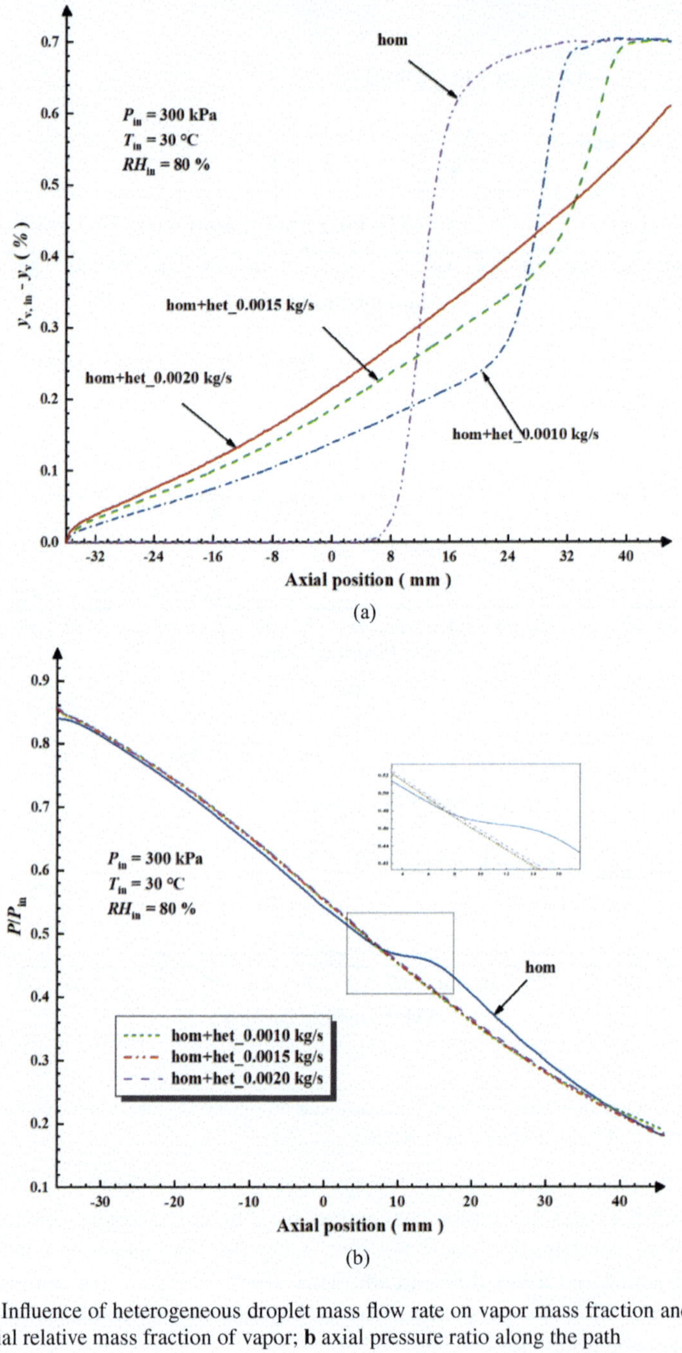

Fig. 7.11 Influence of heterogeneous droplet mass flow rate on vapor mass fraction and pressure ratio: **a** axial relative mass fraction of vapor; **b** axial pressure ratio along the path

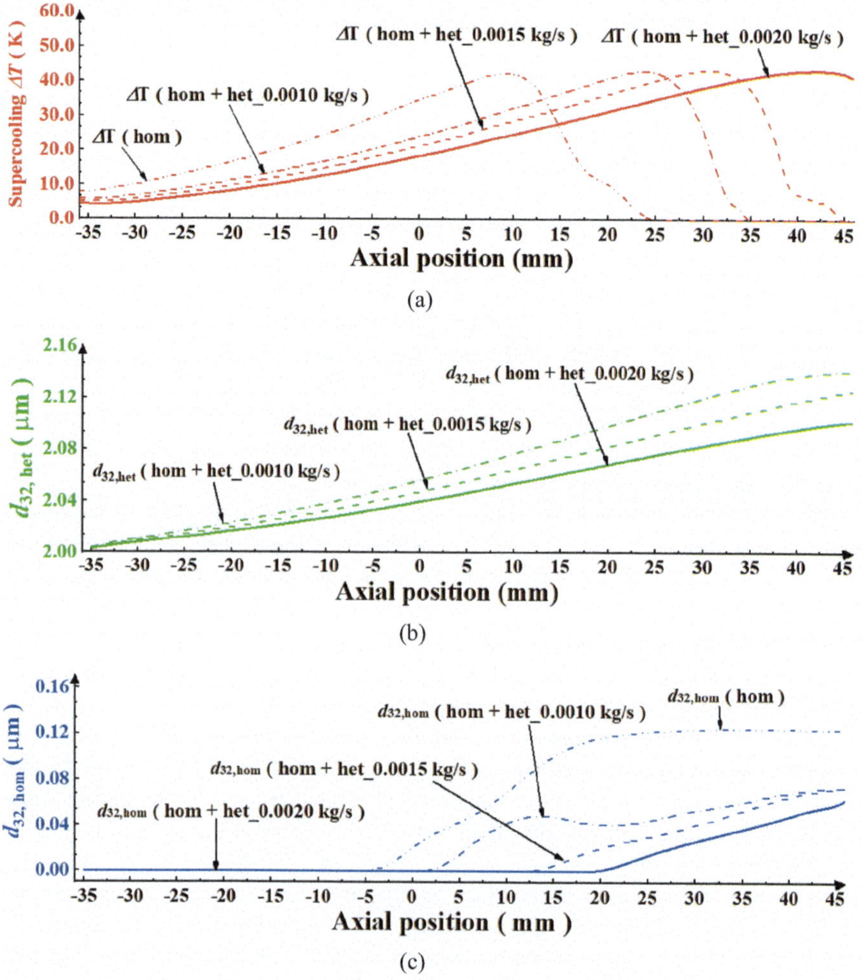

Fig. 7.12 Influence of heterogeneous droplet mass flow rate on subcooling and droplet Sauter diameter: **a** subcooling ΔT; **b** $d_{32,\mathrm{het}}$; **c** $d_{32,\mathrm{hom}}$

7.5 Characteristic Analysis of Gas, Droplet, and Liquid Film

This section presents and analyzes the flow characteristics of gas, droplets, and liquid films to better illustrate the complex flow field characteristics inside the supersonic separator.

7.5.1 Gas Behavior in Supersonic Swirling Flow

Creating a high-speed swirl field is crucial for achieving effective gas–liquid separation [4]. Figure 7.13a displays the pressure and Mach number distributions along the section at $y = 0$ mm in supersonic separator. As gas flows through the convergent section of supersonic nozzle, the temperature and pressure of the fluid decrease due to the reduction in the nozzle's cross-sectional area. This decrease in temperature and pressure causes the fluid to expand, thereby increasing its velocity. At the nozzle throat, the gas reaches supersonic speeds, characterized by a Mach number (Ma) of 1. As the gas continues to flow through divergent section, it continues to expand, further increasing its velocity. Just before the gas pressure rises in the diffuser, the pressure can drop to a minimum of 20 kPa, while the maximum Mach number can reach as high as 2.3. This detailed analysis of the pressure and Mach number distributions highlights the importance of creating a high-speed swirl field to enhance the separation process within the supersonic separator.

The significant reduction in the pressure downstream of the nozzle throat results in the low-temperature environment, as depicted in Fig. 7.13b. This temperature drop, which is approximately the 60 K, facilitates the condensation of vapor. As a result, the mass fraction of the vapor decreases since the vapor mass is transferred to condensed droplets, thereby enabling effective gas purification.

Moreover, within the diffuser, there is an observed increase in the vapor mass fraction due to the creation of an evaporative environment, which is a consequence of rising temperature and pressure conditions. Despite this, under the conditions described in case 1, the outlet vapor mass fraction is still lower than the inlet vapor mass fraction, indicating that the separation process is effectively removing vapor from the gas stream. This outcome demonstrates efficiency of the separation process under tested conditions.

High swirl strength reduces the Mach number and increases the static temperature inside the nozzle, thereby disrupting the expansion characteristics of the supersonic nozzle [27]. Figure 7.14 illustrates the swirling gas flow within a supersonic separator.

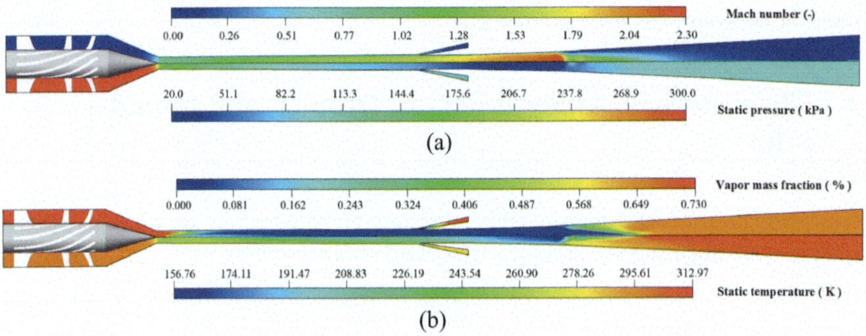

Fig. 7.13 Generation and evolution of supersonic gas for case 1: **a** the change of pressure and velocity in the gas phase; **b** the change of temperature and vapor mass fraction in the gas phase

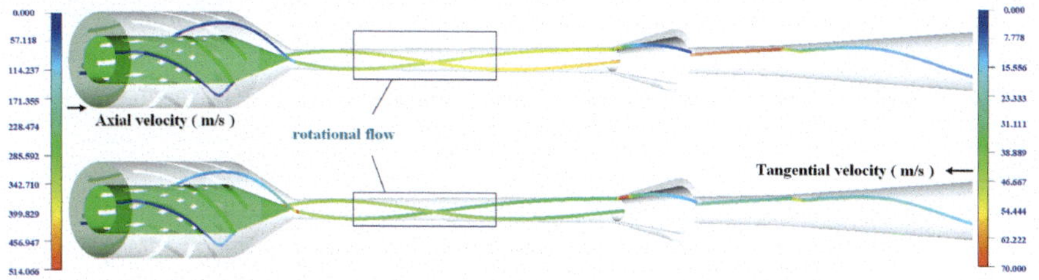

Fig. 7.14 Tangential and axial velocity of swirling gas for case 1

The swirl strength is quantified by the ratio of tangential velocity to axial velocity at a specific point within the system.

In this chapter, the supersonic separator demonstrates a maximum swirl strength of 20.6% for case 1, which is characterized by a peak tangential velocity of 70 m/s. This indicates that at the point of maximum swirl strength, the tangential velocity reaches 70 m/s, and the swirl strength represents 20.6% of the axial velocity. This significant swirl effect profoundly alters the flow dynamics within the nozzle, impacting the overall performance of the supersonic separator. The increased swirl strength leads to changes in the flow behavior, which in turn affects the efficiency and effectiveness of the separation process within the nozzle.

7.5.2 *Dynamic Evolution of Droplet and Film*

Figure 7.15 illustrates the detailed trajectories of both heterogeneous and homogeneous droplets in case 1. Due to the presence of swirling flow within the system, heterogeneous droplets frequently collide with one another and subsequently coalesce, resulting in the formation of larger droplets upstream of the supersonic nozzle. Under the specific influence of a swirl strength measured at 20.6%, all heterogeneous droplets are eventually deposited in the central region of the nozzle's divergent section.

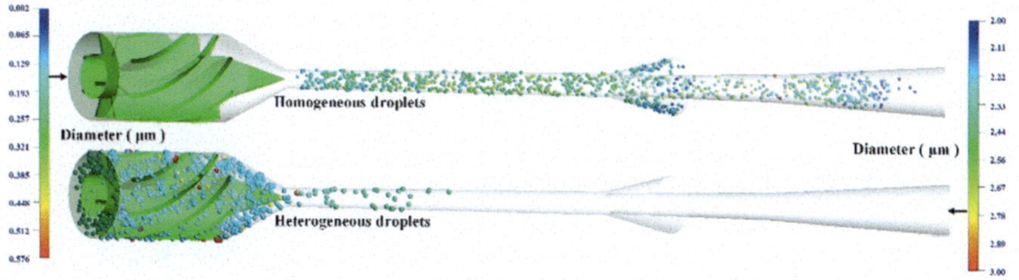

Fig. 7.15 Generation and evolution of droplets for case 1

However, because of the inherently limited vapor content available, homogeneous droplets remain relatively small in size, with the maximum observed diameter reaching only 0.576 μm. Consequently, the swirl strength of 20.6% proves to be insufficient to achieve the complete separation of all homogeneous droplets. As a result, some of these smaller droplets manage to escape into the diffuser section and eventually re-evaporate into vapor during the subsequent rewarming process. This behavior underscores the challenge in achieving total separation of fine homogeneous droplets under the given swirl conditions, highlighting the need for further optimization to improve separation efficiency.

Figure 7.16 illustrates the axial variations in the Sauter diameter of condensed droplets and the extent of subcooling within the nozzle. At the nozzle throat, the Sauter diameter of heterogeneous droplets, $d_{32,\mathrm{het}}$, has increased to 2.33 μm, which surpasses the initial diameter $d_{\mathrm{p}}^{\mathrm{in}}$ of 2.2 μm. This increase confirms that heterogeneous condensation has occurred upstream of the throat. The subsequent deposition of droplets is responsible for the observed decline in the $d_{32,\mathrm{het}}$ curve.

As subcooling intensifies downstream of the throat, homogeneous droplets begin to form. After reaching its peak, the degree of subcooling starts to diminish due to the high rate of condensation and growth of these homogeneous droplets. At an axial position of approximately 200 mm, the condensation rate decreases, leading to a gradual increase in the Sauter diameter of homogeneous droplets, $d_{32,\mathrm{hom}}$, which ultimately reaches 0.262 μm at the drain outlet.

Figure 7.17 presents the distributions of axial and radial slip velocities along the radial direction at different cross-sections for case 1. The axial slip velocity typically decreases progressively along the flow direction due to the drag force exerted by the flow field. In contrast, the radial slip velocity plays a crucial role in the separation of droplets from the primary gas stream.

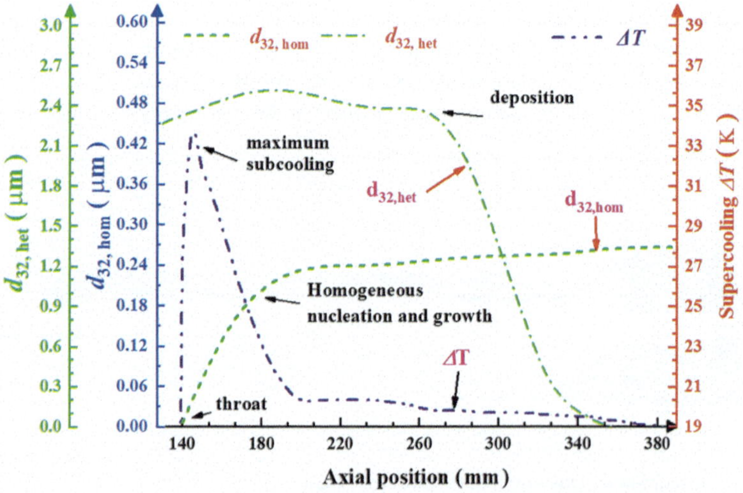

Fig. 7.16 Sauter diameter and subcooling distributions of the supersonic nozzle for case 1

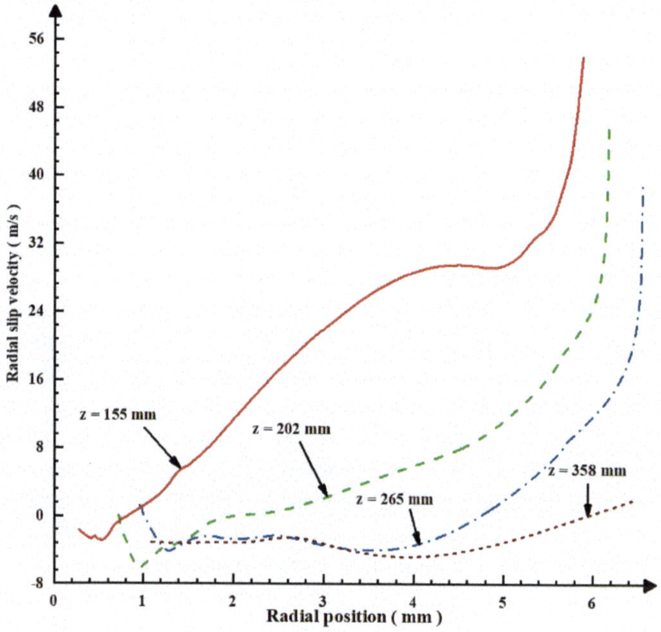

(a)

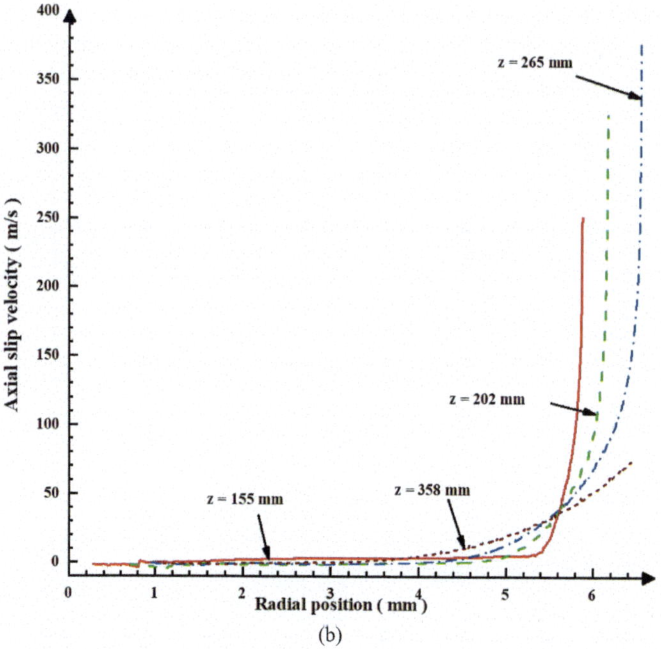

(b)

Fig. 7.17 Slip velocity in different axial positions for case 1: **a** radial slip velocity; **b** axial slip velocity

As the radial position increases from 0 to 7 mm, the radial velocity of the droplets initially lags behind the radial velocity of the gas. Eventually, this lag diminishes, and the radial velocity of the droplets exceeds that of the gas. The magnitude of this lag becomes more pronounced as the axial position moves further downstream. At specific axial positions, the radial velocities that correspond to zero radial slip velocity are as follows: 0.825 mm at $z = 155$ mm, 2.015 mm at $z = 202$ mm, 4.729 mm at $z = 265$ mm, and 5.902 mm at $z = 358$ mm. This distribution highlights the varying behavior of slip velocities throughout the nozzle and underscores the impact of radial and axial positions on droplet separation efficiency.

The larger droplets are propelled to the wall by the swirling centrifugation, leading to the formation of a thin liquid film on the wall. Figure 7.18 illustrates the axial distribution of deposited d_{32} along the supersonic nozzle for case 1. In the convergent section, the deposition strength due to the swirling flow surpasses that of condensation, causing larger droplets to be continuously absorbed by the liquid film and resulting in a drop of about 0.5 μm in deposited $d_{32,\text{het}}$. This stage corresponds to the initial formation of the liquid film, as confirmed in Fig. 7.19. Around $z = 110$ mm, the number of large droplets decreases, slowing the reduction of $d_{32,\text{het}}$. Additionally, as shown in Fig. 7.19, the liquid film's velocity increases at a much slower rate than the gas velocity. The strong shear force between the film and gas leads to mass transfer from the liquid film to the gas and droplets, reducing the film thickness and increasing the diameter of the entrained droplets.

In the area of intense condensation downstream of the throat, the continual conversion of vapor into droplets leads to a rise of approximately 0.38 μm in the deposited $d_{32,\text{het}}$. During this phase, the deposition of homogeneous droplets also begins. As the

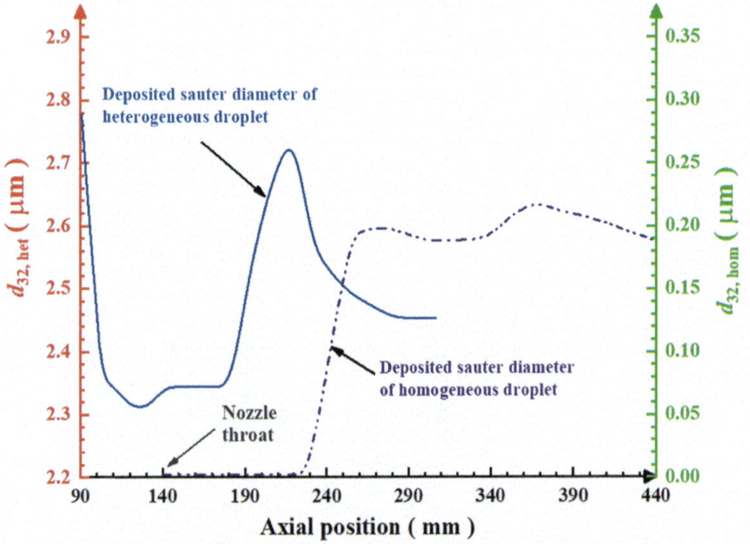

Fig. 7.18 Axial deposited Sauter diameter of droplets for case 1

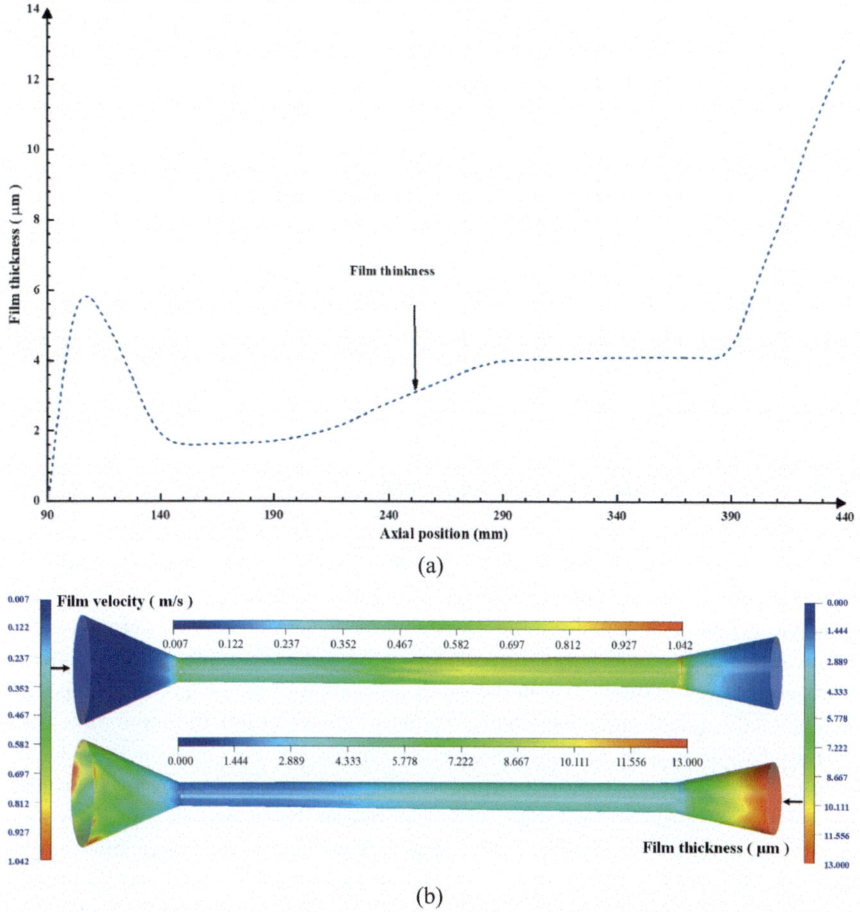

Fig. 7.19 Generation and evolution of liquid film for case 1: **a** film thickness distribution line diagram; **b** cloud map of film velocity distribution and film thickness distribution

diameters of the remaining foreign cores become sufficiently large, heterogeneous droplets lose their capacity to serve as condensation nuclei. Consequently, their diameters stabilize, but the deposition process continues, resulting in a decrease of about 0.26 μm in deposited $d_{32,\text{het}}$. This reduction in the diameter of deposited $d_{32,\text{het}}$ facilitates the deposition of homogeneous droplets. As a result, the deposited $d_{32,\text{hom}}$ increases to 0.2 μm and remains at this level until reaching the liquid outlet.

In Fig. 7.19, it can be observed that the film thickness progressively increases along the flow direction within the divergent section of the nozzle. This growth is influenced by the integrated effects of deposition, entrainment, evaporation, and condensation. As the high-speed gas separates downstream of the drain outlet, the entrainment strength notably decreases, leading to a sharp increase in the liquid film thickness. Ultimately, this thickness reaches 12.74 μm at the liquid outlet. This behavior indicates that the dynamics within the nozzle contribute significantly to the development and final thickness of the liquid film, highlighting the complex interplay of various factors affecting the separation process.

7.6 Sensitivity Analysis of the Inlet Characteristics

In another paper using Euler–Lagrange–Euler Model by Ding et al. [7], the effects of inlet heterogeneous droplet diameter d_p^{in}, inlet heterogeneous droplet concentration q_p^{in}, and inlet pressure p_{in} on separation performance are discussed in detail. This will be introduced in this section.

7.6.1 *Effect of Inlet Droplet Diameter*

In the first group of tests, the inlet droplet diameter was systematically varied to observe its effects on the system. Figures 7.20 and 7.21 provide detailed depictions of the vapor mole fraction along the $y = 0$ mm cross-section and the corresponding droplet trajectories for different droplet diameters. As the inlet droplet diameter increases, there is a notable rise in the vapor mole fraction at the gas outlet. Specifically, when the inlet droplet diameter d_p^{in} is set to 2.0 and 2.1 μm, the vapor mole fraction initially shows a decreasing trend before increasing. This phenomenon occurs because some droplets are not centrifuged effectively to the liquid outlet; instead, they enter the diffuser and subsequently evaporate in the secondary diffuser, which leads to an increase in the vapor mole fraction. Conversely, for larger inlet droplet diameters of $d_p^{in} = 2.2$ μm and $d_p^{in} = 4.0$ μm, the vapor mole fraction demonstrates a steady and consistent decrease along the axial direction of the system. This observed decrease is indicative of an absence of droplet evaporation within the supersonic separator, thereby signifying the complete and efficient separation of all droplets from the vapor phase. This thorough separation ensures that the larger droplets are entirely removed from the vapor content, thereby maintaining both the stability and high efficiency of the separator system.

Additionally, to thoroughly examine the distribution of the liquid film, axial distribution data for film thickness across different inlet diameters were systematically gathered and plotted in Fig. 7.22. The data reveal that as the inlet droplet diameter d_p^{in} increases, the overall film thickness correspondingly becomes greater, with the film thickness at the liquid outlet also rising. Moreover, with an increase in the inlet particle size, the onset position of the film development stage shifts closer to the nozzle throat. This shift occurs because larger droplets are collected by the wall of the separator sooner, leading to an earlier deposition onset location nearer to the nozzle throat.

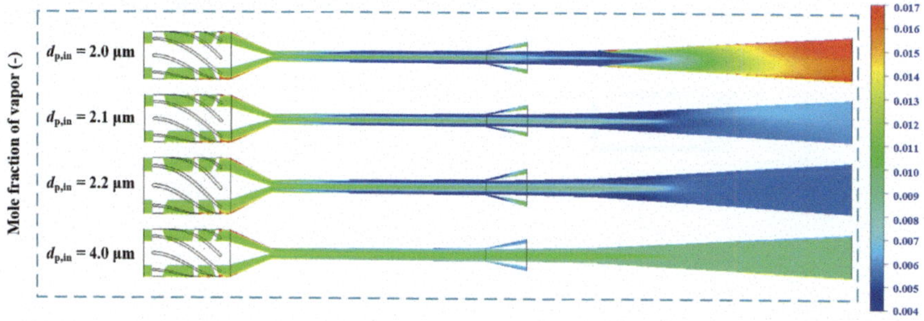

Fig. 7.20 Mole fraction of vapor of the first group when $q_{\mathrm{p}}^{\mathrm{in}} = 0.001$ kg/s and $p_{\mathrm{in}} = 3$ atm

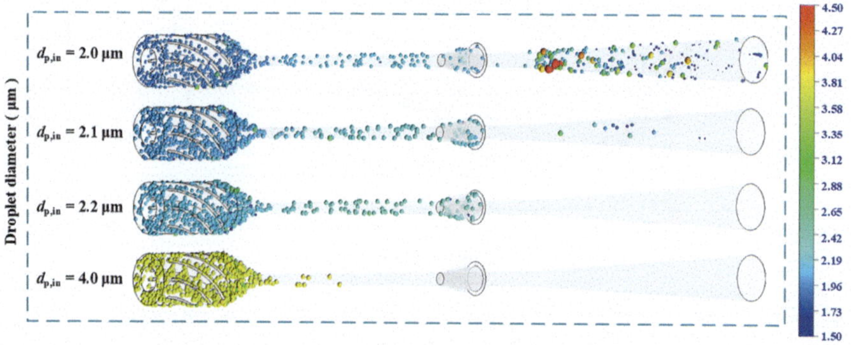

Fig. 7.21 Droplet trajectory of the first group when $q_{\mathrm{p}}^{\mathrm{in}} = 0.001$ kg/s and $p_{\mathrm{in}} = 3$ atm

However, in this set of tests, where the inlet concentration is kept constant, once all droplets are effectively separated—specifically when $d_{\mathrm{p}}^{\mathrm{in}}$ exceeds 2.2 μm—the film thickness at the liquid outlet remains relatively stable. It maintains an approximate thickness of 85.2 μm, indicating that beyond this diameter threshold, further increases in inlet droplet size do not significantly affect the final film thickness at the outlet.

7.6.2 Effect of Inlet Droplet Concentration

In the second set of tests, the inlet droplet diameter $d_{\mathrm{p}}^{\mathrm{in}}$ was maintained at a constant value of 2.2 μm, and the inlet pressure p_{in} was kept steady at 3 atm. Under these controlled conditions, the inlet droplet mass flow rate $q_{\mathrm{p}}^{\mathrm{in}}$ was varied to observe its effects on the system's performance. The vapor mole fraction at the cross-section $y = 0$ mm was measured and is presented in Fig. 7.23. It was observed that when $q_{\mathrm{p}}^{\mathrm{in}} = 0.0020$ kg/s and $q_{\mathrm{p}}^{\mathrm{in}} = 0.0050$ kg/s, droplet evaporation occurs within the diffuser. The distribution of droplet diameters for different values of $q_{\mathrm{p}}^{\mathrm{in}}$ is illustrated

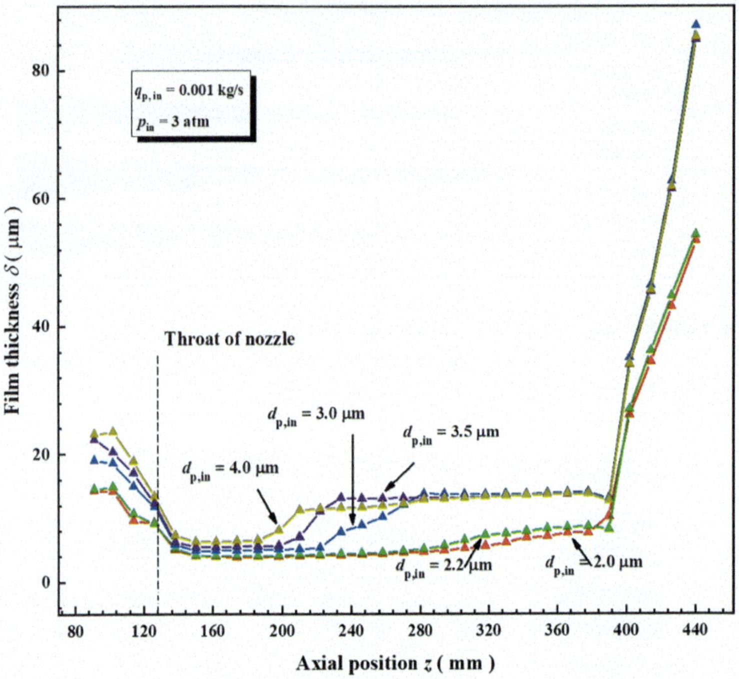

Fig. 7.22 Film thickness distributions of the first group when $q_p^{in} = 0.001$ kg/s and $p_{in} = 3$ atm

in Fig. 7.24. At the higher mass flow rate of $q_p^{in} = 0.0050$ kg/s, a substantial number of droplets manage to enter the diffuser and subsequently evaporate. Conversely, when the mass flow rate is reduced to $q_p^{in} = 0.0020$ kg/s, although fewer droplets enter the diffuser, they still cannot be entirely separated from the gas stream, indicating an incomplete separation process. These observations lead to the conclusion that optimizing the inlet droplet concentration by carefully adjusting the mass flow rate can significantly enhance the overall performance of the supersonic separator. By ensuring more efficient droplet separation and minimizing evaporation within the diffuser, the system can achieve better operational stability and effectiveness.

The variations in the liquid film thickness observed during the second set of tests are thoroughly illustrated in Fig. 7.25. The liquid film within the supersonic separator undergoes a series of distinct stages: initial formation, acceleration, development, and a sharp increase in thickness. As the inlet droplet mass flow rate $q_{p,in}$ increases, there is a noticeable corresponding rise in the overall thickness of the liquid film. During the deceleration stage, it becomes evident that a higher mass flow rate leads to a more significant increase in the growth range of the liquid film. Specifically, the film thicknesses measured at the liquid outlet for different values of $q_{p,in}$ are as follows: 31.2 μm for $q_p^{in} = 0.0001$ kg/s, 62.6 μm for $q_p^{in} = 0.0015$ kg/s, 79.5 μm for $q_p^{in} =$

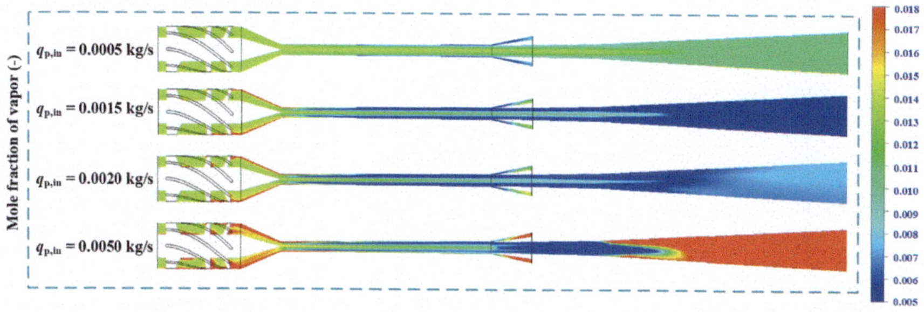

Fig. 7.23 Mole fraction of vapor of the second group when $d_p^{in} = 2.2$ μm and $p_{in} = 3$ atm

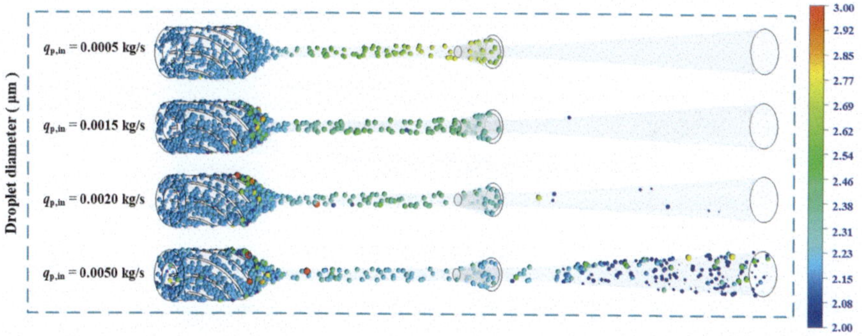

Fig. 7.24 Droplet trajectory of the second group when $d_p^{in} = 2.2$ μm and $p_{in} = 3$ atm

0.0020 kg/s, 101.6 μm for $q_p^{in} = 0.0035$ kg/s, and 139.7 μm for $q_p^{in} = 0.0050$ kg/s. These measurements indicate that as the mass flow rate increases, the thickness of the liquid film at the outlet also increases, reflecting a more substantial accumulation of liquid.

7.6.3 Effect of Inlet Pressure

The third set of tests was designed to investigate the effect of varying inlet pressure p_{in} on system performance. Figures 7.26, 7.27, and 7.28 provide detailed illustrations of the variations observed in vapor mole fraction, droplet trajectories, and liquid film thickness across different inlet pressures. The results indicate that higher inlet pressures are associated with a reduced evaporation range and improved efficiency in droplet removal.

Specifically, at lower inlet pressure values of 2.5 and 2.0 atm, shock waves become evident at the drain outlet. These shock waves cause some of the droplets to enter the diffuser and subsequently evaporate, leading to an increase in the mole fraction of

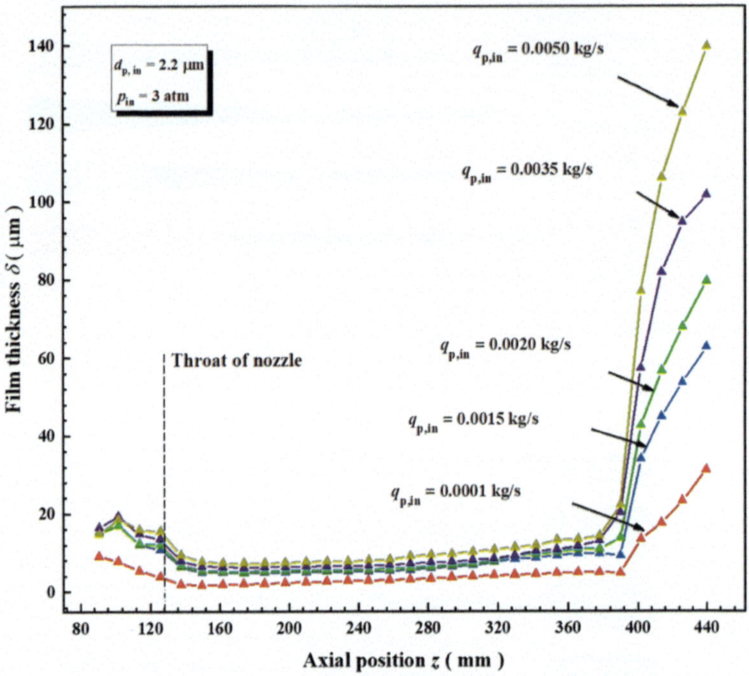

Fig. 7.25 Film thickness distributions of the second group when $d_p^{in} = 2.2$ μm and $p_{in} = 3$ atm

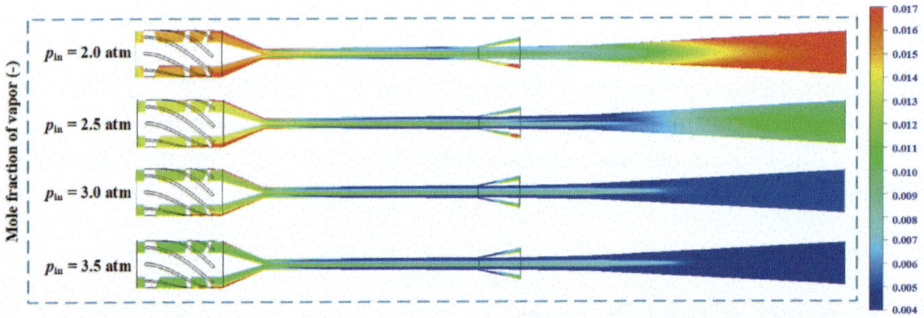

Fig. 7.26 Mole fraction of vapor of the third group when $d_p^{in} = 2.2$ μm and $q_p^{in} = 0.0015$ kg/s

vapor at the gas outlet. Furthermore, when p_{in} is set to 2.0 atm, a significant portion of the droplets fails to separate effectively. This inefficiency results in a higher mole fraction of vapor at the outlet compared to the inlet, highlighting the challenges in droplet removal under these conditions.

In contrast, as the inlet pressure increases to 3.0, 3.5, and 4.0 atm, the thickness of the liquid film at the outlet remains relatively stable, showing minimal variation across these higher-pressure values. However, due to the presence of shock waves

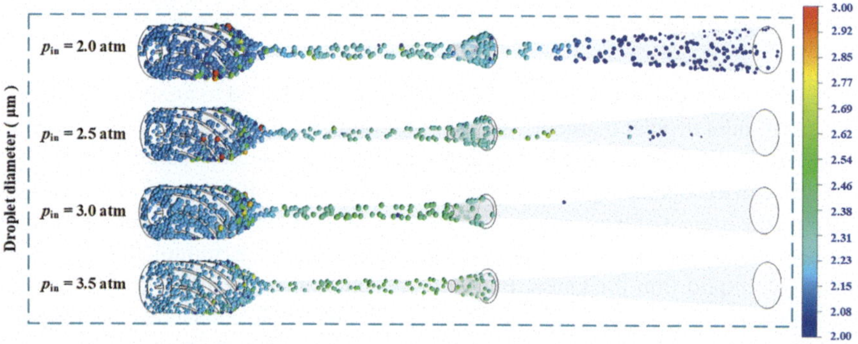

Fig. 7.27 Droplet trajectory of the third group when $d_{\mathrm{p}}^{\mathrm{in}} = 2.2$ μm and $q_{\mathrm{p}}^{\mathrm{in}} = 0.0015$ kg/s

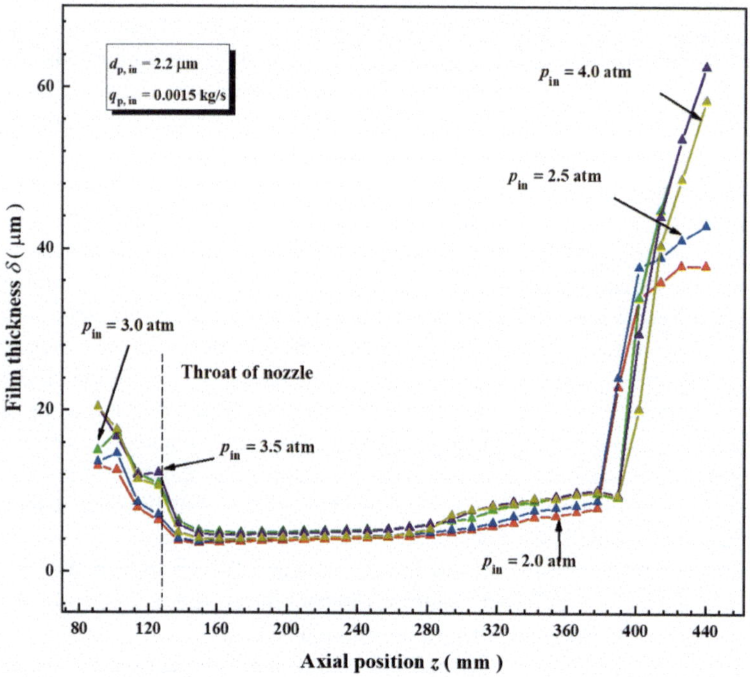

Fig. 7.28 Film thickness distributions of the third group when $d_{\mathrm{p}}^{\mathrm{in}} = 2.2$ μm and $q_{\mathrm{p}}^{\mathrm{in}} = 0.0015$ kg/s

at the drain outlet, the film thickness for lower inlet pressures of 2.0 and 2.5 atm is notably reduced. This observation underscores the impact of inlet pressure on both droplet removal efficiency and film thickness, emphasizing the importance of optimizing pressure conditions to enhance system performance.

7.6.4 *Optimization Analysis of Separation Efficiency*

In this section, a comprehensive investigation was conducted to analyze the effects of varying inlet droplet diameter, inlet droplet mass flow rate, and inlet pressure on the dynamics of droplets and liquid film within the supersonic separator. This analysis was carried out through three distinct sets of tests, each designed to isolate and examine the impact of these parameters on system performance. The findings reveal that the inlet droplet mass flow rate plays a critical role in determining the maximum achievable liquid film thickness. Furthermore, the combined influence of the inlet droplet diameter, mass flow rate, and pressure significantly impacts the overall efficiency of both droplet and vapor removal processes. Consequently, this section will focus on a detailed analysis aimed at optimizing the separation efficiency of the supersonic separator by systematically varying and adjusting these inlet conditions.

Specifically, the first set of tests was conducted to explore how different inlet droplet diameters affect the formation and behavior of the liquid film within the supersonic separator. These tests provided insights into how variations in droplet size influence the dynamics of the liquid film, including its thickness and distribution. The second set of tests was designed to investigate the impact of varying the inlet droplet mass flow rate on the liquid film. This set of experiments demonstrated the significant effect that changes in the mass flow rate of droplets have on the maximum achievable thickness of the liquid film. It highlighted the relationship between droplet mass flow rate and the resulting film characteristics. Finally, the third set of tests focused on examining the effects of varying inlet pressure on the system. These tests illustrated how fluctuations in inlet pressure can alter the trajectories of the droplets and affect their evaporation rates. As a result, the performance of the supersonic separator is influenced by these pressure variations, impacting its overall separation efficiency.

By comprehensively analyzing the data obtained from these tests, we aim to develop a deeper and more detailed understanding of the complex dynamics occurring within the supersonic separator. This thorough understanding will serve as a foundation for optimizing various operational parameters, with the goal of achieving higher separation efficiency. Specifically, we seek to ensure that both droplets and vapor are effectively removed from the gas stream. To accomplish this, we will carefully evaluate how different conditions impact the separator's performance and use these insights to fine-tune operational strategies. Consequently, this section will provide valuable insights into the optimal conditions for operating the supersonic separator, ultimately enhancing its overall performance, efficiency, and reliability.

Figure 7.29 vividly demonstrates the impact of varying inlet droplet diameters on the separation efficiency of the supersonic separator, with the assumption that all other conditions remain constant. The data show that as the inlet droplet diameter $d_{\mathrm{p}}^{\mathrm{in}}$ increases, the droplet removal rate (η_{p}) correspondingly improves. The droplet removal rate η_{p} refers to the ratio of the mass flow rate of droplets separated at the liquid outlet to the mass flow rate at the liquid inlet, and is also an important indicator for measuring the performance of supersonic separators. Its calculation formula is:

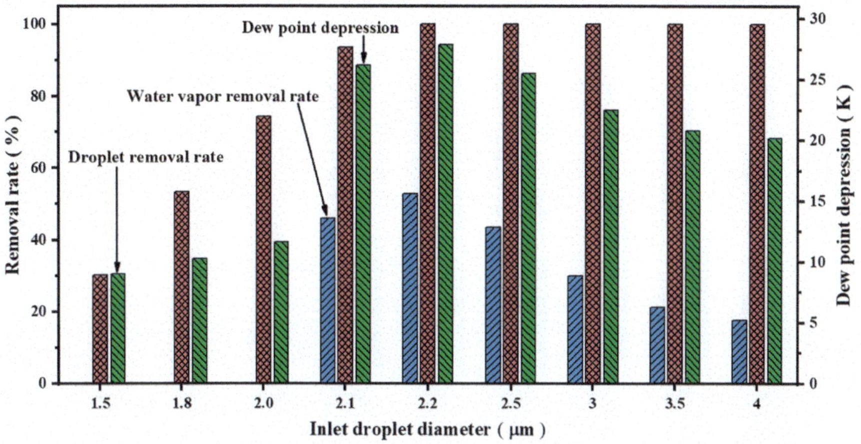

Fig. 7.29 The separation efficiency of the supersonic separator with different inlet droplet diameters when $q_{\mathrm{p}}^{\mathrm{in}} = 0.001$ kg/s and $p_{\mathrm{in}} = 3$ atm

$$\eta_{\mathrm{p}} = \frac{q_{\mathrm{p,out}}}{q_{\mathrm{p,in}}} \times 100\% \tag{7.28}$$

Remarkably, the droplet removal rate reaches an impressive 100% when the inlet droplet diameter is 2.2 μm.

To comprehensively evaluate the separation efficiency, two key metrics are used: the vapor removal rate (η_{v}) and the dew point depression (ΔT_{d}). Higher values of these metrics indicate better performance of the separator. Specifically, an inlet droplet diameter of 2.2 μm results in a vapor removal rate (η_{v}) of 57.4% and a dew point depression (ΔT_{d}) of 27.9 K. These values are considered optimal for achieving high separation efficiency.

The findings highlight that optimizing the inlet droplet diameter is crucial for enhancing the performance of the supersonic separator. By achieving a diameter of 2.2 μm, the system not only maximizes droplet removal but also significantly improves the overall efficiency of vapor removal and dew point depression. This optimal diameter ensures that the separator operates at peak performance, effectively removing both droplets and vapor from the gas stream.

Figure 7.30 effectively illustrates the separation efficiency of the supersonic separator at varying inlet mass flow rates. The data clearly demonstrate that an inlet mass flow rate of 0.0015 kg/s optimally removes both droplets and vapor, thereby significantly enhancing the overall separation efficiency of the supersonic separator.

At this specific mass flow rate, the separator achieves a balanced and efficient performance, effectively managing the separation of droplets from the gas stream while also ensuring the removal of vapor. This optimal rate indicates that the system can handle the gas–liquid mixture with maximum efficiency, preventing the escape of droplets and reducing the vapor content to desirable levels, effectively.

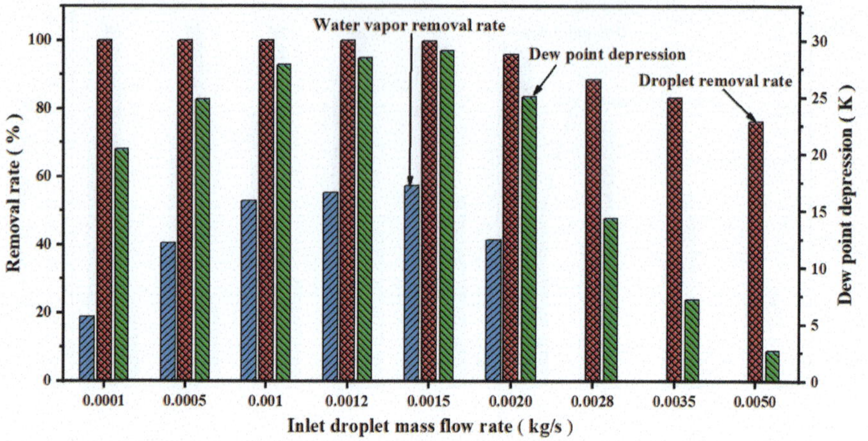

Fig. 7.30 The separation efficiency of the supersonic separator with different inlet droplet mass flow rates when $d_p^{in} = 2.2$ μm and $p_{in} = 3$ atm

Thus, maintaining an inlet mass flow rate of 0.0015 kg/s is crucial for achieving peak operational efficiency in the supersonic separator. This rate not only optimizes the separation process but also contributes to the reliable and stable functioning of the separator, ensuring consistent performance across various operating conditions.

Regarding the impact of inlet pressure on separation efficiency, Fig. 7.31 provides a clear illustration of the observed trend. The data presented in the figure demonstrate that increasing the inlet pressure generally leads to an enhancement in the overall separation efficiency of the supersonic separator. This improvement occurs because higher inlet pressures contribute to more effective droplet separation and vapor removal. However, it is also evident from the data that once the inlet pressure exceeds 3 atm, further increases in pressure result in a diminishing effect on the separator's performance. This suggests that there is a threshold beyond which additional pressure does not contribute significantly to the separation efficiency and may even lead to diminishing returns.

This trend suggests that while higher inlet pressures initially contribute to improved separation efficiency, the benefits of further increasing the pressure become less significant beyond a certain point. Therefore, optimizing the inlet pressure up to approximately 3 atm appears to be effective for maximizing separation performance. Beyond this threshold, further adjustments to the pressure may yield only marginal improvements, indicating that there is an optimal pressure range where the separator operates most efficiently.

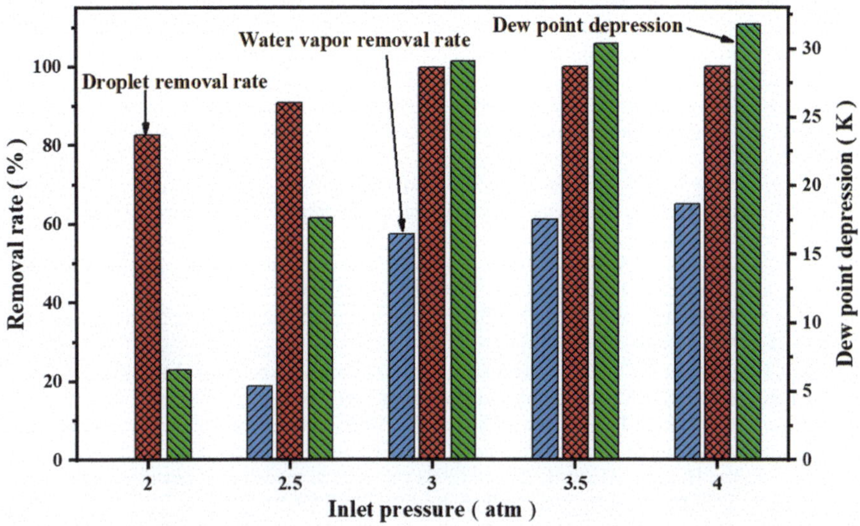

Fig. 7.31 The separation efficiency of the supersonic separator with different inlet pressures when $d_{\mathrm{p}}^{\mathrm{in}} = 2.2\ \mu\mathrm{m}$ and $q_{\mathrm{p}}^{\mathrm{in}} = 0.0015$ kg/s

In conclusion, selecting an inlet droplet diameter ($d_{\mathrm{p}}^{\mathrm{in}}$) of 2.2 μm, an inlet droplet mass flow rate ($q_{\mathrm{p}}^{\mathrm{in}}$) of 0.0015 kg/s, and an inlet pressure (p_{in}) of 3 atm results in optimal separation efficiency under the given conditions. This configuration achieves a droplet removal rate (η_{p}) of 100%, a vapor removal rate (η_{v}) of 57.4%, and a dew point depression (ΔT_{d}) of 29.1 K, ensuring the best performance for the supersonic separator.

References

1. Adechy D, Issa RI (2004) Modelling of annular flow through pipes and T-junctions. Comput Fluids 33(2):289–313
2. Bai B, Li X, Li S (2016) Computation of supersonic branching flow with aerosol particle separation. AIAA J 54(7):2069–2076
3. Baker N, Kelly G, O'Sullivan PD (2020) A grid convergence index study of mesh style effect on the accuracy of the numerical results for an indoor airflow profile. Int J Vent 19(4):300–314
4. Bian J, Cao X, Teng L, Sun Y, Gao S (2019) Effects of inlet parameters on the supersonic condensation and swirling characteristics of binary natural gas mixture. Energy 188:116082
5. Deng Y, Zhang L, Hou H, Yu B, Sun D (2019) Modeling and simulation of the gas-liquid separation process in an axial flow cyclone based on the Eulerian–Lagrangian approach and surface film model. Powder Technol 353:473–488
6. Ding H, Sun C, Wang C, Wen C, Tian Y (2020) Prediction of dehydration performance of supersonic separator based on a multi-fluid model with heterogeneous condensation. Appl Therm Eng 171:115074

7. Ding H, Sun C, Wen C, Liang Z (2022a) The droplets and film behaviors in supersonic separator by using three-field two-fluid model with heterogenous condensation. Int J Heat Mass Transf 184:122315
8. Ding H, Zhang Y, Sun C, Yang Y, Wen C (2022b) Numerical simulation of supersonic condensation flows using Eulerian–Lagrangian and Eulerian wall film models. Energy 258:124833
9. Ding H, Zhang Y, Yang Y, Wen C (2023) A modified Euler–Lagrange–Euler approach for modelling homogeneous and heterogeneous condensing droplets and films in supersonic flows. Int J Heat Mass Transf 200:123537
10. Dykas S, Wróblewski W (2012) Numerical modelling of steam condensing flow in low and high-pressure nozzles. Int J Heat Mass Transf 55(21–22):6191–6199
11. Dykas S, Wróblewski W (2013) Two-fluid model for prediction of wet steam transonic flow. Int J Heat Mass Transf 60:88–94
12. Han J, Feng J, Hou T, Chen W, Peng X (2021) Numerical and experimental study on gas-water separators for a PEMFC system. Int J Green Energy 18(5):490–502
13. Hoekstra AJ, Derksen JJ, Van Den Akker HEA (1999) An experimental and numerical study of turbulent swirling flow in gas cyclones. Chem Eng Sci 54(13–14):2055–2065
14. Huang Z, Yan H, Liu L, Gong H, Zhou P (2021) Numerical modeling of reactive bubbly flows based on Euler–Lagrange approach. Chem Eng Sci 239:116640
15. Kobashi Y, Zama Y, Kuboyama T (2020) Modeling wall film formation and vaporization of a gasoline surrogate fuel. Int J Heat Mass Transf 147:119035
16. Liu Y, Cui J, Li WZ (2011) A two-phase, two-component model for vertical upward gas–liquid annular flow. Int J Heat Fluid Flow 32(4):796–804
17. Liu X, Liu Z, Li Y (2014) Investigation on separation efficiency in supersonic separator with gas-droplet flow based on DPM approach. Sep Sci Technol 49(17):2603–2612
18. Liu Y, Cao X, Yang J, Li Y, Bian J (2021) Energy separation and condensation effects in pressure energy recovery process of natural gas supersonic dehydration. Energy Convers Manage 245:114557
19. Liu Z, Tao C, Yuan S, Wang W, Tamaddon M, Ng L et al (2022) Eulerian wall film model for predicting dynamic cell culture process to evaluate scaffold design in a perfusion bioreactor. Med Nov Technol Devices 13:100104
20. O'Rourke PJ (1981) Collective drop effects on vaporizing liquid spray. Ph.D. thesis, Princeton University
21. Ounis H, Ahmadi G, McLaughlin JB (1991) Brownian diffusion of submicrometer particles in the viscous sublayer. J Colloid Interface Sci 143(1):266–277
22. Pađen I, Petranović Z, Edelbauer W, Vujanović M (2021) Numerical modeling of spray secondary atomization with the Euler–Eulerian multi-fluid approach. Comput Fluids 222:104919
23. Shooshtari SR, Shahsavand A (2018) Optimal operation of refrigeration oriented supersonic separators for natural gas dehydration via heterogeneous condensation. Appl Therm Eng 139:76–86
24. Springer TE, Zawodzinski TA, Gottesfeld S (1991) Polymer electrolyte fuel cell model. J Electrochem Soc 138(8):2334
25. Wang X, Chang H, Corradini M, Cong T, Wang J (2016) Prediction of falling film evaporation on the AP1000 passive containment cooling system using ANSYS FLUENT code. Ann Nucl Energy 95:168–175
26. Wang Y, Yu Y, Hu D, Xu D, Yi L, Zhang Y, Zhang S (2020) Improvement of drainage structure and numerical investigation of droplets trajectories and separation efficiency for supersonic separators. Chem Eng Process Process Intensif 151:107844
27. Wen C, Cao X, Yang Y, Zhang J (2011) Swirling effects on the performance of supersonic separators for natural gas separation. Chem Eng Technol 34(9):1575–1580
28. Wen C, Cao X, Yang Y, Zhang J (2012) Evaluation of natural gas dehydration in supersonic swirling separators applying the discrete particle method. Adv Powder Technol 23(2):228–233

29. Wen C, Ding H, Yang Y (2020) Optimisation study of a supersonic separator considering nonequilibrium condensation behaviour. Energy Convers Manage 222:113210
30. Yang Y, Wen C (2017) CFD modeling of particle behavior in supersonic flows with strong swirls for gas separation. Sep Purif Technol 174:22–28
31. Yoo YL, Han DH, Hong JS, Sung HG (2017) A large eddy simulation of the breakup and atomization of a liquid jet into a cross turbulent flow at various spray conditions. Int J Heat Mass Transf 112:97–112
32. Yuan S, Fan Y, Chen B, Li J, Gao L, Zhang S (2020) Forming and stripping of the wall film and the influence on gas–liquid separation. Asia-Pac J Chem Eng 15(3):e2447
33. Yue T, Chen J, Song J, Chen X, Wang Y, Jia Z, Xu R (2019) Experimental and numerical study of upper swirling liquid film (USLF) among gas-liquid cylindrical cyclones (GLCC). Chem Eng J 358:806–820

Chapter 8
Joule-Thomson Effect and Structural Parameter Impact for Energy-Efficient Supersonic Dehydration of Natural Gas

8.1 Introduction

The supersonic separator leverages a unique mechanism for the dehydration of high-pressure natural gas. The dehydration process initiates with the natural gas mixture (CH_4–H_2O) being swirled through the swirl generator, equipped with static vanes characterized by specific parameters such as outlet angle, thickness, and chord length [3]. Subsequently, the water vapor within the natural gas mixture condenses into droplets as it traverses through the supersonic nozzle, which is defined by specific parameters, including inlet diameter, throat diameter, convergent section profile, length, and expansion ratio. These droplets then compact against the shroud wall, forming a liquid film, and undergo separation in the drainage section, characterized by specific parameters of outer angle and discharge length. Following this separation process, the resulting purified natural gas undergoes pressure energy recovery within the diffuser before being discharged to subsequent processes. The diffuser parameters, such as length and expansion ratio, directly influence this recovery process [10]. Consequently, it is evident that the structural parameters of the swirl generator, supersonic nozzle, drainage section, and diffuser hold critical importance in determining the separation performance and pressure energy loss of the supersonic separator. Thus, a comprehensive understanding of how these parameters influence flow behaviors and separation performance is critical for supersonic separator structure and performance optimization. Wen et al. conducted a survey on the impact of three different diffusers on the flow characteristics of natural gas using an Euler single-phase model. The findings revealed that the conical diffuser performed better than the other diffusers [18]. Hu et al. [7] introduced a supersonic separator featuring a reflux channel and analyzed its flow characteristics and separation performance using an Euler single-phase model. They found that the separation efficiency was significantly enhanced when the designed structural parameters of a 7 mm reflux channel diameter and an insertion depth of 30 mm [16]. Then both the Euler–Euler (E-E) approach and Euler–Lagrange (E-L) method, coupling with condensation effects,

H. Ding et al., *Supersonic Separators for Sustainable Gas Purification*, Green Energy and Technology, https://doi.org/10.1007/978-3-031-85213-8_8

have been employed to deeply study the sensitivity of structural parameters on separation performance [11]. Wang et al. compared the effects of different drainage structures on the separation performance of a specific supersonic separator design incorporating reflux using the E-L model [17].

In a supersonic separator, the Laval nozzle and swirl generator are the critical two components for energy-efficient dehydration. For Laval nozzle, the main function is to accelerate the fluid to supersonic speed and provide a low-temperature environment suitable for condensation. Cao et al. introduced a novel structure featuring an inner body with an annular nozzle size of 3.94 mm and assessed its separation performance using an Euler single-phase model [4]. Jassim et al. examined the impact of a supersonic nozzle's structural parameters on the flow characteristics under a high-pressure environment using an Euler single-phase model [9]. Bian et al. optimized the supersonic nozzle by modifying the length and angle of the divergent section using an Euler single-phase model, resulting in a 47.5% reduction in energy loss and a notable improvement in separation performance [1]. Moreover, Shooshtari et al. investigated the influence of expansion angles of the supersonic nozzle on flow field characteristics using a formulated E-L method. The results indicated maximal removal rates of water vapor and droplets at an expansion angle of 3° and an injected particle diameter of 5 μm [15].

For swirl generator, the main function is to produce a swirling flow field, which can be quantified by swirl strength. This parameter, a critical parameter in this process, plays a pivotal role in determining the separation performance and the efficiency of energy conversion within the separator. Majidi et al. investigated the impact of the cyclonic part and drainage section on the separation performance and pressure energy loss of a supersonic separator using an Euler single-phase model [11]. Then, Wen et al. studied the impact of swirl vane structures on flow characteristics and droplet removal efficiency in a supersonic separator using the E-L model [19]. The findings suggested that a vane-to-pipe diameter ratio of 0.67 significantly improved separation performance, achieving a droplet separation efficiency of 70%. Similarly, Yang et al. examined the influence of static vane parameters on droplet behavior and separator performance using the E-L mode [23]. They found that a 45–60° swirl angle and 8–16 vanes contribute positively to enhancing the separation performance. Therefore, varying levels of swirl strength can result in distinct predictions for the separation effectiveness and energy conversion rates, with observable trends changing as the inlet pressure of the natural gas fluctuates. This highlights the importance of optimizing the swirl strength to achieve the most favorable conditions for efficient water removal and energy recovery in the context of supersonic separation technology. The swirling number S_w is used to describe the swirl strength and can be calculated using Eq. (8.1).

$$S_w = \frac{u_{tan}}{u_{mix}} \tag{8.1}$$

where u_{tan} is the tangential velocity, u_{mix} is the axial velocity of the gas–liquid mixture.

Meanwhile, in the energy-efficient supersonic dehydration of natural gas, the flow complexity necessitates a detailed understanding of the thermodynamic properties of the CH_4–H_2O mixture, as these properties significantly affect the evaluation of flow behavior and the separation efficiency of the supersonic separator. Existing studies have analyzed the effects of structural parameters and thermodynamic properties on flow characteristics and separation performance by different numerical methods.

This chapter analysis the impacts of multiple coupled parameters of Laval nozzle and swirl generator on supersonic flow behaviors, droplet separation efficiency, dehydration efficiency, dew-point depression, and pressure energy loss of the supersonic separator under different thermodynamic properties based on the E-L-E two-phase three-phase model.

8.2 Thermodynamic Properties Effect

The thermodynamic properties of natural gas mixture are crucial [5]. The ideal gas model (IGM) and the real gas model (RGM) have completely different prediction results for Wilson point [6] and separation performance [8]. In non-swirling supersonic flows, Bian et al. analyzed the condensation characteristics of natural gas based on the RGM [2]. Yang et al. [22] and Wen et al. [21] compared the prediction results of condensation and shock waves by the IGM and four RGMs in a Laval nozzle. It was proved that the IGM overpredicts the condensation effect than other RGMs, and the RGMs (Redlich–Kwong RGM, Redlich–Kwong–Soave RGM, Redlich–Kwong–Aungier RGM, and Peng–Robinson RGM) are predicted almost similar results. The NIST RGM can be used for performance evaluations of high-pressure supersonic separations in natural gas processing [20].

The IGM is expressed by Eq. (8.1)

$$p = \rho R_{\mathrm{v}} T_{\mathrm{g}} \tag{8.2}$$

The Redlich–Kwong RGM is described as:

$$p = \frac{RT}{V - b} - \frac{T^{-0.5} a}{V^2 + \lambda_1 V} \tag{8.3}$$

where V is volume. For the natural gas mixture of methane and water vapor, a, and b are calculated by Eqs. (8.3a) and (8.3b).

$$a = \sum_{i=1}^{2} \sum_{j=1}^{2} x_i x_j \sqrt{a_i a_j} \left(1 - k_{ij}\right) \tag{8.3a}$$

$$b = \sum_{i=1}^{2} x_i b_i \tag{8.3b}$$

where x is the mole fraction, k_{ij} is the interaction coefficient.

The NIST RGM is built upon REFPROP v9.1, developed and supported by the National Institute of Standards and Technology (NIST). For natural gas mixture, the thermodynamic properties of each component are based on the modified Benedict–Webb–Rubin equation, Helmholtz-energy equation, and extended corresponding states. Based on the GERG-2008 equation of state, the Helmholtz free energy α_{mix} is calculated by Eq. (8.4):

$$\alpha^{\text{mix}} = \alpha^{\text{o}}(\rho, T) + \alpha^{\text{r}}(\delta, \tau) \tag{8.4}$$

where the reduced mixture density $\delta = \rho/\rho_{\text{c}}$, and the inverse reduced mixture temperature $\tau = T_{\text{c}}/T$. The subscript c means critical. The Helmholtz free energy for the ideal part $\alpha_{\text{id}}(\delta, \tau)$ is given by:

$$\alpha^{\text{o}}(\rho, T) = \sum_{i}^{n} x_i\left[\alpha^{\text{r}}_{\text{oi}}(\rho, T) + \ln x_{\text{i}}\right] \tag{8.5}$$

where n is the number of components, $\alpha^{\text{r}}_{\text{oi}}(\rho, T)$ is the dimensionless Helmholtz free energy for ideal part, x_{i} is the mole fraction.

The residual part $\alpha^{\text{r}}(\delta, \tau)$ is expressed by:

$$\alpha^{\text{r}}(\delta, \tau) = \sum_{i}^{n} x_{\text{i}}\alpha^{\text{r}}_{\text{oi}}(\delta, \tau) + \sum_{i=1}^{n-1}\sum_{j=i+1}^{n} x_i x_j F_{ij}\alpha^{\text{r}}_{ij}(\delta, \tau) \tag{8.6}$$

where $\alpha^{\text{r}}_{\text{oi}}(\delta, \tau)$ is the residual part of the reduced Helmholtz free energy of component i, F_{ij} is the adjustable factor.

8.2.1 Two-Phase Flow Behaviors

Different inlet foreign droplets diameters and concentrations induce various heterogeneous condensation, which is expected to influence the separation performance. Therefore, based on the IGM, Redlich–Kwong RGM, and NIST RGM, the impact of the foreign droplets' diameter and concentration on the high-pressure offshore natural gas dehydration processing was analyzed with various p_{in} of 6, 30, 60, 90, and 120 bar. The operation conditions are shown in Table 8.1.

The natural gas dehydration processing in the SS mainly includes two complex processes: condensation phase change and supersonic swirling flows. There are complex mass and heat transfer phenomena during the processing. The flow field parameters are critical for analyzing the mechanism of heat and mass transfer and

Table 8.1 Operation conditions

p_{in} (bar)	p_{out} (bar)	T_{in} (K)	d_p^{in} (μm)	q_p^{in} (kg s^{-1})
6	2	300.0	1.4–1.9	0.0008–0.0013
30	10	300.0	1.9–2.4	0.0012–0.0018
60	20	300.0	2.4–2.9	0.0021–0.0029
90	30	300.0	2.8–3.3	0.0023–0.0036
120	40	300.0	3.1–3.6	0.0024–0.0037

quantifying the separation performance of SS. The comparison of flow field characteristics by the three thermodynamic models of IGM, NIST RGM, and Redlich–Kwong RGM are shown in Figs. 8.1, 8.2, 8.3, 8.4, 8.5 and 8.6, with $p_{in} = 120$ bar, $q_p^{in} = 0.0026$ kg s^{-1}, $d_p^{in} = 3.3$ μm, $p_{out} = 40$ bar, and $T = 300$ K.

As shown in Fig. 8.1, in the Laval nozzle divergence section, the NIST RGM and Redlich–Kwong RGM predict similar pressure ratio p_r variation results, while the IGM predicts a value 8.1% higher than the other two. The p_r indicated by NIST RGM is lower, about a maximum of 17.0%, than Redlich–Kwong RGM, which has the same value as IGM in the cyclone separation section. It is also found that the shockwave location predicted by NIST RGM is 3 mm later than that of Redlich–Kwong RGM and IGM. Moreover, as shown in Fig. 8.2, the NIST RGM indicates a maximum of 10.6 and 3.4% lower static temperature values than IGM and Redlich–Kwong RGM in the cyclone separation section. While the temperature obtained by IGM is about 13.1 and 12.4% larger than NIST RGM and Redlich–Kwong RGM, i.e., the NIST RGM

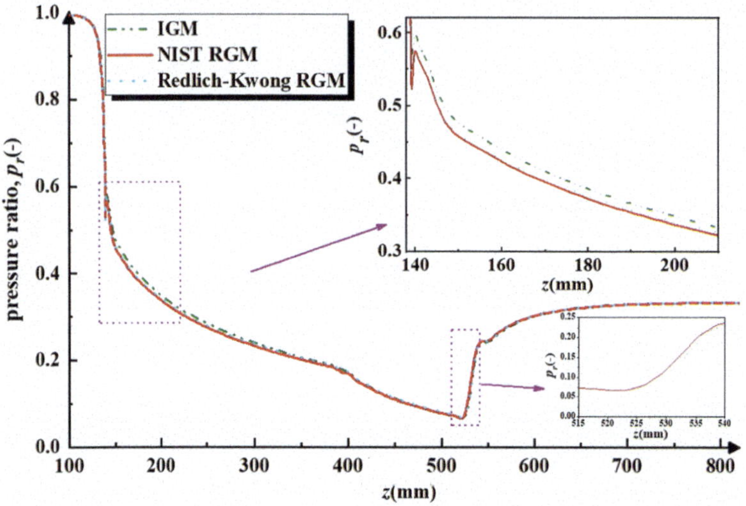

Fig. 8.1 Pressure ratio distributions

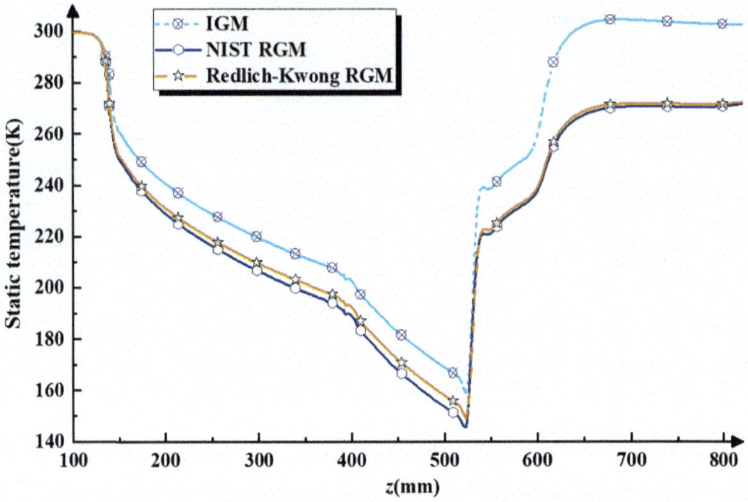

Fig. 8.2 Static temperature distributions

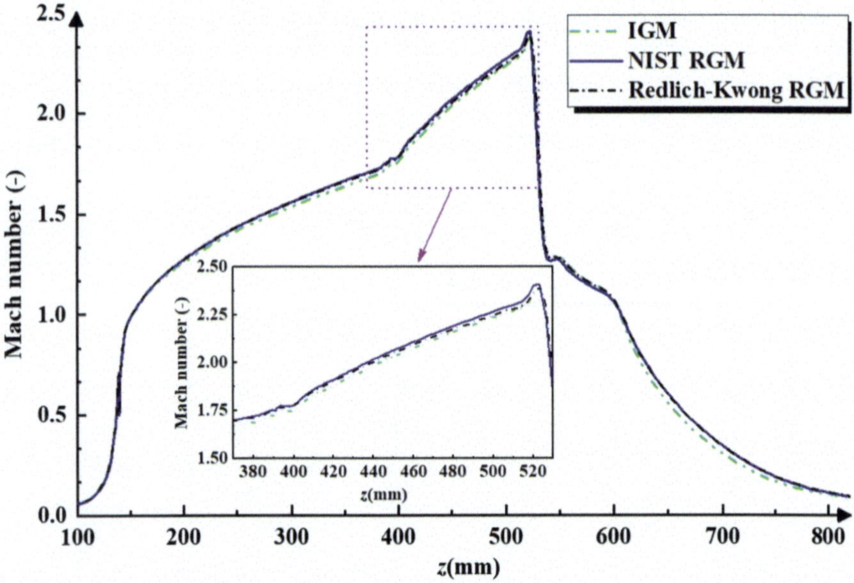

Fig. 8.3 Mach number distributions

gets the lowest static temperature at the dry gas outlet. This indicates that the IGM and Redlich–Kwong RGM overpredict the condensation phase change processing of the water vapor than NIST RGM, which will lead to a different prediction of the separation performance in SS.

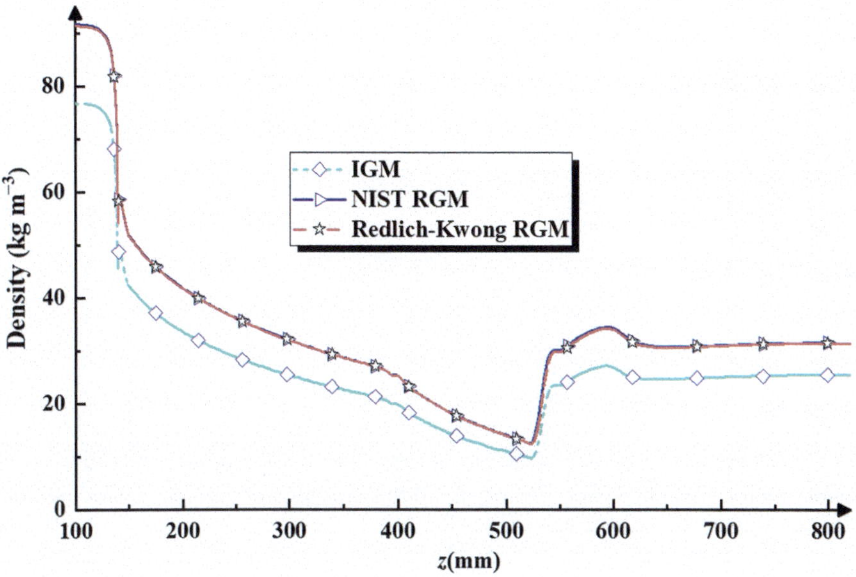

Fig. 8.4 Gas density distributions

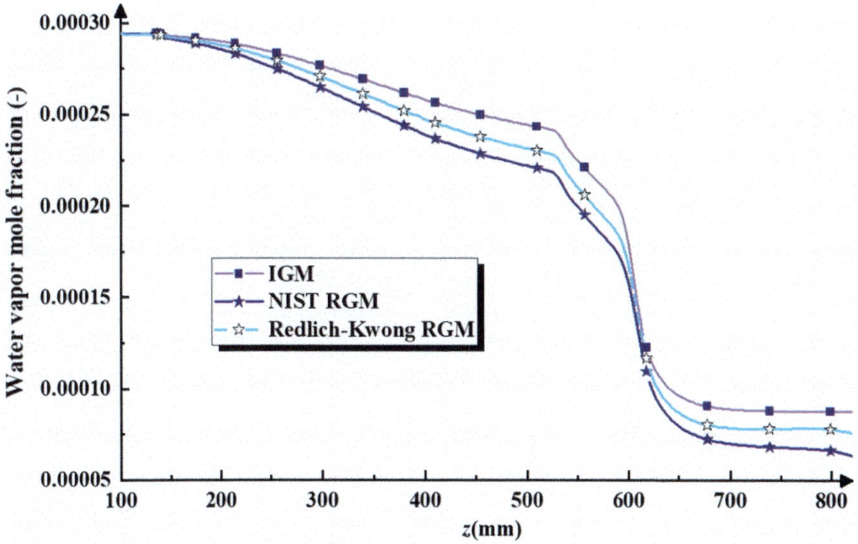

Fig. 8.5 Water vapor mole fraction distributions

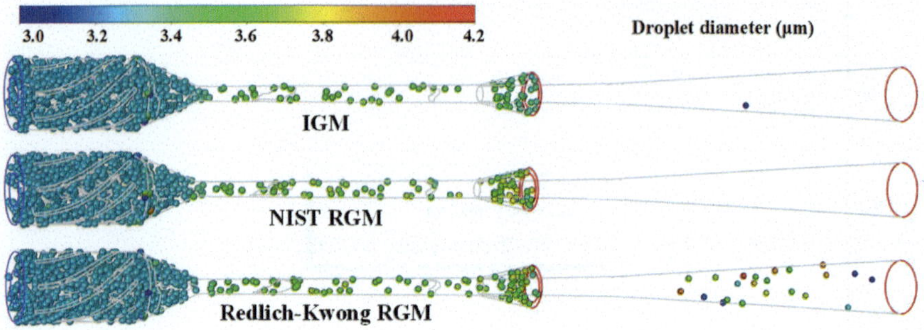

Fig. 8.6 Droplet diameter distributions

Figure 8.3 shows the comparison results of the Mach number predicted by the three thermodynamic models. One can see that the IGM exhibits a slightly lower Mach number than the other two models in the cyclone separation section, and the opposite is true in the diffuser section. The NIST RGM and Redlich–Kwong RGM almost predict the same distribution. The predicted maximum Mach numbers are 2.40, 2.39, and 2.38 for NIST RGM, Redlich–Kwong RGM, and IGM. The Mach number deviation obtained between IGM and NIST RGM is 1.9% in the cyclone separation section and 13.4% in the diffuser.

Besides, from Figs. 8.1, 8.2, 8.3, 8.4, 8.5 and 8.6, one can see that the IGM, NIST RGM, and Redlich–Kwong RGM predict the same p_r, static temperature, and Mach number of the natural gas mixture in the section between the inlet and the throat of Laval nozzle. As the Mach number decreases rapidly to less than 1.00 downstream of the shockwaves in the diffuser, the static pressure recovery rates are the same. However, the NIST RGM predicts the lowest static temperature recovery. That is to say, most of the cold energy and kinetic energy are converted back to pressure energy, and the NIST RGM obtains the highest energy conversation efficiency. This will effectively improve the energy efficiency of the SS.

Figure 8.4 illustrates the gas density diagram obtained by the IGM, NIST RGM, and Redlich–Kwong RGM. The NIST RGM and Redlich–Kwong RGM calculate similar results. In comparison, there is a significant difference in gas density between the IGM and the other two, with the maximum relative error of about 37.5% in the cyclone separation section. Thus, the difference in gas density will lead to apparent errors in flow field parameters of high-pressure offshore natural gas supersonic dehydration through a SS. More important, the three thermodynamic models compute different characteristics of water vapor mole fractions in the SS, as shown in Fig. 8.5. In the high-pressure offshore natural gas supersonic dehydration process, the water vapor mole fractions predicted by IGM, NIST RGM, and Redlich–Kwong RGM are 0.000088, 0.000064 and 0.000076 at the dry gas outlet, respectively. The IGM obtains higher water vapor mole fraction by about 37.5 and 15.8% compared to the NIST RGM and Redlich–Kwong RGM. It demonstrates that the IGM underestimates the water vapor removal rate of the SS more than the other two models under the same operation condition, and the NIST RGM obtains the best in the high-pressure offshore natural gas supersonic dehydration process.

Figure 8.6 shows the droplet diameter evolution with the three thermodynamic models in the SS. One can see that the droplet diameter and dynamic behaviors predicted by the three thermodynamic models are different. The maximum droplet diameters are 3.7 μm for IGM, 3.9 μm for NIST RGM, and 4.1 μm for Redlich–Kwong RGM, respectively. The relative errors are about 5.4 and 10.8%. The droplets predicted by NIST RGM are separated in the wet gas outlet, while the droplets estimated by IGM and Redlich–Kwong RGM are not. The escaped particle concentrations are 7.22×10^{-5} and 6.56×10^{-4} kg m^{-3} predicted by IGM and Redlich–Kwong RGM. This indicates that droplets condensation and dynamic behavior vary with different thermodynamic models with the same operation condition. Therefore, for an energy-efficient separation performance in the SS, attention must be paid to the influences of thermodynamic properties on high-pressure offshore natural gas dehydration process with different p_{in} and foreign droplets parameters.

8.2.2 Separation Performance

The maximum separation performances predicted by the three thermodynamic models with different operation conditions are shown in Fig. 8.7, considering the $p_{\text{in}} = 6$, 30, 60, 90, and 120 bar. Figure 8.7a implies that the NIST RGM obtains the highest η_{w} than others, with the maximum relative deviation of 10.9 and 10.2% than IGM and Redlich–Kwong RGM when $p_{\text{in}} = 6$ bar. For an average relative deviation at the five p_{in}, the NIST RGM maintains a higher value of 7.7 and 5.0% than the IGM and Redlich–Kwong RGM. It can be seen in Fig. 8.7b that the NIST RGM also predicts the highest ΔT_{d} than others, with the maximum relative deviation of 5.9 and 5.8% than the IGM and Redlich–Kwong RGM when $p_{\text{in}} = 6$ bar. For an average relative deviation, the NIST RGM predicted ΔT_{d} is 5.0% higher than IGM, and 4.5% higher than Redlich–Kwong RGM. This means that for the high-pressure offshore natural gas dehydration process of the SS, the NIST RGM can obtain the most increased separation performance than the other two models.

Under NIST RGM, as p_{in} rises from 6 to 120 bar, the η_{w} increases from 67.05 to 81.75% with 21.78%, the ΔT_{d} rises from 32.19 to 37.65 K with 16.81%. This implies that increasing the inlet pressure has a positive impact on enhancing separation performance.

8.2.3 Energy Conversion

In the SS, the water vapor condensed and separated by the energy conversion with the conversion of enthalpy into kinetic energy. The energy conversion efficiency has a

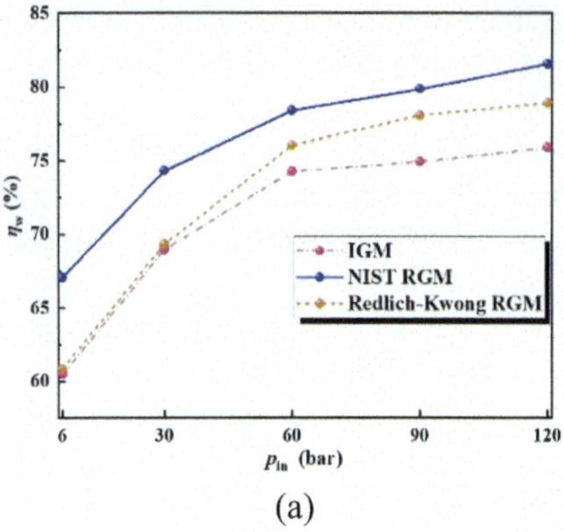

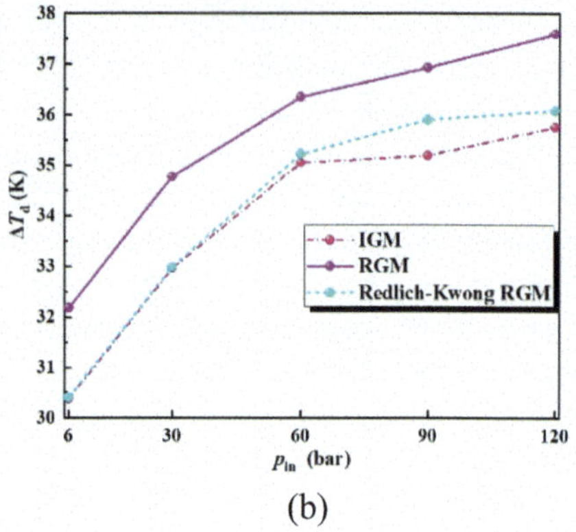

Fig. 8.7 Maximum separation performance with the three thermodynamic models under different p_{in}: **a** the comparison of η_w; **b** the comparison of ΔT_d

critical effect on the separation performance from the view of this aspect. Therefore, the temperature drop ΔT and entropy generation ΔE were studied based on the three thermodynamic models. It is noted that the temperature rise caused by the latent heat of the condensation phase change was fully taken into account in this process.

Figure 8.8 illustrates the comparison results of maximum ΔT with the three thermodynamic models. One can see that the IGM does not predict the ΔT. This means that the IGM under-predicts the cold energy generation due to the conversion of pressure energy, which also confirms the fact that the IGM has a low prediction on the separation performance of the SS in the high-pressure environment. However, ΔT becomes more significant with p_{in}. For the same pressure energy loss, the NIST RGM predicted ΔT increases from 5 to 29 K, and that obtained by Redlich–Kwong RGM rises from − 1 to 28.5 K. This means that the Redlich–Kwong RGM and NIST RGM get the Joule–Thomson effect of the SS. The energy conversion efficiency is enlarged with p_{in}, and the NIST RGM obtains higher energy conversion efficiency. It also corresponds to the previous results that the higher the p_{in}, the better the separation performance predicted by NIST RGM.

Furthermore, the three thermodynamic models predict different ΔE under the optimum separation performance of the SS with different p_{in}, as shown in Fig. 8.9. One can see that with p_{in} increasing, the IGM estimated ΔE shows an increasing trend from 627.9 to 788.7 J kg^{-1} K^{-1} with a relative growth rate of 25.3%. On the contrary, the ΔE predicted by NIST RGM decreases from 573.4 to 514.3 J kg^{-1} K^{-1} with 10.3%. The Redlich–Kwong RGM predicted ΔE decreases from 615.5 to 525.1 J kg^{-1} K^{-1} with 14.7%. Under the same p_{in}, the NIST RGM obtains the lowest ΔE, and the maximum relative deviation is 53.4% for IGM and 7.3% for Redlich–Kwong RGM. This means that the NIST RGM predicts the highest energy

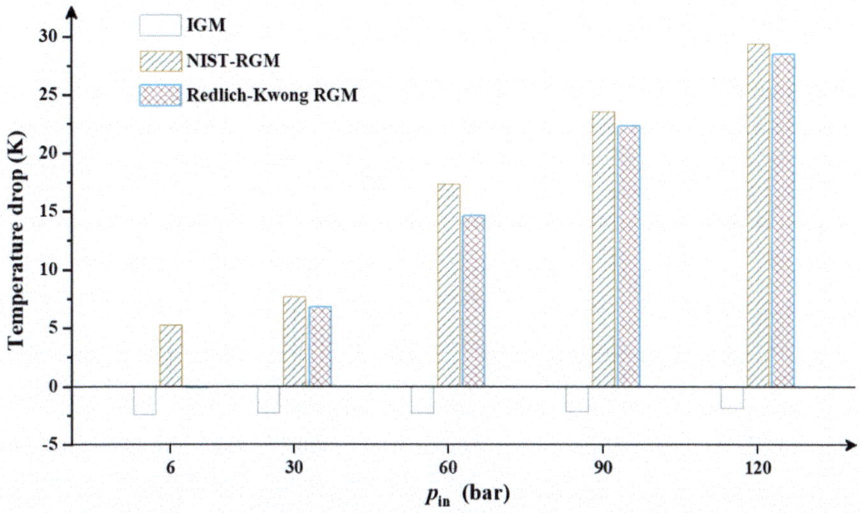

Fig. 8.8 Maximum ΔT with the three thermodynamic models

conversion efficiency. Therefore, for the high-pressure offshore natural gas dehydration process of the SS, the NIST RGM needs to be used for separation performance quantification. p_{in} has a positive effect on improving separation performance and energy efficiency.

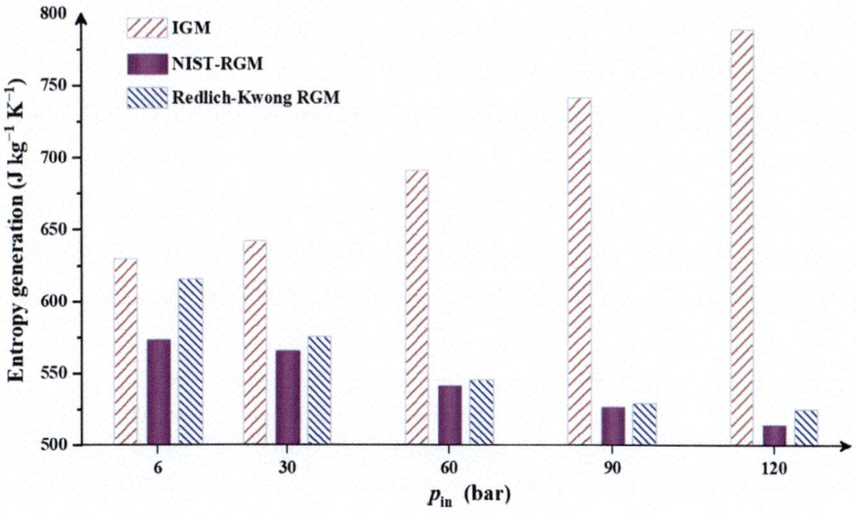

Fig. 8.9 Maximum ΔE with the three thermodynamic models

8.3 Joule–Thomson Effect

In thermodynamics, the Joule–Thomson effect describes the temperature change of a real gas when it is forced by a pressure difference through a valve while keeping it adiabatic [14]. This procedure is called the Joule–Thomson process. The effect is purely due to deviation from ideality, as any ideal gas has no Joule–Thomson effect. This effect is very important in practical applications,for example, it is widely used in gas refrigeration and liquefaction processes. The liquefaction process of gases such as nitrogen, oxygen, air, and natural gas uses the Joule–Thomson effect to reduce temperature [12].

For the supersonic separator, based on the NIST RGM, the total temperature drop ($\Delta T_{\text{J-T}} = T_{\text{in}} - T_{\text{dry}}$) and entropy generation between the inlet and dry gas outlet are shown in Table 8.2 with $S_{\text{w}} = 0.636$ under the five inlet pressures of $p_{\text{in}} = 6$, 30, 60, 90, and 120 bar. The released latent heat (T_{con}) by the condensation of water vapor is taken into account when calculating the temperature depression. With inlet pressure increases, the temperature drop becomes more significant. That is to say, a larger temperature drop is caused by a higher loss of pressure, i.e., the corresponding energy conversion efficiency from pressure energy to cold energy is enlarged. For instance, the $\Delta T_{\text{J-T}}$ increases from − 0.1 K of $p_{\text{in}} = 6$ bar to 28.5 K of $p_{\text{in}} = 120$ bar. The Joule–Thomson effect of high-pressure natural gas supersonic dehydration can influence the production of cold energy from pressure energy. The produced cold energy can be reused to improve the dehydration efficiency of the SS by heat exchange between outlet dry natural gas and inlet source natural gas, which can provide a superior low-temperature environment for condensation will be provided by the Laval nozzle section in the supersonic separator. And it also can be recycled to generate power by the Organic-fluid Rankine Cycle (ORC) system in the dehydration technology of oil–gas fields.

The Joule–Thomson coefficient ($u_{\text{J-T}}$) is used to illustrate the Joule–Thomson effect, which can be calculated by:

$$u_{\text{J-T}} = \left(\frac{\partial T}{\partial p}\right)_{\text{H}} \tag{8.7}$$

where ∂T and ∂p are the rate of change of temperature and pressure.

Table 8.2 Temperature drop and entropy generation with different inlet pressure

p_{in} (bar)	p_{out} (bar)	T_{in} (K)	T_{dry} (K)	T_{con} (K)	$\Delta T_{\text{J-T}}$ (K)
6	2	300.0	301.1	6.4	5.3
30	10	300.0	293.6	1.3	7.7
60	20	300.0	283.2	0.6	17.4
90	30	300.0	276.8	0.4	23.6
120	40	300.0	270.9	0.3	29.4

In the supersonic separator, the Joule–Thomson coefficient along the z-axis direction is varied with temperature and pressure variation with RGM. The evolution law of the Joule–Thomson coefficient along the z-axial direction is depicted in Fig. 8.10 when $p_{in} = 120$ bar. The Joule–Thomson coefficient remains almost a constant value of about 0.28 K bar^{-1} in the convergent section of the Laval nozzle because both temperature and pressure are tended to decrease with the same variation in the Laval nozzle due to the expansion effect. That is to say, a very small part of the pressure energy converts to cold energy in this section. As the natural gas flows through the throat, a significant jump of the Joule–Thomson coefficient is appeared with the increment from 0.28 to 0.55 K bar^{-1} due to the influence of the heterogeneous condensation by water vapor around the throat. Then the Joule–Thomson coefficient will increase to 1.74 K bar^{-1} in the cyclone separation section before the shockwave location. It can reflect that more pressure energy is converted into cold energy, and the generated cold energy reaches the maximum before shockwave. When the natural gas flows through the aerodynamic shockwave location, the Joule–Thomson coefficient drops abruptly. The aerodynamic shockwave can cause a harsh environment, that is, the sudden increment of temperature and pressure with different amplitudes in the SS. Then the Joule–Thomson coefficient fluctuates slightly and eventually becomes constant of 0.48 K bar^{-1} at the diffuser section. Most of the cold energy is converted into pressure energy again in this section.

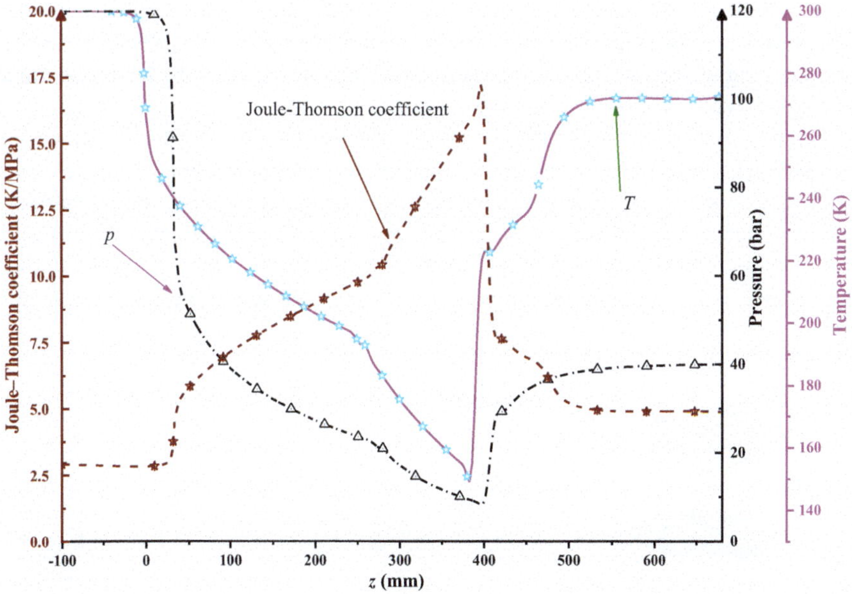

Fig. 8.10 Joule–Thomson coefficient distribution along the z-axis when $p_{in} = 120$ bar

8.4 Swirl Strength Effect

In the process of supersonic natural gas dehydration, swirl strength is one of the key factors affecting separation performance and energy conversion rate. It greatly affects the evaluation of flow characteristics and separation performance in the supersonic separation process of high-pressure natural gas. Notably, different swirl strengths result in distinct predictions for separation performance and energy conversion, showing various trends with changes in inlet pressure.

8.4.1 Separation Performance

Three swirl generators are designed to qualitatively analyze the impact of swirl strength on supersonic separation performance. Each swirl generator has a unique outlet angle β: a weak swirl strength with $\beta = 20°$, a normal swirl strength with $\beta = 35°$, and a strong swirl strength with $\beta = 50°$. The corresponding swirl numbers are $S_w = 0.485$, 0.636, and 0.885, respectively. Figure 8.11 displays the droplet diameter distributions for different swirl strengths, with $p_{in} = 120$ bar, $q_p^{in} = 0.0026$ kg s^{-1}, $d_p^{in} = 3.3$ μm, $p_{out} = 40$ bar, and $T = 300$ K. A higher number of foreign droplets escape toward the diffuser and undergo evaporation by the shock wave when $S_w = 485$. With $S_w = 0.885$, all foreign droplets impact the wall in the expansion section of the Laval nozzle. Conversely, with $S_w = 0.636$, all droplets are successfully collected by the liquid outlet section. This suggests that different swirl strengths lead to distinct droplet behaviors within the supersonic separator.

The distributions of film thickness with different swirl strengths are shown in Fig. 8.12. The liquid film forms in the convergence section of the Laval nozzle and finally accumulates and separates in the drainage section of the supersonic separator. The maximum film thickness at the wet gas outlet for $S_w = 0.485$, $S_w = 0.636$, and $S_w = 0.885$ are 26.96 μm, 48.41 μm, and 50.44 μm, respectively. It is evident that swirl strength also significantly influences the formation and evolution of the liquid film.

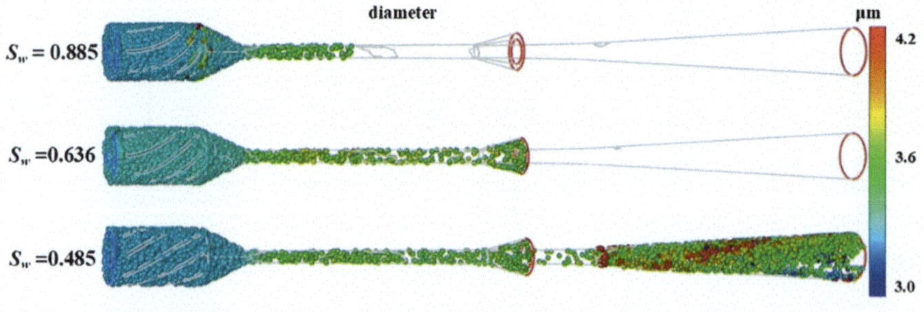

Fig. 8.11 Droplet diameter distributions with different swirl strengths

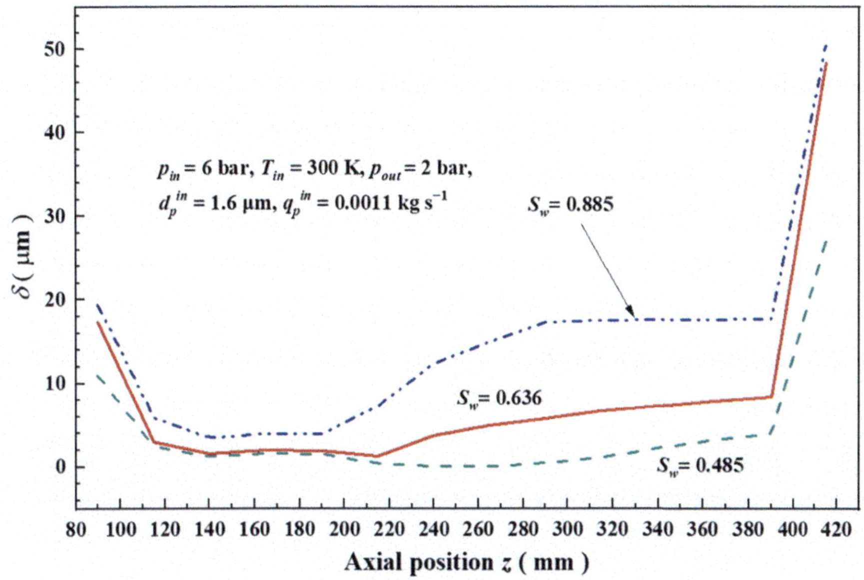

Fig. 8.12 Film thickness distributions with different swirl strengths

Figure 8.13 illustrates the separation performance of the supersonic separator under different swirl strengths.

When $S = 0.485$, $\eta_p = 45.41\%$, whereas for other cases, it reaches 100%. The η_w are 81.75% and 69.26% when $S_w = 0.636$ and $S_w = 0.885$, respectively, while

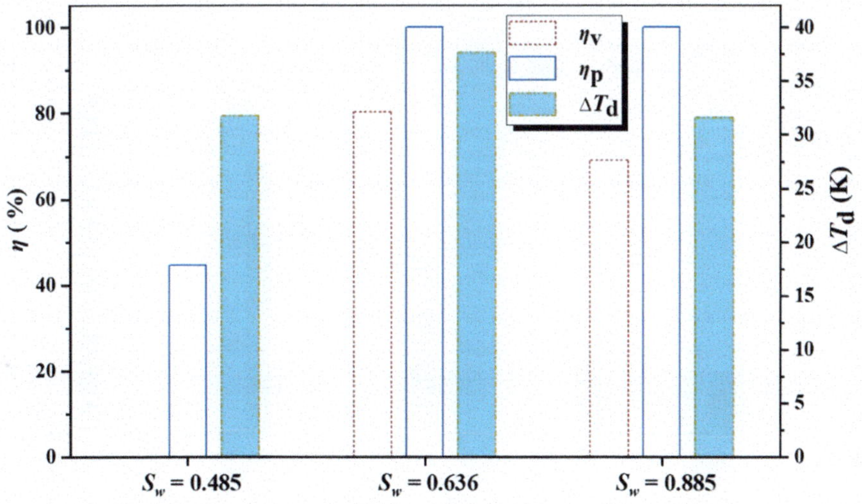

Fig. 8.13 Separation performance with different swirl strengths

η_v is 0 with $S_w = 0.485$. The ΔT_d is 37.65 K with $S_w = 0.636$, exhibiting a higher value of 19.08 and 18.73% compared with $S_w = 0.485$ and $S_w = 0.885$. This implies that while strong swirl strength effectively separates all foreign droplets, it does not significantly enhance the overall separation performance of the supersonic separator. On the other hand, weak swirl strength not only fails to achieve complete droplet separation but also impedes dehydration efficiency. In contrast, normal swirl strength is the optimal choice for maximizing the separation performance of the supersonic separator.

8.4.2 Energy Conversion

Water vapor undergoes condensation and separation through the conversion of enthalpy into kinetic energy in the high-pressure supersonic dehydration. The efficiency of this energy conversion plays a crucial role in determining the separation performance from this perspective. The temperature drop (ΔT_{J-T}) and entropy generation (ΔE) at different swirl intensities and specific inlet pressures are important indicators reflecting energy conversion.

The comparison results of ΔT_{J-T} under different swirl strengths are illustrated in Fig. 8.14.

As p_{in} rises from 6 to 120 bar, ΔT_{J-T} increases from 7.15 to 31.68 K when $S_w = 0.485$. Similarly, with $S_w = 0.636$, ΔT rises from 5.29 to 29.42 K. For $S_w = 0.885$, ΔT increases from 3.17 to 33.10 K. This implies that the inlet pressure positively influences the enhancement of energy conversion efficiency from pressure energy to

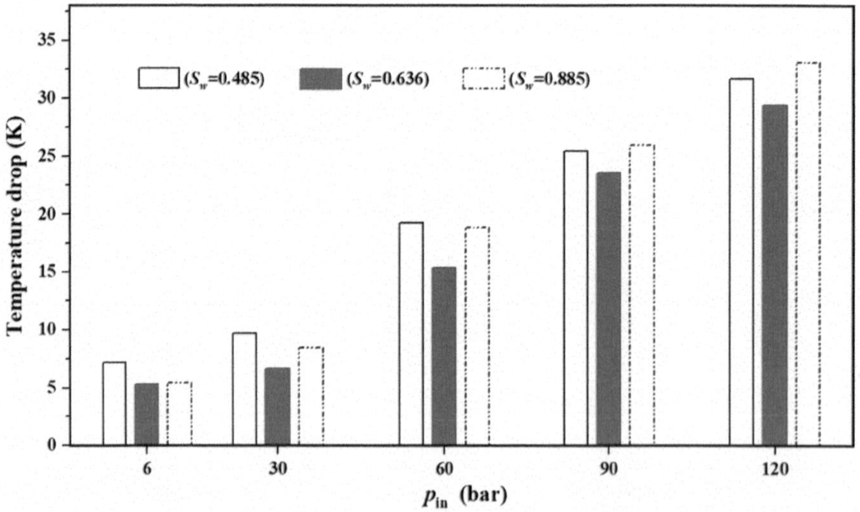

Fig. 8.14 Comparisons of ΔT_{J-T} with different swirl strengths under different p_{in}

cold energy, which can be considered to improve the separation performance of the supersonic separator. Under all conditions, ΔT_{J-T} value for $S_w = 0.636$ consistently maintains the lowest value, exhibiting maximum deviations of 20.70% and 12.51% when compared with $S_w = 0.485$ and $S_w = 0.885$, respectively. This shows that both weak and strong swirl strengths can achieve a higher temperature drop than normal swirl strength. This occurs because weak swirl strength cannot completely separate the foreign droplets, and the secondary evaporation of the droplets in the diffuser causes the gas temperature to decrease.

Figure 8.15 shows the comparisons of ΔE attributed to different swirl strengths under five p_{in}. As p_{in} increases from 6 to 120 bar, for $S_w = 0.485$, ΔE exhibits a decline from 585.5 to 493.1 J kg^{-1} K^{-1}. Similarly, for $S_w = 0.636$, ΔE demonstrates a decreasing pattern from 573.4 to 514.3 J kg^{-1} K^{-1}. Additionally, when $S_w = 0.885$, ΔE decrease from 583.5 to 479.6 J kg^{-1} K^{-1}. This feature considerably indicates that increasing the inlet pressure leads to decreased entropy generation of the SS, regardless of the swirl strength. This is because the Joule − Thomson effect is predicted by the NIST RGM in the high-pressure supersonic dehydration processing of the natural gas as described in Sect. 8.2.

Moreover, when $p_{in} = 6$ bar, the ΔE for $S_w = 0.636$ is lower than that for $S_w = 0.485$ and $S_w = 0.885$, with deviations of 2.07% and 1.73%, respectively. However, as $p_{in} \geq 30$ bar, the ΔE for $S_w = 0.636$ displays maximum deviations of 4.12% and 6.75% compared with $S_w = 0.485$ and $S_w = 0.885$, respectively. These observations suggest that, under the same p_{in}, the swirl strength is also a factor driving changes in entropy generation. The impact of swirl strength on entropy generation shows a gradually increasing trend as the inlet pressure rises.

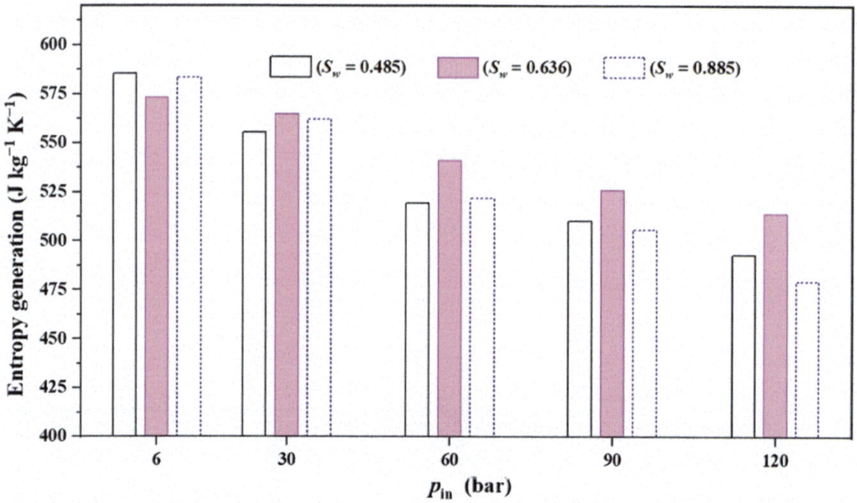

Fig. 8.15 Comparisons of ΔE with different swirl strengths under different p_{in}

8.5 Laval Nozzle Parameter Sensitivity Analysis

The Laval structural parameters are crucial in determining both separation performance and pressure energy loss. However, a comprehensive understanding of how them influence flow behaviors and performance is hindered by their coupling effects and the limitations of numerical methods. Notably, different parameters of the Laval nozzle significantly impact separation performance and pressure energy loss.

Figure 8.16 details the structural diagram of the straight inlet section and supersonic nozzle. The inlet and throat flow areas remain constant with a specific flow rate, as shown in Table 8.3. Eight static vanes are uniformly positioned within the annular flow channel in the straight inlet section, featuring a fixed outlet angle of 35°. The features of the shroud wall and the inner body determine the structural parameters of the annular nozzle. Specifically, the effective size of the inlet is governed by the inlet radius of the inner body ($R_{\mathrm{hub,in}}$), while the effective size of the throat is dictated by the throat radius of the inner body ($R_{\mathrm{hub,th}}$). As shown in Fig. 8.16, the corresponding structural parameters and their respective ranges of variation are provided in Table 8.3. One can see that the z-axis is defined as the rotation axis, with the center point of the throat serving as the origin of the coordinates, denoted as $z = 0$ mm. The expansion ratio of the divergent section (ERDS) is defined as $\mathrm{Er} = A_{\mathrm{out}}/A_{\mathrm{th}}$, correlates with its length (L_5). The convergent section is designed with quintic curves to ensure a uniform flow field, as described by:

$$\frac{R_z - R_{\mathrm{th}}}{R_{\mathrm{eff,in}} - R_{\mathrm{th}}} = 1 - 10\left(\frac{z}{L_3}\right)^3 + 15\left(\frac{z}{L_3}\right)^4 - 6\left(\frac{z}{L_3}\right)^5 \tag{8.8}$$

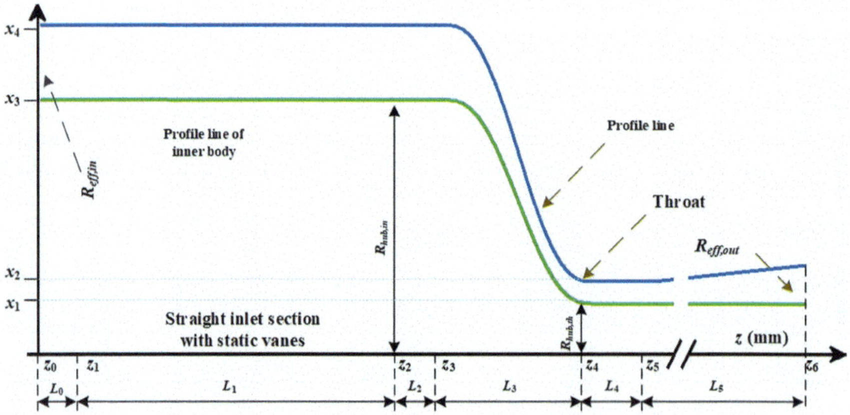

Fig. 8.16 Geometric structure of the straight inlet section and supersonic nozzle

Table 8.3 Physical dimensions of the straight inlet section and supersonic nozzle

Dimensions of structure	Value
Straight inlet section of supersonic separator (L_0)	10.00 mm
Straight inlet section with static vanes (L_1)	80.00 mm
Straight inlet section of the supersonic nozzle (L_2)	10.00 mm
Length of the convergent section (LCS, L_3)	20.00–100.00 mm
Straight section of the throat (L_4)	15.00 mm
Length of the divergent section (L_5)	200.00–300.00 mm
Inlet radius of the inner body (IRIB, $R_{\text{hub,in}}$)	15.00–45.00 mm
Throat radius of the inner body (TRIB, $R_{\text{hub,th}}$)	0.50–13.50 mm
Effective inlet diameter ($d_{\text{eff,in}}$)	$d_{\text{eff,in}} = 2 \times \sqrt{(x_4^2 - x_3^2)}$ mm
Effective throat diameter ($d_{\text{eff,th}}$)	$d_{\text{eff,th}} = 2 \times \sqrt{(x_2^2 - x_1^2)}$ mm
Effective flow area of the inlet (A_{in})	1717.67×10^{-6} mm^2
Effective flow area of the throat (A_{th})	105.07×10^{-6} mm^2

where $R_{\text{eff,in}}$ signifies the inlet radius determined by $R_{\text{hub,in}}$. $R_{\text{eff,th}}$ represents the throat radius of the supersonic nozzle, calculated using $R_{\text{hub,th}}$. R_z denotes the cross-sectional radius at any given position z. L_3 represents the length of the convergent section (LCS).

It can be concluded that in a supersonic nozzle, the parameters such as $R_{\text{hub,in}}$, $R_{\text{hub,th}}$, L_3 and Er are pivotal parameters that need to be analyzed in this paper.

8.5.1 Expansion Ratio of Divergent Section (ERDS) Effect

During the dehydration process of the CH_4–H_2O gas mixture in a supersonic separator, the water vapor accelerates along with the supersonic gas mixture and subsequently impacts to the shroud wall in the divergent section due to swirling motion. Figure 8.17a shows the mass fraction of water vapor and droplet distributions at varying ERDS values, specifically from Er = 1.0 to 1.6, with dimensions L_3 = 60 mm, $R_{\text{hub,in}}$ = 15 mm, and $R_{\text{hub,th}}$ = 6.75 mm. It is evident that heterogeneous droplets migrate toward the shroud wall downstream of the throat due to centrifugal forces, resulting in a significantly higher concentration of droplets near the shroud wall compared to the inner body. Additionally, the water vapor mass fraction exhibits radial inhomogeneity across a specific cross-section, gradually decreasing from the inner body toward the shroud wall. Significant differences are observed among Er values of 1.0, 1.3, and 1.6. As Er increases from 1.0 to 1.6, the mean droplets concentration decreases from 0.17 to 0 kg m^{-3}, and the water vapor mass fraction declines from 0.51 to 0.37%. These results indicate that the ERDS plays a critical role in influencing the distribution of each phase and in determining the behavior of heterogeneous droplets. Furthermore, variations in swirl strength and centrifugal acceleration are also crucial in this process.

Figure 8.17b shows the variations in swirl strength and centrifugal acceleration along the z-axis. Swirl strength is quantified using a dimensionless parameter S_w, which can be defined as the ratio of rotational momentum to axial momentum in the natural gas mixture flow. The average value of S_w on the annular flow channel cross-section at the current coordinates serves as the evaluation index. Centrifugal

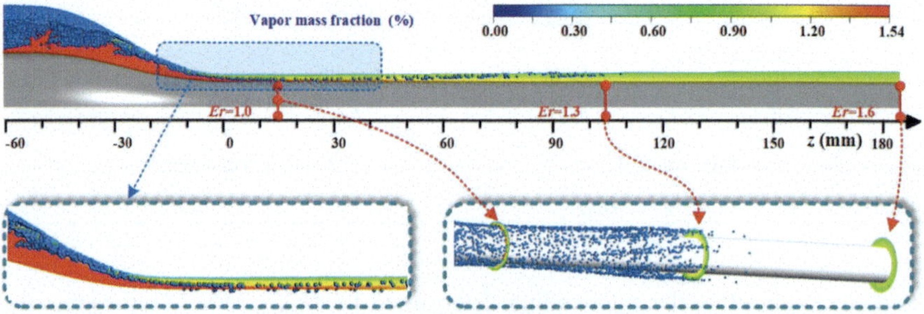

(a)

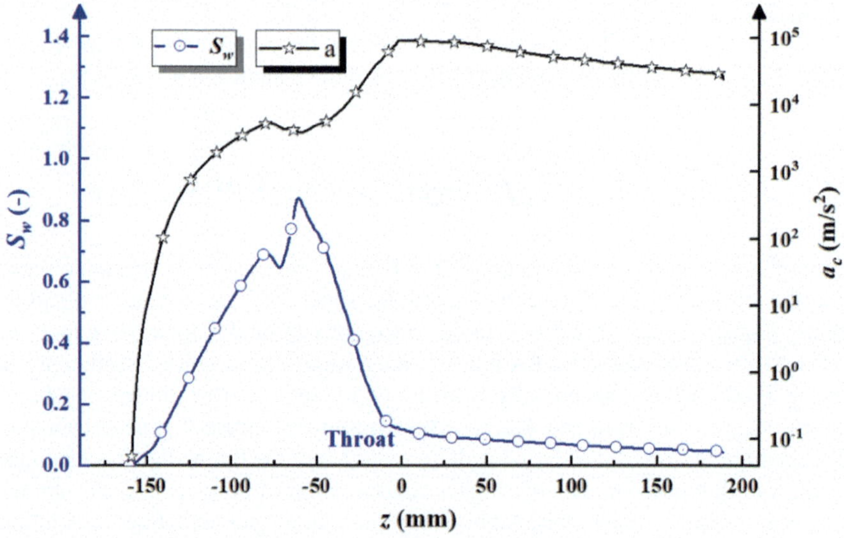

(b)

Fig. 8.17 Flow behaviors with different ERDS values: **a** distributions of water vapor mass fraction and droplet; **b** distribution of the swirl strength and centrifugal acceleration

acceleration a_c is utilized to characterize the centrifugal force within the flow field, equal to the ratio of the square of the airflow's rotational speed to its radial position. Similarly, the average value of a_c on the annular flow channel cross-section is used for analysis in Fig. 8.17b.

As illustrated in Fig. 8.17b, a_c reaches its maximum value of 9.14×10^4 m s^{-2} at Er = 1.0, and subsequently decreases with increasing Er. At Er = 1.3 and Er = 1.6, the a_c values are 4.73×10^4 m s^{-2} and 2.41×10^4 m s^{-2}, respectively, making a relative decrease is approximately 73.63%. This indicates that in the divergent section of the supersonic nozzle, with an increase in ERDS, the flow field's capacity to centrifuge droplets to the shroud wall diminishes. Figure 8.17b clearly shows that the dimensionless parameter S_w reaches its pick value of 0.914 at the supersonic nozzle inlet ($z_4 = -60.00$ mm) and decreases to 0.118 at the throat ($z_5 = 0.00$ mm). Subsequently, it undergoes further reduction with the increasing Er, with S_w values of 0.102, 0.067, and 0.044 observed at Er = 1.0, 1.3, and 1.6, respectively. Through data fitting analysis conducted within the divergent section, it is found that S_w follows Eq. (8.9):

$$S_w = S_{w,0} + S_{w,th} e^{-\zeta \frac{(z - z_5)}{d_{eff,th}}} \tag{8.9}$$

where $S_{w,0}$ represents the fitting constant, set as -0.0103, ζ denotes the attenuation coefficient, equal to -0.0318. $S_{w,th}$ and $d_{eff,th}$ denote the swirl strength at the throat and the effective diameter of the throat, respectively. Figure 8.18 illustrates the fitting accuracy of swirl strength. It is evident that the simulation data predominantly fall within the $\pm$ 5% error band of the fitted curve, with an R^2 value of 0.986. This demonstrates a high level of accuracy in the fitted curve.

To illustrate the sensitivity of ERDS on separation performance, consider a structure with dimensions $L_3 = 60.00$ mm, $R_{hub,in} = 27.50$ mm, and $R_{hub,th} = 0.50$ mm.

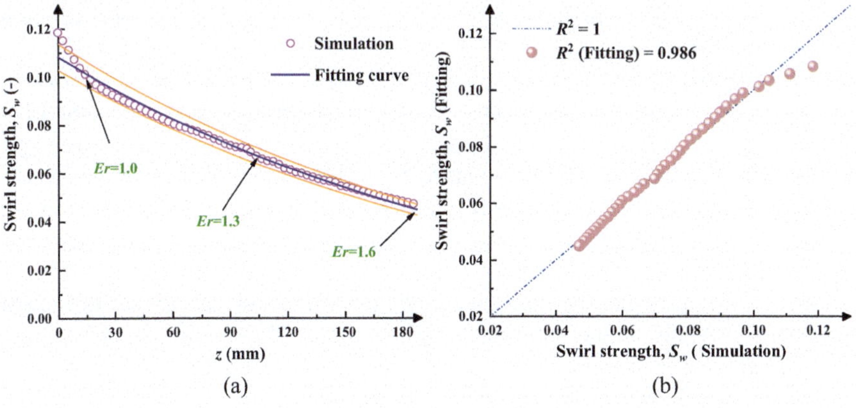

Fig. 8.18 Comparison of simulation results and fitting results of swirl strength: **a** fitting curve and simulation results; **b** the comparison of R^2

The flow behavior and separation performance are illustrated in Fig. 8.19. Analysis of Fig. 8.19a reveals notable trends: as the Er increases from 1.0 to 1.6, η_w first increases from 89.93 to 95.09% at Er = 1.5, and then decrease to 93.56% at Er = 1.6. ΔT_d expands from 12.70 to 25.28 K at Er = 1.5, then drops to 22.32 K. γ rises from 0.12 to 0.47 Er = 1.5, then decreases to 0.45 at Er = 1.6. In summary, as ERDS increases, the separation performance will first increase to a maximum at a specific Er, then begin to decrease, and will not increase indefinitely with further increases in ERDS. The main reason is the aerodynamic shock waves. As shown in Fig. 8.19c, it is evident that, at z = 250.00 mm, the pressure, static temperature, and sound speed reach their minimum values of 0.71 atm, 230.81 K, and 395.78 m s^{-1}, respectively, while resulting that the Mach number reaches its peak value of 1.56. Then as the ERDS increases to Er = 1.6, the pressure, static temperature, and sound speed rise to 0.92 atm, 248.26 K, and 410.15 m s^{-1}, respectively, while the Mach number decays to 1.35. These changing patterns of the axis profiles also verify the above conclusions.

These results emphasize the critical role of ERDS in impacting the separation performance of supersonic separator. It should be noted that this influence comes with varied pressure energy loss as ERDS values rise. Therefore, it is crucial to carefully balance the conflicting demands of separation performance and pressure energy loss during structural design and optimization to achieve an optimal compromise.

8.5.2 Length of Convergent Section (LCS) Effect

Figure 8.20 illustrates the flow characteristics, separation performance, and pressure energy loss of the supersonic separator under three LCS values: L_3 = 20.00, 50.00, and 100.00 mm. This analysis is conducted under the condition of Er = 1.2, $R_{hub,in}$ = 27.50 mm, $R_{hub,th}$ = 6.75 mm. Figure 8.20a, b illustrate the distributions of swirl strength and centrifugal acceleration. It is evident that the swirl strength values at the throat are S_w = 0.157, 0.150, and 0.142, respectively. Correspondingly, the centrifugal acceleration values at the throat are $a_c = 1.28 \times 10^5$ m s^{-2}, 1.31×10^5 m s^{-2} and 1.36×10^5 m s^{-2}, respectively. It is evident that as the LCS increases, the swirl strength at the throat gradually decreases while the centrifugal acceleration steadily increases by about 6.25%. This trend indicates an increase in the rotational speed of the airflow and a concurrent reduction in the disparity between the rotational speed and axial velocity due to the acceleration effect induced by the supersonic nozzle.

Figure 8.20c presents the cloud diagrams that show several critical flow field parameters of static temperature and Mach number. These help visualize phase changes and the distribution of flow field parameters, which are crucial for analyzing the performance and efficiency of supersonic separators.

Figure 8.20d describes the droplet distribution, pressure, mass fraction of water vapor dehydration efficiency, dew-point depression, and pressure loss ratio across different LCS values of L_3 = 20.00, 50.00, and 100.00 mm. As the LCS value

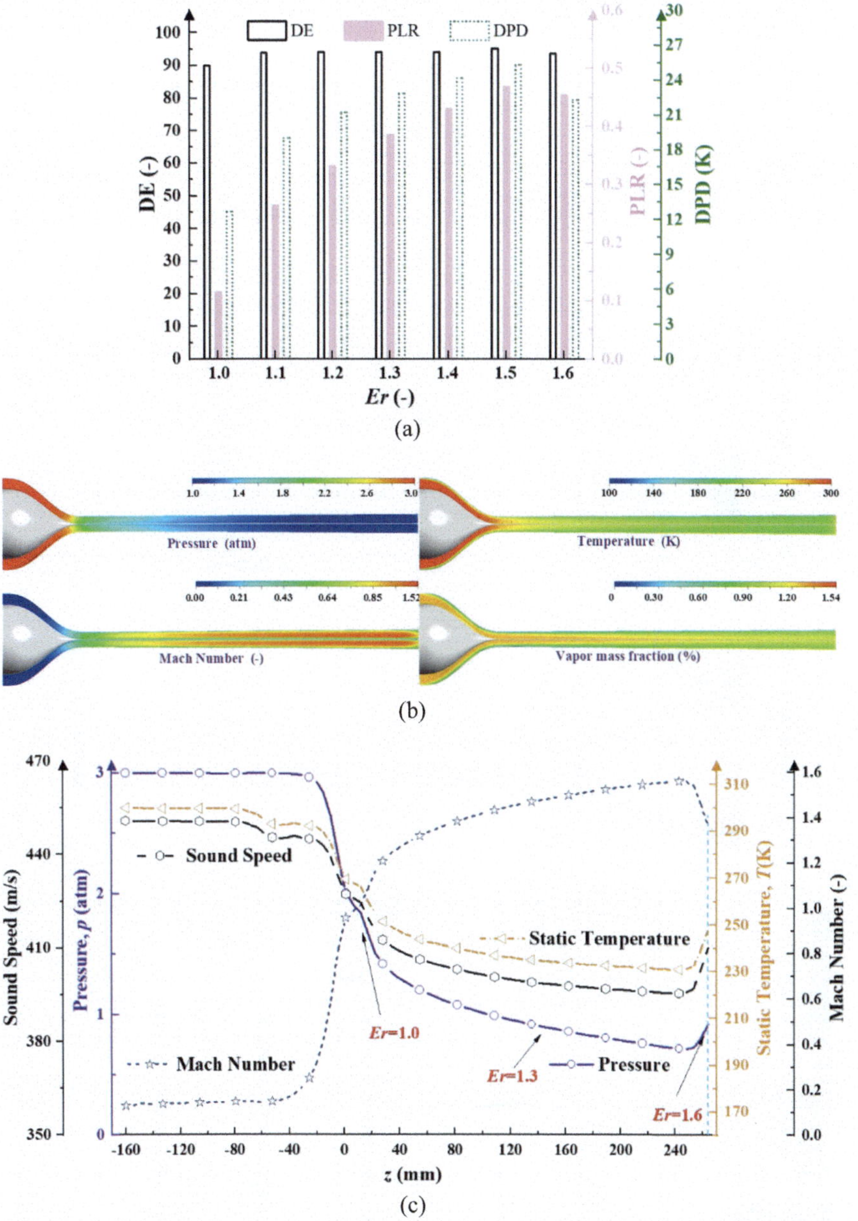

Fig. 8.19 Separation performance and axial profiles with different ERDS values: **a** separation performance and pressure energy loss; **b** cloud diagram; **c** axial profiles of flow variables

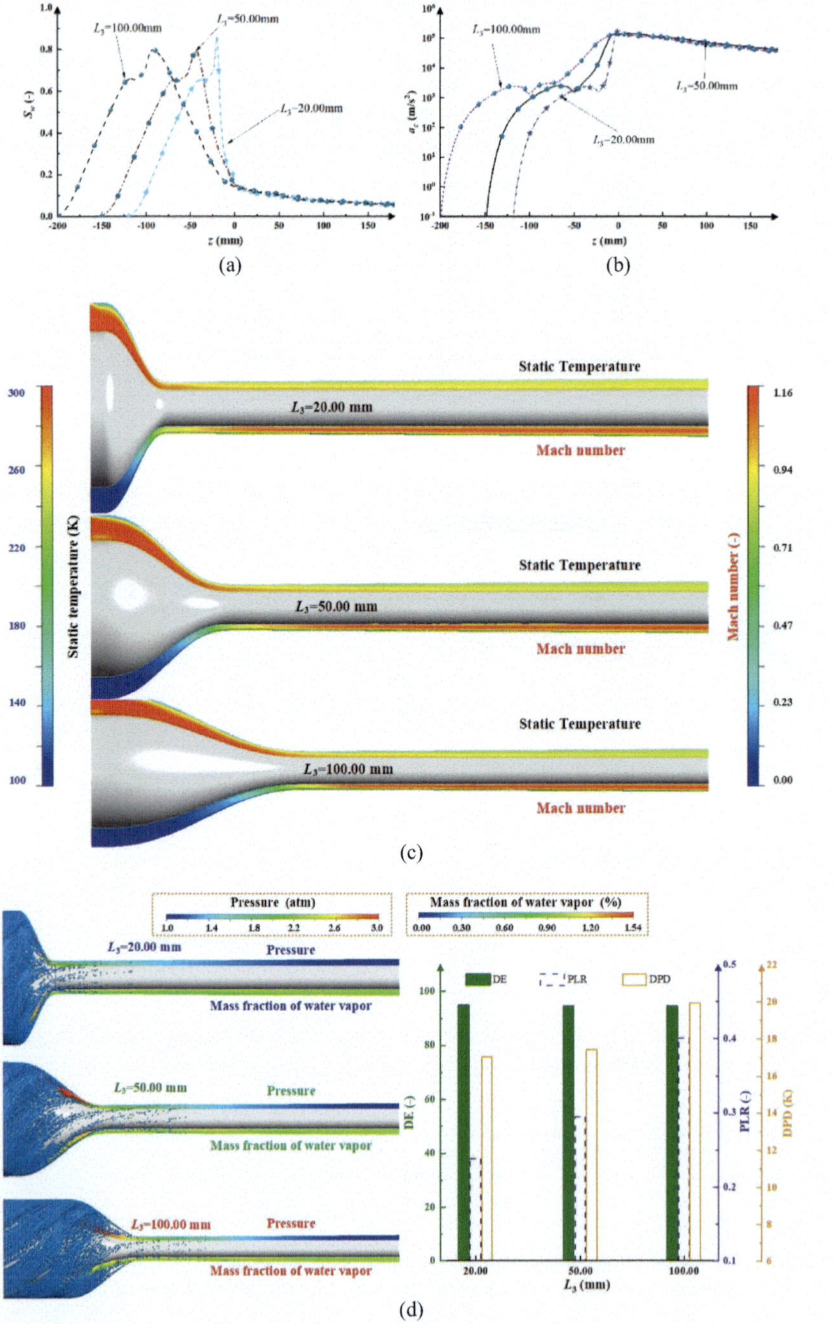

Fig. 8.20 Flow characteristics and separation performance with different LCS values: **a** distribution of the swirl strength; **b** distribution of the centrifugal acceleration; **c** cloud diagram of flow field parameters; **d** separation performance and pressure energy loss

increases, the droplet migration time in the flow field lengthens. More important, the maximum values of η_w are observed to be 94.95%, 94.50%, and 94.60%, respectively. The corresponding dew-point depressions ΔT_d are 17.05 K, 17.44 K, and 19.96 K, respectively. Additionally, the corresponding pressure loss ratios γ are 0.24, 0.29, and 0.40, respectively. These results suggest that as the LCS increases, the maximum dehydration performance initially decreases and gradually increases, though less pronouncedly. However, LCS significantly impacts pressure energy loss, with longer LCS values corresponding to higher pressure energy losses when maximum separation performance is achieved.

Figure 8.21 illustrates the axial profiles of pressure, static temperature, sound speed, Mach number, and the mass fraction of water vapor for the three LCS values. As the supersonic nozzle expands, the pressure, temperature, and sound speed of the natural gas mixture progressively decrease, resulting in a continuous increase in the Mach number. However, when the LCS value reaches $L_3 = 100.00$ mm, the static temperature and sound speed begin to increase beyond $z = 120$ mm, which negatively impacts separation performance. In addition, the sensitivity of the above LCS value to the separation performance can be further verified from Fig. 8.21e.

8.5.3 *Inlet Radius of Inner Body (IRIB) Effect*

Figure 8.22 illustrates the flow characteristics, maximum separation performance, and pressure energy loss of the supersonic separator for three distinct IRIB values: $R_{hub,in} = 15.00$, 27.50, and 45.00 mm. This analysis keeps constant values of $L_3 = 60$ mm, $R_{hub,th} = 6.75$ mm, and $A_{in} = 1717.67$ mm^2. As illustrated in Fig. 8.22a, b, at throat, $S_w = 0.119, 0.150$ and 0.194. The centrifugal accelerations at the throat are calculated as $a_c = 8.56 \times 10^4$, 1.46×10^5 and 2.58×10^5 m s^{-2}. Clearly, both the swirl strength and the centrifugal acceleration at the throat increase with the IRIB. The findings suggest that the IRIB has the potential to enhance the swirling capability of the supersonic flow field. Notably, before reaching $z = -30$ mm, a larger IRIB is associated with lower centrifugal acceleration. However, after the flow is accelerated by the supersonic nozzle, this trend reverses, and a larger IRIB corresponds to higher centrifugal acceleration. Understanding this relationship is crucial for optimizing the design and performance of the supersonic separator.

Figure 8.22c presents cloud diagrams showing the distributions of static temperature and Mach number for various IRIB values. These diagrams demonstrate the influence of various IRIB configurations on key flow parameters in the supersonic separator, providing qualitative insights into IRIB's impact on flow dynamics and aiding structural sensitivity analysis. Figure 8.22d describes the droplet trajectory, pressure, water vapor mass fraction, dehydration efficiency, dew-point depression, and pressure loss ratio with varying IRIB values: $R_{hub,in} = 15.00$, 27.50, and 45.00 mm. It can be seen that the droplets are entirely separated from the natural gas mixture at Er

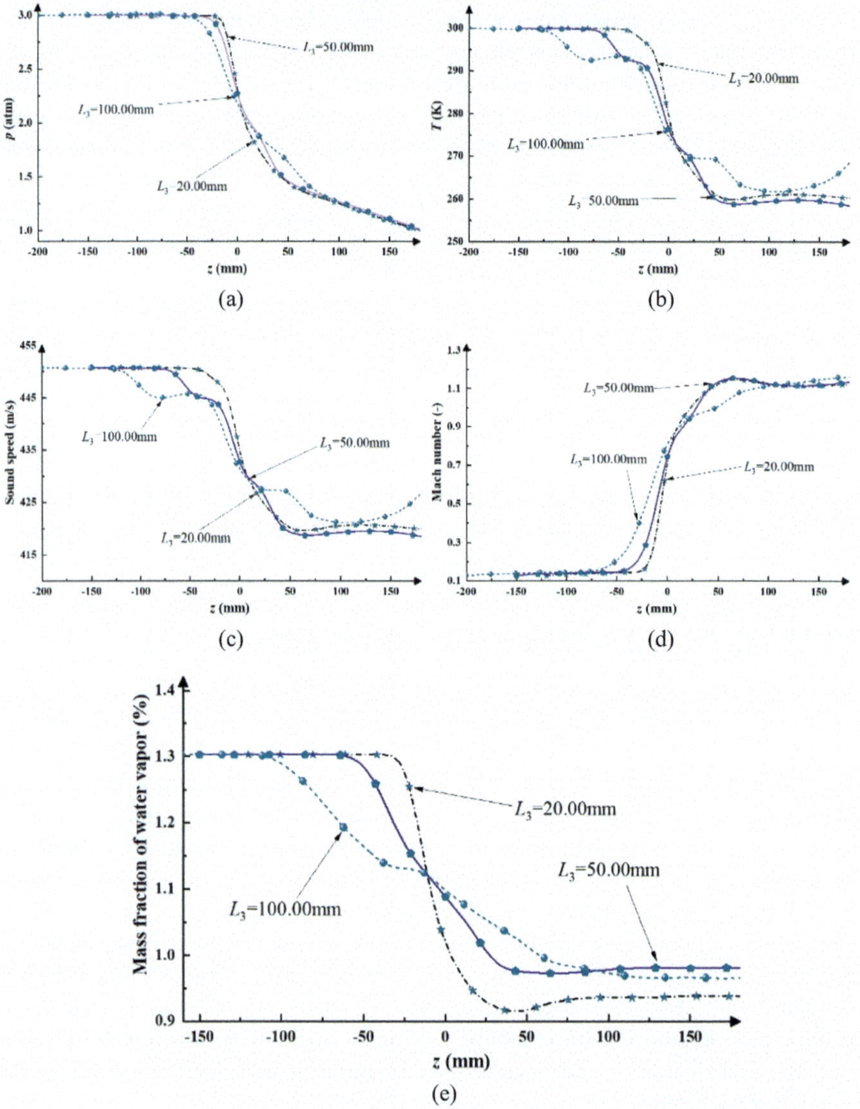

Fig. 8.21 Axial profiles with different LCS values: **a** pressure; **b** static temperature; **c** sound speed; **d** Mach number; **e** mass fraction of water vapor

= 1.6. When $R_{\text{hub,in}} = 15.00$ mm, the droplet migration time in the flow field is the longest. Meanwhile, the maximum values of η_w are 95.09, 94.47, and 93.97%. The corresponding dew-point depression values of ΔT_d are 20.99, 19.57, and 17.82 K, Additionally, the corresponding values of γ are noted as 0.36, 0.37, and 0.38.

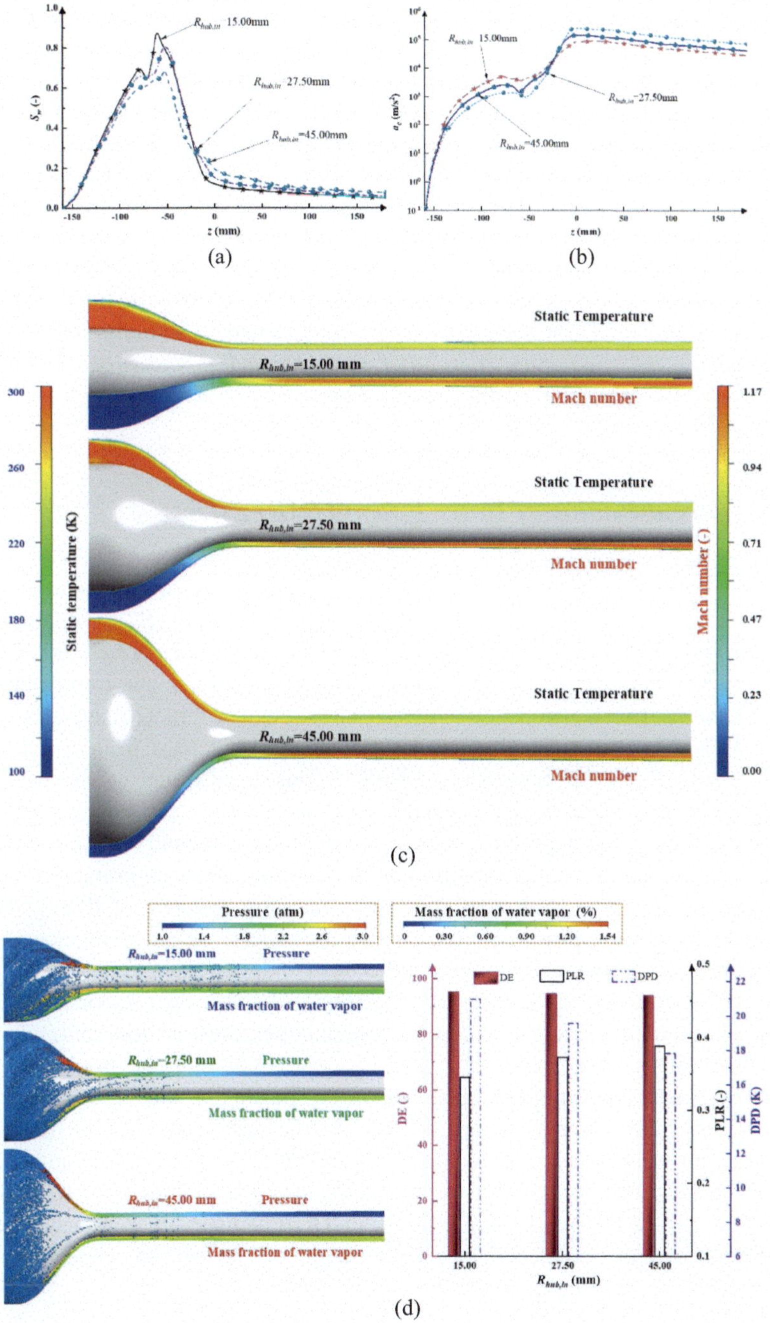

Fig. 8.22 Flow characteristics and separation performance with different IRIB values: **a** distribution of the swirl strength; **b** distribution of the centrifugal acceleration; **c** cloud diagram of flow field parameters; **d** separation performance and pressure energy loss

It is apparent that as IRIB increases, the maximum separation performance decreases while the pressure energy loss continues to rise. These findings demonstrate that in a supersonic separator, IRIB variations significantly affect the separation performance, pressure energy loss, and flow characteristics, underscoring the vital importance of accurate IRIB selection to enhance efficiency.

Figure 8.23a–e illustrates the axial profiles of pressure, static temperature, sound speed, Mach number, and the water vapor mass fraction for the three IRIB values. These provide insights into how varying the IRIB affects the flow characteristics along the axis of the supersonic separator, which is critical for evaluating the separation performance and pressure energy loss of the supersonic separator. In addition, these axial profiles also verify the sensitivity of IRIB values to the separation performance obtained in Fig. 8.22. In particular, it can be verified from Fig. 8.23e that when $R_{\mathrm{hub,in}} = 15.00$ mm, the separation performance is obviously optimal.

8.5.4 *Throat Radius of Inner Body (TRIB) Effect*

Figure 8.24 presents the flow characteristics, separation performance, and pressure loss ratio of the supersonic separator with varying TRIB values: $R_{\mathrm{hub,th}} = 0.50$ mm, 7.50 mm, and 13.50 mm. This analysis maintains constant values of $L_3 = 60.00$ mm, $R_{\mathrm{hub,in}} = 27.50$ mm and $A_{\mathrm{th}} = 105.07$ mm^2. From Fig. 8.24a, b, one can see that, as $R_{\mathrm{hub,th}}$ increases from 0.50 to 13.50 mm, the swirl strength at throat decays from $S_{\mathrm{w}} = 0.35$ to 0.08, and the centrifugal acceleration at the throat decrease from $a_{\mathrm{c}} = 1.97 \times 10^6$ to 1.85×10^4 m s^{-2} by approximate 99.06%. This suggests that an increase in the TRIB significantly hampers the enhancement of the swirling capability, ultimately influencing the separation performance. Figure 8.24c shows cloud diagrams of static temperature and Mach number for different TRIB values. These diagrams offer qualitative insights into how TRIB configurations affect flow dynamics. Different TRIB values lead to varying flow behaviors, which are crucial for analyzing separation performance and pressure energy loss in the supersonic separator.

Figure 8.24d describes the droplet trajectory, pressure, water vapor mass fraction, dehydration efficiency, dew-point depression, and pressure loss ratio with the three TRIB values. One can see that when $R_{\mathrm{hub,th}} = 13.50$ mm, the droplets are not entirely separated from the natural gas mixture. The smaller the TRIB values, the shorter the migration time of the droplet. Meanwhile, across the three TRIB values, the maximum values of η_{w} are 94.10, 94.50, and 90.37%. The corresponding ΔT_{d} are 21.23 K, 20.53 K, and 22.18 K. Additionally, the corresponding γ are 0.33, 0.44, and 0.49. It can be inferred that an increase in TRIB hampers the complete separation of heterogeneous droplets due to the weakening of the swirling capability. The dehydration efficiency initially rises to a peak value before decreasing. The dew-point depression initially decreases and then increases while the pressure energy loss continues to rise. Therefore, excessively large and small TRIB values negatively impact the separation performance, while the pressure energy loss increases with increasing TRIB values.

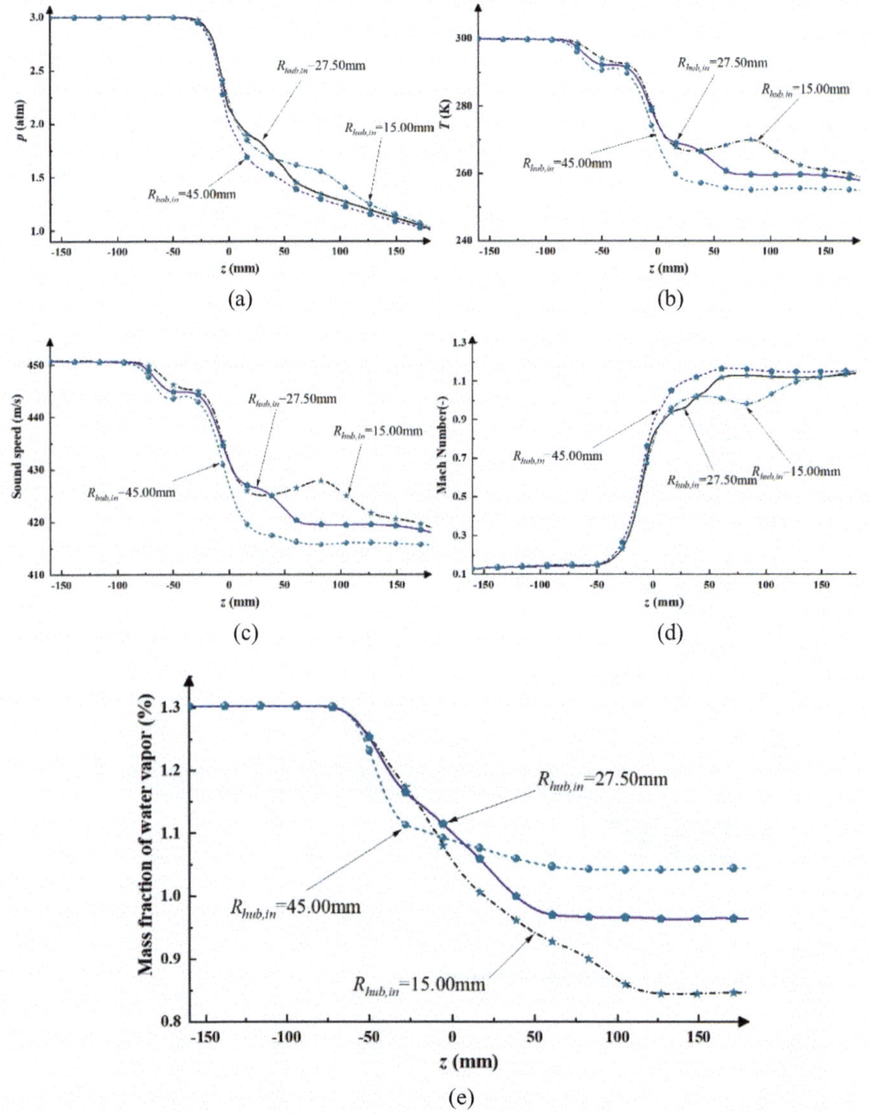

Fig. 8.23 Axial profiles with different IRIB values: **a** pressure; **b** static temperature; **c** sound speed; **d** Mach number; **e** mass fraction of water vapor

Figure 8.25 illustrates the axial profiles of pressure, static temperature, sound speed, Mach number, and the mass fraction of water vapor for the three TRIB values. The figures reveal distinct trends among these parameters based on varying TRIB configurations. When $R_{\mathrm{hub,th}} = 0.50$ mm, Fig. 8.25a–d show that pressure and static temperature are at their lowest values, while the Mach number reaches its highest. However, as depicted in Fig. 8.25e, after $z = 30.00$ mm, the corresponding mass

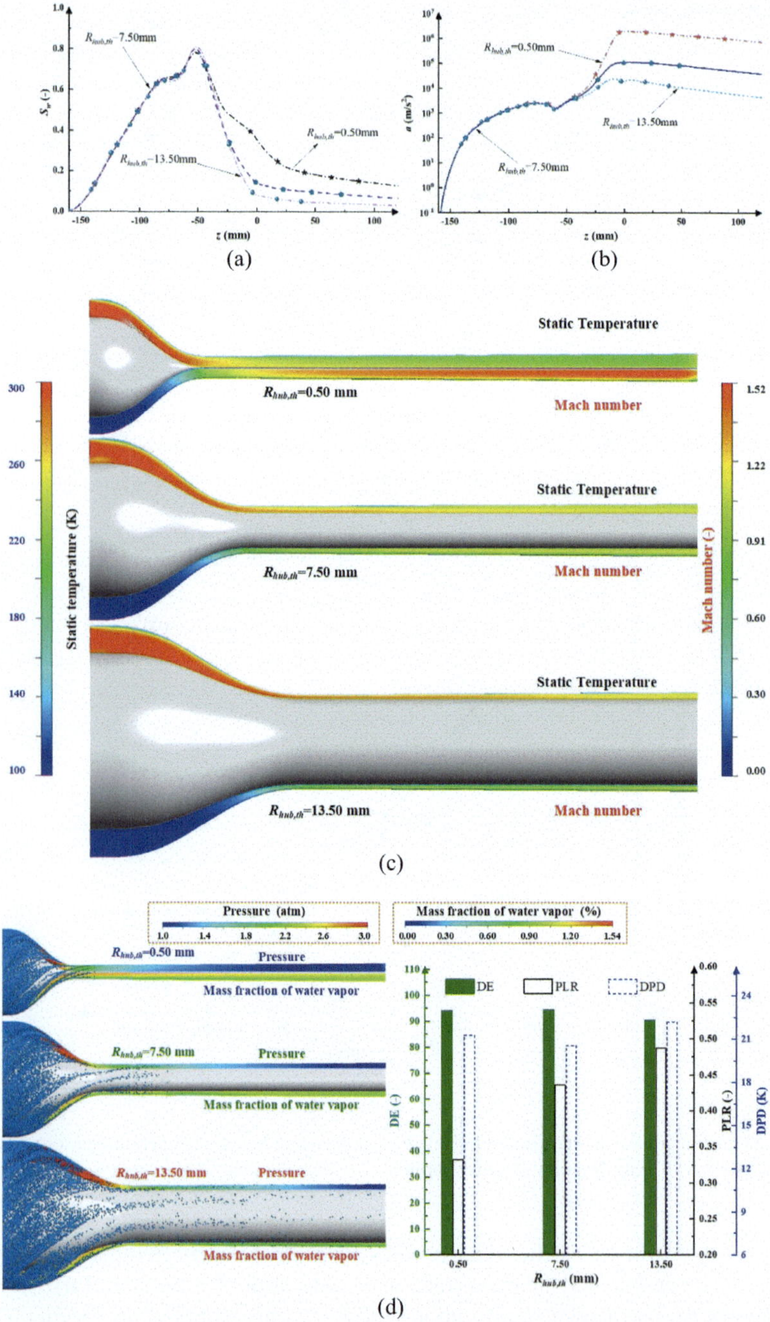

Fig. 8.24 Flow characteristics and separation performance with different TRIB values: **a** distribution of the swirl strength; **b** distribution of the centrifugal acceleration; **c** cloud diagram of flow field parameters; **d** separation performance and pressure energy loss

fraction of water vapor is the highest under these conditions. This is due to the short migration time of the droplets, which reduces the condensation effect within the flow field, thus hindering overall separation performance. Such conditions do not improve the overall separation performance. Simultaneously, Fig. 8.25e also indicates that when $R_{\mathrm{hub,th}} = 13.50$ mm, the mass fraction of water vapor at the outlet remains the lowest compared to the other two conditions. Despite this lower mass fraction of water vapor, the separation performance is not maximized because heterogeneous droplets are not fully separated. This observation emphasizes that achieving optimal separation performance goes beyond merely reducing the mass fraction of water vapor; it also requires careful consideration of the separation of heterogeneous droplets.

8.6 Optimum Foreign Droplet Parameters

For the high-pressure natural gas supersonic dehydration process enhanced by heterogeneous condensation, the parameters of foreign droplets are crucial to the condensation effect and separation performance. Adjustment of the parameters of the foreign droplets enables active control of them. As obtained in Sect. 8.3, the normal swirl strength ($S_{\mathrm{w}} = 0.636$) is the optimal choice for maximizing the separation performance of the supersonic separator.

Therefore, when $S_{\mathrm{w}} = 0.636$ and $p_{\mathrm{in}} = 120$ bar, with the NIST RGM the impact of inlet foreign droplets parameters on condensed droplet diameter and water vapor mole fraction is illustrated in Fig. 8.26. As shown in Fig. 8.26a, when $d_{\mathrm{p}}^{\mathrm{in}} = 3.1$ μm, $q_{\mathrm{p}}^{\mathrm{in}} = 0.0020$ kg s^{-1}, a significant quantity of water droplets remains unseparated, escaping from the dry gas outlet alongside the supersonic natural gas. The corresponding water vapor mole fraction at the dry gas outlet is a relatively high value of about 0.000074. When $q_{\mathrm{p}}^{\mathrm{in}}$ remains constant and $d_{\mathrm{p}}^{\mathrm{in}}$ increases to 3.3 μm, all the water droplets are separated from the wet gas outlet. The corresponding water vapor mole fraction at the dry gas outlet is about 0.000064. While $d_{\mathrm{p}}^{\mathrm{in}}$ increases to 3.5 μm, although the water droplets are separated, the water vapor mole fraction at the dry gas outlet increases to 0.000069. This phenomenon occurs due to the fact that smaller droplet sizes undergo a diminished centrifugal force, insufficient to impact the wall and create a liquid film. In contrast, larger particle sizes experience a heightened centrifugal force. Nevertheless, premature wall collisions impede the establishment of satisfactory cores for heterogeneous condensation. It is worthy that this rule also still applies to the IGM and Redlich–Kwong RGM.

Figure 8.26b illustrates the impact of $q_{\mathrm{p}}^{\mathrm{in}}$ on the separation performance of the SS. When $d_{\mathrm{p}}^{\mathrm{in}} = 3.3$ μm, $q_{\mathrm{p}}^{\mathrm{in}} = 0.0010$ kg s^{-1}, the droplets are totally captured by the drainage section, and the water vapor mole fraction at the dry gas outlet is 0.000089. As the $d_{\mathrm{p}}^{\mathrm{in}}$ remains 3.3 μm, and $q_{\mathrm{p}}^{\mathrm{in}}$ increases to 0.0025 kg s^{-1}, not only do the water droplets separate completely, but the water vapor mole fraction at the dry gas outlet decreases to 0.000059. While $q_{\mathrm{p}}^{\mathrm{in}}$ increases to 0.0040 kg s^{-1}, some water droplets

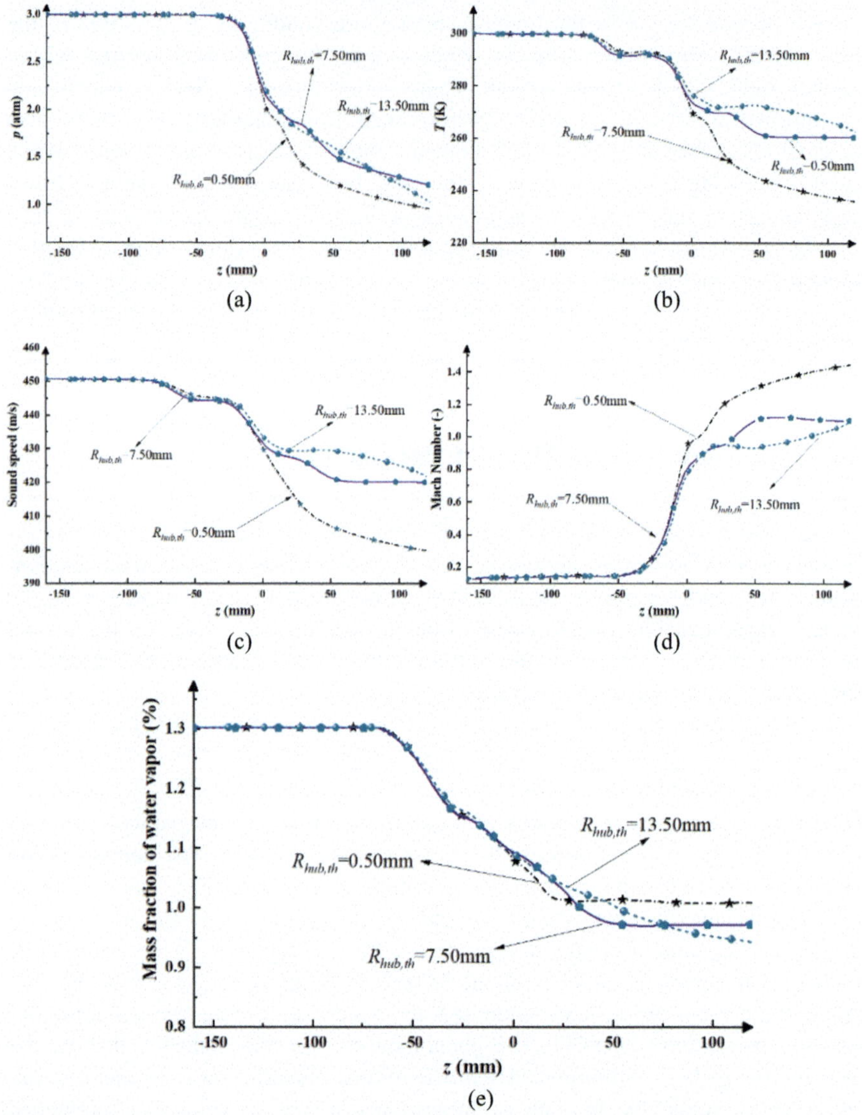

Fig. 8.25 Axial profiles with different TRIB values: **a** pressure; **b** static temperature; **c** sound speed; **d** Mach number; **e** mass fraction of water vapor

flow out of the dry gas outlet with the natural gas, and the water vapor mole fraction at the dry gas outlet is back up to 0.000067. This occurs because a droplet concentration that is too small under specified conditions fails to provide an adequate core for optimal heterogeneous condensation. On the contrary, an excessively high concentration results in some droplets escaping, leading to incomplete separation. Therefore,

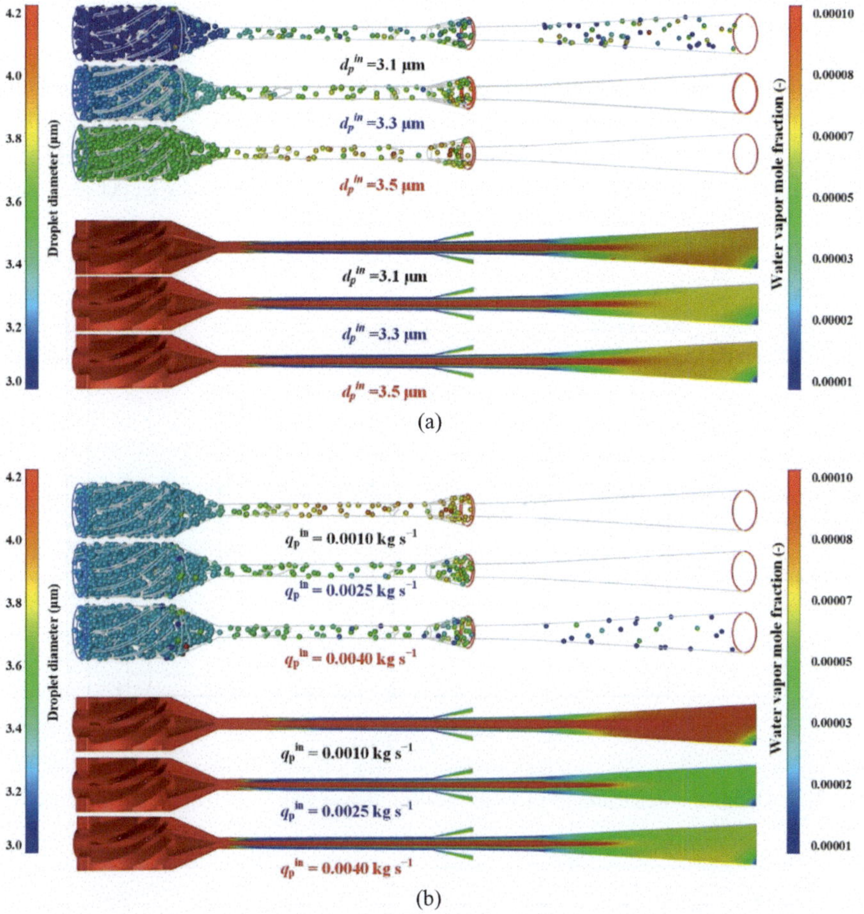

Fig. 8.26 Effect of different foreign droplet parameters on distributions of droplet and water vapor mole fraction with $S_w = 0.636$ when $p_{in} = 120$ bar: **a** different droplet diameter; **b** different droplet concentration

under given p_{in} and d_p^{in} conditions, there exists an optimum foreign water droplet diameter $d_{p,o}^{in}$ and concentration $q_{p,o}^{in}$, that can improve the dehydration performance of the SS with a certain swirl strength. It is worthy that this rule also still applies to the IGM and Redlich–Kwong RGM.

The different optimal foreign droplet parameters predicted by the IGM, NIST RGM, and Redlich–Kwong RGM under different p_{in} are illustrated in Figs. 8.27 and 8.28. It can be observed in Fig. 8.27, that the $d_{p,o}^{in}$ predicted by the three thermodynamic models become more significant with the increase of p_{in}. The NIST RGM indicated $d_{p,o}^{in}$ increases from 1.6 to 3.3 μm with the increment of 106.3%, while the IGM and e Redlich–Kwong RGM predicted $d_{p,o}^{in}$ increases from 1.7 to 3.4 μm with the increasing of 100.0%. The NIST RGM obtains the minimum $d_{p,o}^{in}$ than the other two models under all p_{in}, with the maximum relative deviation of 14.8 and 4.8% than IGM and Redlich–Kwong RGM.

Figure 8.28 shows $q_{\text{p,o}}^{\text{in}}$ predicted by the three thermodynamic models. One can see that with the increasing of p_{in}, the $q_{\text{p,o}}^{\text{in}}$ predicted by NIST RGM is raised from 0.0011 to 0.0026 kg s^{-1} with the increment of 136.4%, the IGM predicted $q_{\text{p,o}}^{\text{in}}$ increases from 0.001 to 0.0035 kg s^{-1} by 250.0%, and the Redlich–Kwong RGM obtained

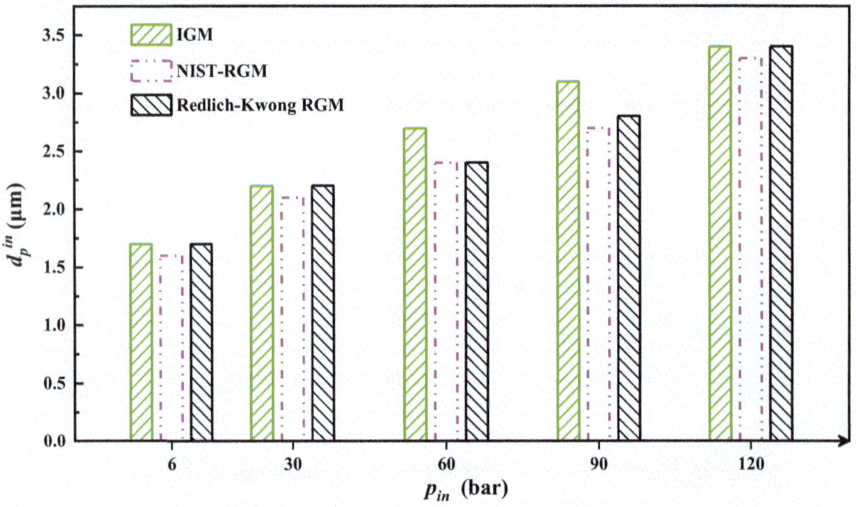

Fig. 8.27 Optimum foreign droplet diameters under different with $S_{\text{w}} = 0.636$ under different p_{in}

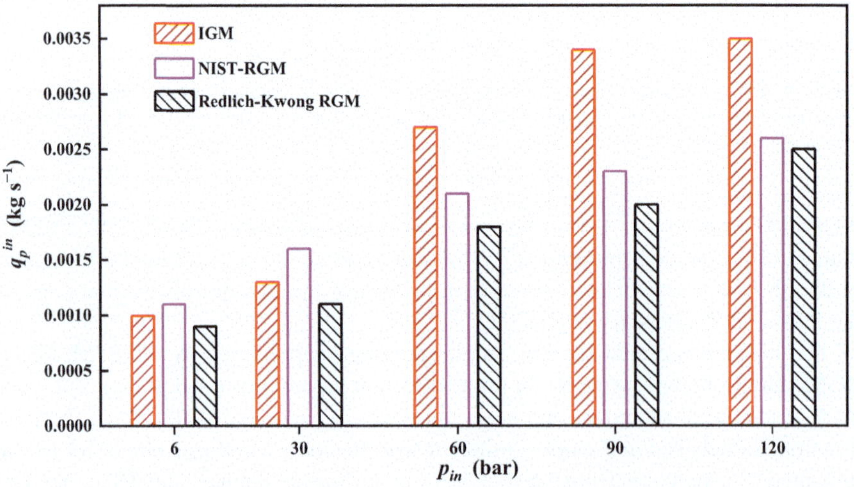

Fig. 8.28 Optimum foreign droplet concentrations under different with $S_{\text{w}} = 0.636$ under different p_{in}

$q_{p,o}^{in}$ increases from 0.0009 to 0.0025 kg s^{-1} by 177.8%. The Redlich–Kwong RGM predicted $q_{p,o}^{in}$ is lower than the other two models with a maximum relative deviation is 45.5% than NIST RGM and 70.0% than IGM. Besides, the IGM under-predicts the $q_{p,o}^{in}$ with a maximum relative deviation of 14.3% than NIST RGM when p_{in} = 6 and 30 bar. However, when p_{in} = 60, 90, and 120 bar, as an opposite result, the IGM overestimates the $q_{p,o}^{in}$ with a maximum relative deviation of 47.8% than NIST RGM. This indicates that IGM has a significant uncertainty in predicting the $q_{p,o}^{in}$ and $d_{p,o}^{in}$ in natural gas dehydration of the SS.

8.7 Economic Analysis

Following the TEG dehydration process [13], the supersonic separator can be strategically placed at the wellhead. Subsequently, the outlet of the supersonic separator serves as the inlet for the TEG process, establishing a novel SS + TEG dehydration process. The parameters for the natural gas dehydration process are listed in Table 8.4.

Assuming that the inlet pressure of the SS is 120 bar, and the outlet pressure and temperature are maintained to meet the required conditions for the TEG process. Eight SSs are interconnected in parallel to fulfill the designated processing capacity, with each individual SS having a flow rate of 5232 Nm^3/day. According to Sect. 8.3, the optimal water vapor removal rate is 81.75%. Thus, 502.97 kg/day of water is pre-separated. As a result, the water absorbed by TEG can be reduced to 112.28 kg/day, leading to a reduction in both TEG cost, steam cost, and fixed capital investment (FCI) within the TEG process. The fixed capital investment (FCI) for the TEG process stands at $2,459,226. In the case of the SS + TEG process, the FCI is calculated as the combination of SS (8 units at $7500 each) and TEG ($508,808).

The total cost of production (TCOP) is calculated by Eq. (8.10)

$$\mathrm{TCOP} = \mathrm{VCOP} + \mathrm{FCOP} + \mathrm{ACC} \tag{8.10}$$

Table 8.4 Parameters of natural gas dehydration process

Properties	Value
Inlet pressure of the SS	120 bar
Inlet pressure of the TEG	41 bar
Temperature	30 °C
Total flow rate	1,000,000 Nm^3/day
Flow rate of a SS	5232 Nm^3/day
Water absorbed by TEG (TEG process)	615.25 kg/day
Water separated by SS (SS + TEG process)	502.97 kg/day
Water absorbed by TEG (SS + TEG process)	112.28 kg/day

Table 8.5 Cost comparison between SS + TEG and TEG

Items	SS + TEG (cost, US $)	TEG (cost, US $)
Variable cost of production (VCOP)	818,007	843,933
• Annual TEG cost (including TEG loss)	1592	8726
• Annual steam cost	3426	18,778
• Annual electrical energy cost	800,000	800,000
• Annual miscellaneous (consumable materials, waste water treatments, waste disposal)	16,429	16,429
Annual capital cost (ACC)	81,409	393,476
Annual fixed cost of production (FCOP)	1,985,292	1,985,292
Total cost of production (TCOP)	2,888,148	3,222,701

where the variable cost of production (VCOP), the fixed cost of production (FCOP) and the annual capital cost (ACC) are determined using the same methodology [13].

As evident from Table 8.5, the annual TCOP for the SS + TEG process is 2,888,148 $/year, whereas it amounts to 3,222,701 $/year for the traditional TEG process. Considering a plant life of 20 years, the SS + TEG process achieves a remarkable cost reduction of $6,691,060. Clearly, the incorporation of SSs into the dehydration process can substantially reduce both investment and production expenses.

References

1. Bian J, Jiang W, Teng L, Liu Y, Wang S, Deng Z (2016) Structure improvements and numerical simulation of supersonic separators. Chem Eng Process Process Intensif 110:214–219
2. Bian J, Cao X, Yang W, Guo D, Xiang C (2020) Prediction of supersonic condensation process of methane gas considering real gas effects. Appl Therm Eng 164:114508
3. Cao X, Bian J (2019) Supersonic separation technology for natural gas processing: a review. Chem Eng Process Process Intensif 136:138–151
4. Cao X, Yang W (2015) The dehydration performance evaluation of a new supersonic swirling separator. J Nat Gas Sci Eng 27:1667–1676
5. Castier M (2014) Modeling and simulation of supersonic gas separations. J Nat Gas Sci Eng 18:304–311
6. Ding H, Wang C, Wang G, Chen C (2015) Analytic equations for the Wilson point in high-pressure steam flow through a nozzle. Int J Heat Mass Transf 91:961–968
7. Hu D, Wang Y, Ma C (2018) Numerical simulation of supersonic separator with axial or tangential outlet in reflow channel. Chem Eng Process Process Intensif 124:109–121
8. Jassim E, Abdi MA, Muzychka Y (2008a) Computational fluid dynamics study for flow of natural gas through high-pressure laval nozzles: part 1. Real gas effects and shockwave. Pet Sci Technol 26(15):1757–1772
9. Jassim E, Abdi MA, Muzychka Y (2008b) Computational fluid dynamics study for flow of natural gas through high-pressure supersonic nozzles: part 2. Nozzle geometry and vorticity. Pet Sci Technol 26(15):1773–1785

10. Liu Y, Cao X, Yang J, Li Y, Bian J (2021) Energy separation and condensation effects in pressure energy recovery process of natural gas supersonic dehydration. Energy Convers Manage 245:114557
11. Majidi D, Farhadi F (2020) Effect of the wet outlet geometry on the shockwave position in supersonic separators. Chem Eng Technol 43(1):126–136
12. Moran MJ, Shapiro HN, Boettner DD, Bailey MB (2010) Fundamentals of engineering thermodynamics. Wiley
13. Neagu M, Cursaru DL (2017) Technical and economic evaluations of the triethylene glycol regeneration processes in natural gas dehydration plants. J Nat Gas Sci Eng 37:327–340
14. Roy BN (2002) Fundamentals of classical and statistical thermodynamics. Wiley
15. Shooshtari SR, Shahsavand A (2018) Numerical investigation of water droplets trajectories during natural gas dehydration inside supersonic separator. J Nat Gas Sci Eng 54:131–142
16. Wang Y, Yu Y, Hu D (2020a) Experimental investigation and numerical analysis of separation performance for supersonic separator with novel drainage structure and reflux channel. Appl Therm Eng 176:115111
17. Wang Y, Yu Y, Hu D, Xu D, Yi L, Zhang Y, Zhang S (2020b) Improvement of drainage structure and numerical investigation of droplets trajectories and separation efficiency for supersonic separators. Chem Eng Process Process Intensif 151:107844
18. Wen C, Cao X, Yang Y, Li W (2012) Numerical simulation of natural gas flows in diffusers for supersonic separators, Energy 37:195–200
19. Wen C, Yang Y, Walther JH, Pang KM, Feng Y (2016) Effect of delta wing on the particle flow in a novel gas supersonic separator. Powder Technol 304:261–267
20. Wen C, Ding H, Yang Y (2021) Numerical simulation of nanodroplet generation of water vapour in high-pressure supersonic flows for the potential of clean natural gas dehydration. Energy Convers Manage 231:113853
21. Wen C, Li B, Ding H, Akrami M, Zhang H, Yang Y (2022) Thermodynamics analysis of CO_2 condensation in supersonic flows for the potential of clean offshore natural gas processing. Appl Energy 310:118523
22. Yang Y, Wen C, Wang S, Feng Y (2014) Numerical simulation of real gas flows in natural gas supersonic separation processing. J Nat Gas Sci Eng 21:829–836
23. Yang Y, Li A, Wen C (2017) Optimization of static vanes in a supersonic separator for gas purification. Fuel Process Technol 156:265–270

Chapter 9
High-Pressure Supersonic Carbon Dioxide (CO_2) Separation

9.1 Introduction

The NECPs and LTS of EU Member States are pivotal for achieving the EU's energy and climate goals, though many still depend on natural gas as a transition fuel [36]. The "REPowerEU" proposal seeks to diversify gas supplies and promote renewable alternatives. In China, targets include peaking emissions by 2030 and achieving carbon neutrality by 2060, emphasizing breakthrough technologies like CCS and CCUS. Natural gas is valued for its lower emissions but faces challenges in CO_2 removal during extraction and storage to prevent pipeline corrosion and improve efficiency. Supersonic separation technology shows promise for CO_2 removal in natural gas but lacks extensive study, especially regarding real gas behaviors at high pressures, crucial for accurate predictions at the Wilson point where CO_2 condenses. For decades, the emission of greenhouse gas carbon dioxide (CO_2) has led to rising atmospheric CO_2 levels and an increase in global average temperatures, posing a potential threat to human existence. Achieving sustainable development in energy, the economy, and the environment is a global priority, with the aim of limiting global temperature rise to 1.5 °C above preindustrial levels [20, 35]. Carbon capture, utilization, and storage (CCUS) is recognized as a crucial climate mitigation technology, capturing CO_2 emissions from fossil power generation and industrial processes for reuse or storage. Traditional decarbonization technologies, such as absorption [34], adsorption [37], membrane [2], cyclone [31], and cryogenics [26, 29], face limitations including corrosiveness, complexity, environmental pollution, and low utilization rates in Fig. 9.1.

Supersonic separation technology, originating from an air-conditioning patent in 1989, offers a revolutionary, cost-effective, and efficient separation method with no pollution [27]. This technology cools mixed gas through a supersonic nozzle, condenses it into droplets, and uses swirling action to separate the liquid film, providing a new approach [4] to decarbonization. Acidic gases such as CO_2 in the

H. Ding et al., *Supersonic Separators for Sustainable Gas Purification*, Green Energy and Technology, https://doi.org/10.1007/978-3-031-85213-8_9

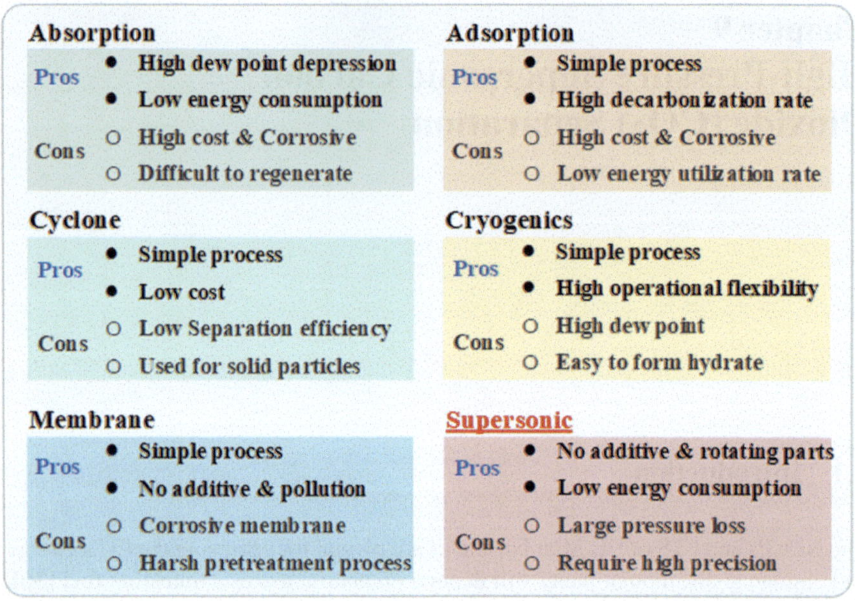

Fig. 9.1 Summary of the advantages and disadvantages of different existing carbon capture, utilization, and storage (CCUS) technologies

natural gas processing process will accelerate pipeline corrosion, so decarbonization technology is particularly critical. CO_2 has complex thermodynamic and kinetic behaviors in supersonic separators. The condensation behavior of CO_2 in supersonic nozzles is crucial for CO_2 separation technologies. Various scholars have conducted significant analyses in this field. Chen et al. [3] developed a CO_2–CH_4 condensation model using computational fluid dynamics under ideal gas assumptions. Jiang et al. [19] assessed CO_2 separation efficiency but lacked a real gas model. Niknam et al. [24] analyzed condensation parameters in supersonic flow numerically, focusing on fluid dynamics within separators. Researchers like dos Santos Cruz et al. [14], Yang et al. [32], and Wen et al. [30] explored real gas properties in supersonic flow, highlighting the limitations of ideal gas models in high-pressure environments.

The condensation position and thermodynamic characteristics of CO_2 gas in a supersonic nozzle, particularly known as the Wilson point where CO_2 begins to condense, are crucial for systematically understanding its condensation behavior. Dobbins [13] proposed an algebraic method to determine the onset of condensation (Wilson line) in low-pressure steam. Huang and Young [18] introduced an analytical solution for the Wilson point in homogeneous nucleation flow. Delale et al. [5] presented an algorithm for the flow properties at the Wilson point. Ding et al. [7] provided an analytical method for the Wilson point in nozzle flow with homogeneous nucleation. Lettieri et al. [21] discussed the Wilson line and its position during the phase change process based on condensation. There is limited research on the Wilson point in the context of CO_2 condensation in supersonic flow, particularly lacking a robust demonstration of Wilson point variations under different inlet conditions (subcritical, critical, and supercritical inlet pressures).

Recent years have seen increased attention on numerical simulations of condensation in supersonic flow. Castier [1] integrated single-phase flow and phase equilibrium processes in their numerical evaluation of thermodynamic properties in natural gas flowing through a supersonic nozzle. Ding et al. [8, 9] focused on a two-fluid model with real gas EOS for studying CO_2 characteristics at the Wilson point in supersonic nozzles. Ding et al. [10] extended their model to supersonic separators, considering gas and liquid interactions but focused on water vapor. These studies advance understanding in CO_2 condensation and separation in supersonic environments, crucial for decarbonization efforts. Understanding the separator's operation hinges on grasping its condensation and swirling processes, illuminated through numerical studies. Liu et al. [23] developed an Euler–Euler model for methane-vapor systems, highlighting pressure energy recovery efficiency (57.1–64.2%) critical for natural gas processing. Shooshtari and Shahsavand [28] introduced an Euler–Euler model considering homogeneous/heterogeneous condensation effects on supersonic separators, while then utilized the Euler–Lagrange method to forecast droplet behavior and separation efficiency within branching flows. Recent advancements include integrating the Euler film model for accurate simulation of liquid film dynamics on walls [16]. Yue et al. [33] explored liquid flow rate sensitivity using the Euler–Euler–Euler model in gas–liquid cylindrical cyclones, showing how higher flow rates impact liquid film distribution uniformity. Deng et al. [6] simulated gas–liquid separation in axial flow cyclones using Euler–Lagrange methods and Euler film models, highlighting progressive liquid film coverage over time. Ding et al. [11] applied these methods to supersonic separators, analyzing condensation and separation performance from continuous and discrete perspectives, identifying key factors affecting efficiency. A combined Euler–Euler–Euler model with NIST real gas EOS has been pivotal in studying supercritical CO_2 capture and mass separation characteristics [12], focusing on CO_2 system condensation and CH_4–CO_2 mass separation. Research on energy loss and entropy analysis remains limited, hindering comprehensive assessment of supersonic separators' decarbonization capabilities [38].

This capture aims to evaluate the effectiveness of non-equilibrium condensation and swirl separation of CO_2 in supersonic decarbonization, using an integrated model that includes entropy analysis, focusing on improving condensation and separation efficiency, and exploring optimization and potential applications, contributing to the development of CCUS by capturing CO_2 more purely and providing a comprehensive numerical basis for this innovative technology.

9.2 Supersonic Condensing Flow of High-Pressure Carbon Dioxide

Based on the real gas state equation, this chapter establishes and verifies a highly sophisticated numerical model, aiming to deeply study the complex physical phenomena in the supersonic decarbonization process under high-pressure environment. This model is not only applicable to pure CO_2 systems, but also compatible with CH_4–CO_2 mixed gas systems, which is closer to actual industrial application scenarios. Through the numerical simulation of carefully designed supersonic nozzles and supersonic separators, this study systematically evaluates the performance of supersonic decarbonization technology in the process of condensation and separation of CO_2. These simulations not only consider the thermodynamic and kinetic characteristics of the gas during the flow process, but also accurately capture the key physical phenomenon of CO_2 condensation near the Wilson point. The core of this chapter's research is to explore the practical application potential of supersonic separation technology under high pressure conditions, and through the careful verification and optimization of the model, it provides a solid scientific basis for the industrial deployment of supersonic decarbonization technology in future. This will not only help promote the further development of carbon capture, utilization, and storage (CCUS) technology, but also provide innovative and sustainable solutions for the global response to climate change and the achievement of sustainable development goals.

9.2.1 *Thermodynamics Modeling of CO_2 Condensation*

To validate the established model and compare the thermodynamic simulations between real gas and ideal gas models, the nozzle structure and operational data from Lettieri et al. [22] were utilized, where the inlet pressure p_{in} is 57.24 bar and inlet temperature T_{in} is 314.78 K. The testing of various nozzles with different expansion rates is facilitated by separating the structural requirements of the test section from the nozzle geometry. A stainless-steel structure retains the windows and provides structural integrity in the high-pressure environment. The nozzle inserts, made of aluminum, are bolted into the main body. This design allows for the investigation of different nozzle geometries and the variation of the location and resolution of pressure measurements. The throat height is selected to minimize the mass flow rate and maximize the blowdown time without significantly altering the stagnation conditions of the charge tank. The test section accommodates 33 pressure transducer ports, but in this study, only 13 pressure transducers were used, focusing the analysis on a single nozzle geometry. The nozzle length is 0.155 m, with a throat area of 20 mm^2. The outlet-to-throat and inlet-to-throat area ratios are 1.3 and 4.1, respectively. The charge tank volume is 400 L.

The condensation behavior was validated using the real gas model, as depicted in Fig. 9.2a. Condensation was observed approximately 2 mm downstream of the nozzle throat, coinciding with a pressure jump caused by latent heat release during condensation. In contrast, the ideal gas model predicted condensation occurring later and less accurately. Quantitative evaluation using mean relative error (MRE) and root mean square error (RMSE) further highlights the differences between the two models. The MRE for the real gas model (0.90%) significantly outperformed that of the ideal gas model (14.37%). Similarly, calculating RMSE yielded 0.01471 for the real gas model and 0.07626 for the ideal gas model in Fig. 9.2b. These results underscore the superior predictive capability of the real gas model in simulating CO_2 condensation in supersonic flows, aligning closely with experimental findings.

$$\mathrm{MRE} = \frac{1}{n}\sum_{i=1}^{n}\frac{(y_i - \hat{y}_i)}{y_i} \tag{9.1}$$

$$\mathrm{RMSE} = \sqrt{\frac{1}{n}\sum_{i=1}^{n}(\hat{y}_i - y_i)^2} \tag{9.2}$$

where $\hat{y}_i$ is the simulation data under real gas model and ideal gas model, y_i is the experiment data respectively, and n is the amount of the data.

Furthermore, Fig. 9.3 illustrates the nucleation rate distribution for both models. The peak nucleation rate, known as the Wilson point, closely matched experimental data for the real gas model. In contrast, the ideal gas model exhibited a nucleation rate peak significantly displaced from the experimental Wilson point. This comparison reaffirms the accuracy and suitability of the real gas model for simulating CO_2 condensation processes in supersonic nozzles.

In the structural composition of the supersonic separator, both the thermodynamic condensation accuracy of the supersonic nozzle and the role of the vortex generator are critical. To align with the separator's working principle, the effect of swirling performance on the nozzle's condensation process was validated using a two-dimensional axisymmetric Laval nozzle in Fig. 9.4. The study set up four different centrifugal accelerations: $1 \times 10^5, 4 \times 10^5, 8 \times 10^5$ m^2/s, and no swirling. As shown in Fig. 9.5a, higher centrifugal acceleration leads to an increased liquid phase fraction along the central axis, causing earlier droplet condensation. Figure 9.5b indicates that higher inlet centrifugal acceleration results in a greater outlet liquid fraction, while zero swirling yields only 3.4%, weakening the condensation effect. Figure 9.5c shows that swirling velocity increases significantly with greater centrifugal acceleration, with the highest velocity near the nozzle wall due to inlet swirling action.

In summary, introducing swirling action enhances CO_2 condensation and advances the condensation position within the supersonic nozzle.

This section delves into the mechanisms of mass and heat transfer through a comparative analysis of condensation characteristics between ideal gas and real gas

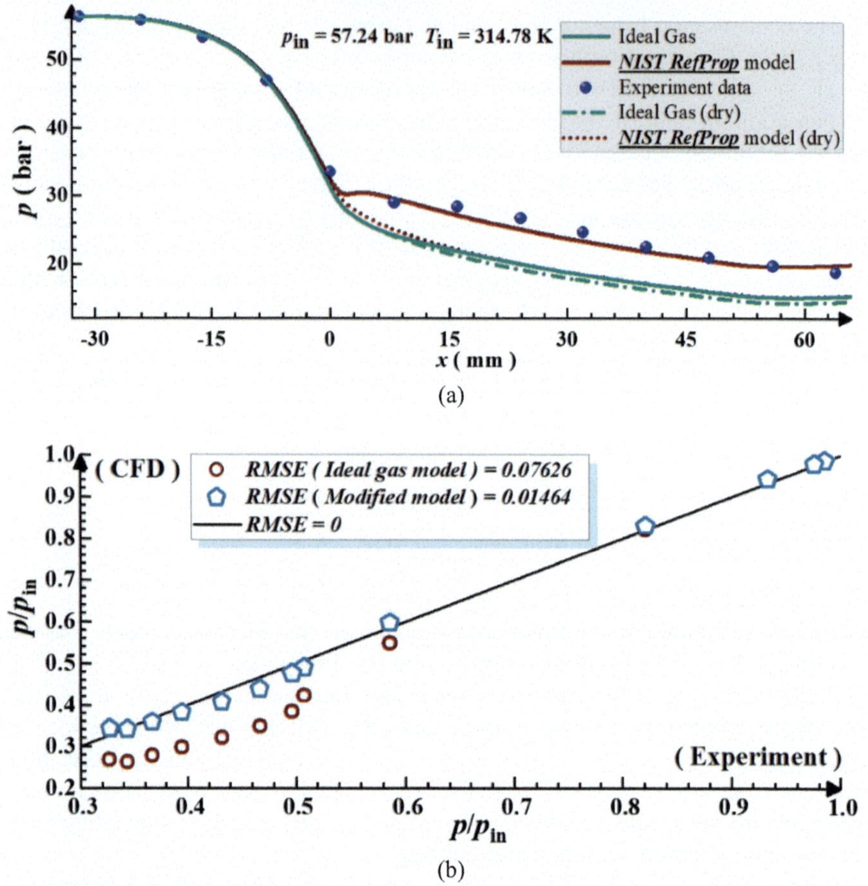

Fig. 9.2 Numerical and experimental results in supersonic nozzle where the inlet pressure is 57.24 bar and the inlet temperature is 314.78 K: **a** the pressure ratio along the axis; **b** the comparison between real gas model and ideal gas model

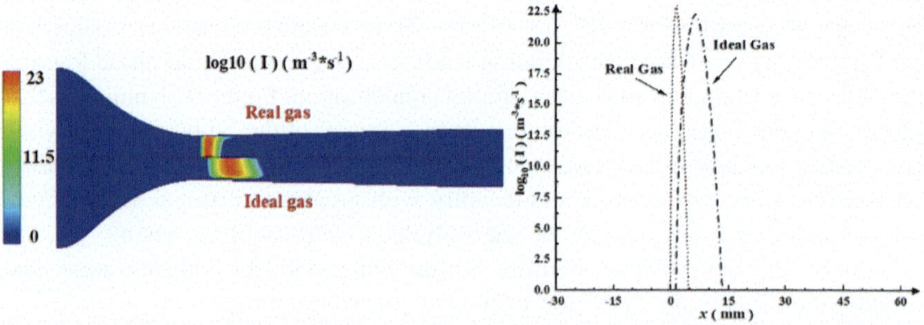

Fig. 9.3 Quantitative comparison of the nucleation rate contour distribution along the axis under certain inlet conditions based on the real gas model and the ideal gas model when the inlet pressure p_{in} is 57.24 bar and inlet temperature T_{in} is 314.78 K, respectively

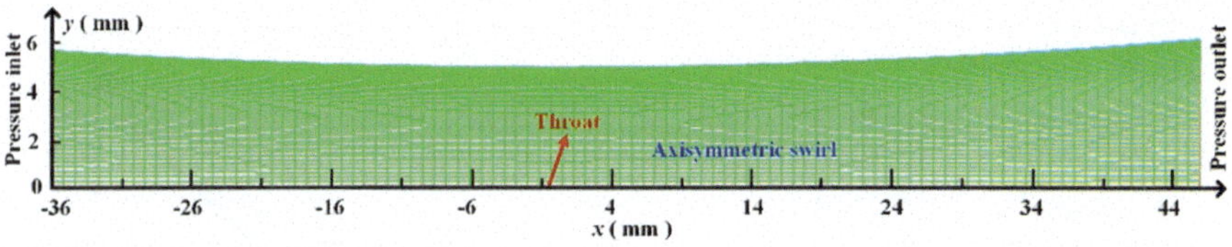

Fig. 9.4 Modeled geometric structure and mesh division of the supersonic large-arc nozzle used for validation

models in a supersonic flow, as illustrated in Fig. 9.6. The primary focus is on examining how these gas models differ in the condensation process under varying inlet conditions. Research reveals that the ideal gas model exhibits greater supercooling $\left(\Delta T_C = T_s - T_{CO_2}\right)$ at the Wilson point compared to the real gas model, with its maximum subcooling occurring further downstream. As inlet pressure increases, supercooling tendencies decrease in both models, yet significant differences persist in the locations of condensation and nucleation regions throughout the transition from critical to supercritical conditions. This comparative analysis sheds light on crucial distinctions that influence the efficiency and effectiveness of condensation processes in supersonic flows.

In terms of mass transfer dynamics, the comparison between the ideal gas and real gas models in Fig. 9.7 reveals significant disparities in how they handle phase transition processes. During nucleation, where phase change begins, the real gas model exhibits a markedly higher rate of mass generation than its ideal gas counterpart. This difference underscores the real gas model's ability to capture more accurately the initial stages of phase transition, where condensation begins. Moving into the droplet growth phase, the discrepancies between the two models continue to be evident. Here, the rate at which mass changes during droplet growth varies noticeably between the ideal gas and real gas models. The real gas model, owing to its consideration of non-ideal behaviors such as molecular interactions and deviations from ideal gas laws, more precisely predicts the evolution of droplet size and mass accumulation during this phase.

Furthermore, the analysis of liquid fraction in Fig. 9.8, which represents the proportion of condensed liquid to gas mixture, reveals another critical distinction between the models. As inlet pressure increases, the real gas model consistently forecasts a significantly higher liquid fraction compared to the ideal gas model. This discrepancy becomes more pronounced under high-pressure conditions, indicating that the ideal gas model tends to underestimate the amount of condensed liquid present. This insight is crucial for applications where accurate prediction of liquid fraction is essential, such as in the design and optimization of supersonic condensation systems for various industrial processes.

In terms of heat transfer characteristics, the ideal gas model and the real gas model exhibit significant differences in predicting subcooling at the Wilson point and the latent heat released at the onset of phase change as shown in Fig. 9.9. The ideal gas model tends to overestimate the subcooling at the Wilson point, which refers to the

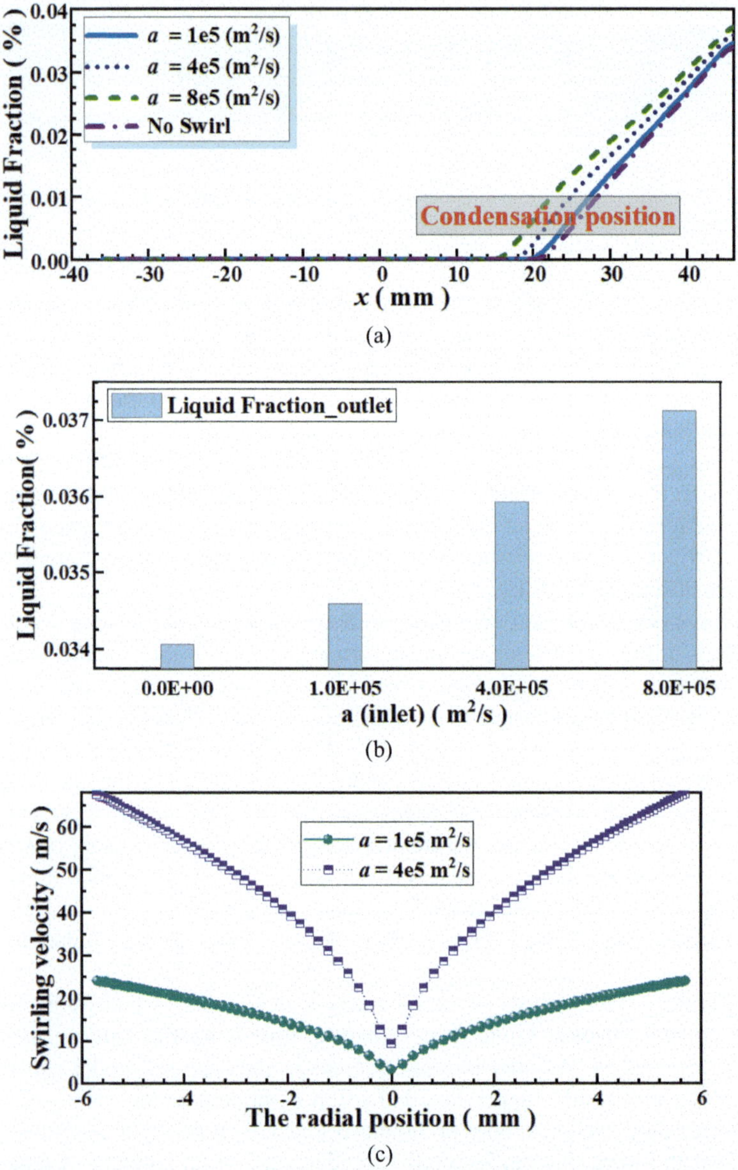

Fig. 9.5 Effects of swirl intensity (m^2/s) of 1e5, 4e5, 8e5, and no swirl on condensation: **a** liquid fraction along the axis; **b** variation trend of condensation intensity with swirl intensity; **c** quantitative analysis of swirl velocity along the axis

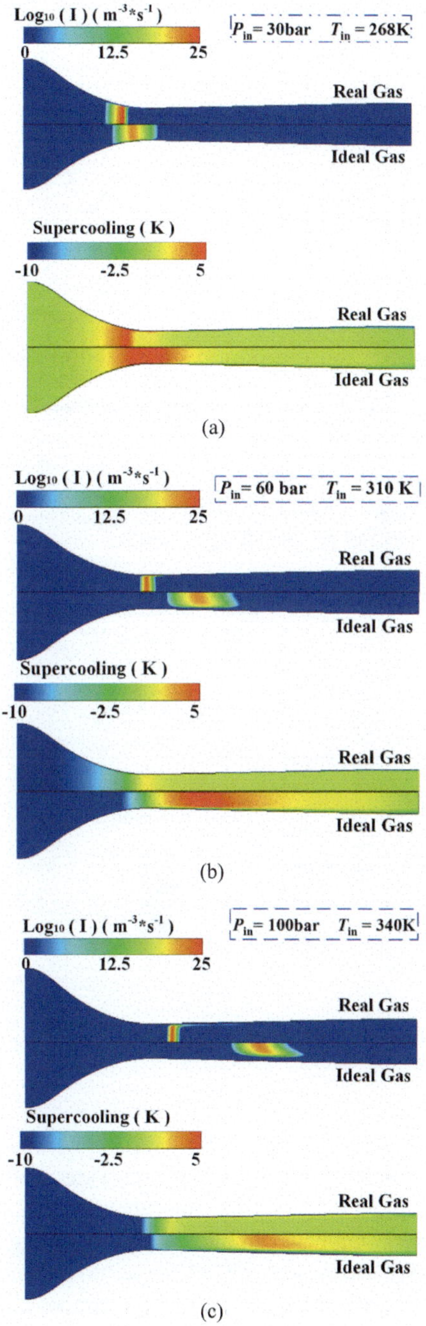

Fig. 9.6 Contours comparison of nucleation rate and supercooling of real gas and ideal gas under different inlet conditions: **a** subcritical $p_{in} = 30$ bar; **b** critical $p_{in} = 60$ bar; **c** supercritical $p_{in} = 100$ bar

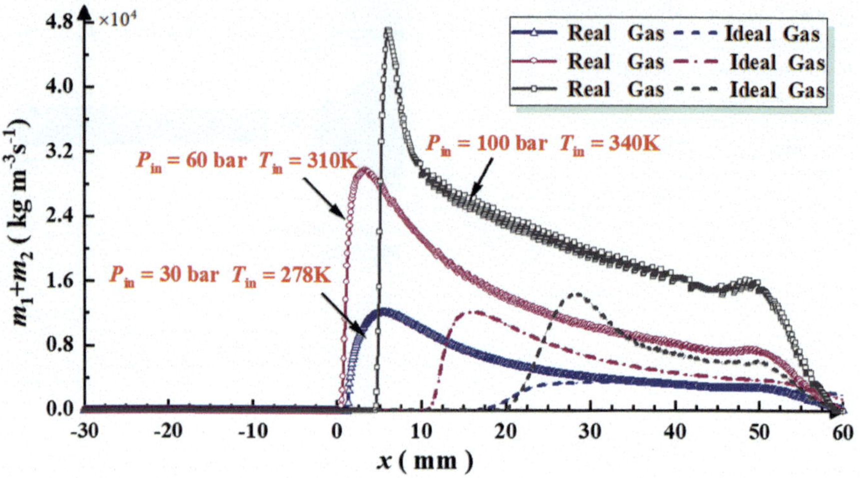

Fig. 9.7 Comparison of the local mass change rate distribution of CO_2 when condensation occurs under real gas and ideal gas models

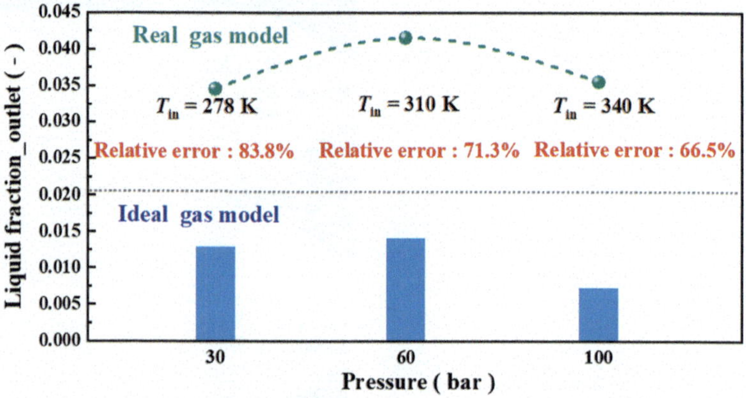

Fig. 9.8 Comparison of liquid fraction of CO_2 at the outlet of the nozzle, distinguishing between real gas and ideal gas properties

difference between gas temperature and the condensation temperature of CO_2. In contrast, the real gas model provides a more accurate representation of supercooling in practical scenarios; the values are listed in Table 9.1. Additionally, there is notable deviation between the two models in predicting the latent heat released at the onset of phase change. The ideal gas model tends to overestimate the amount of latent heat released, whereas the real gas model aligns more closely with actual heat release patterns. These discrepancies become particularly pronounced under high-pressure conditions. In such environments, the relative error between the ideal gas and real gas models significantly increases, highlighting the potential for larger inaccuracies

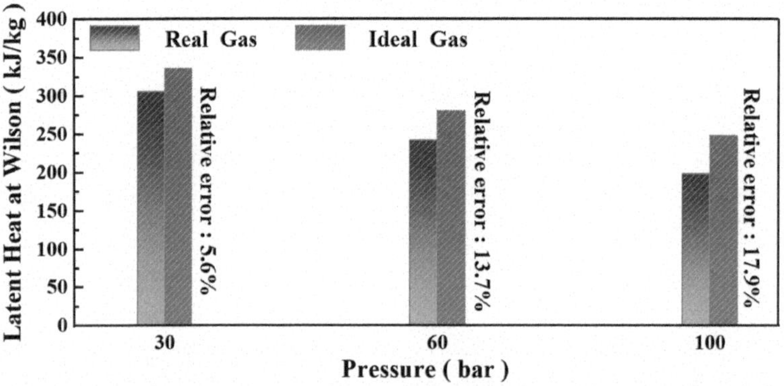

Fig. 9.9 Error analysis of latent heat at Wilson point of real gas and ideal gas

Table 9.1 Value of degree of supercooling at Wilson point of real gas and ideal gas

ΔT_C (K)	$p_{in} = 30$ bar $T_{in} = 268$ K	$p_{in} = 60$ bar $T_{in} = 305$ K	$p_{in} = 100$ bar $T_{in} = 340$ K
Real gas	4.99	1.92	1.22
Ideal gas	5.99	3.44	3.25

in the ideal gas model during complex phase change processes. Therefore, accurate prediction of subcooling and latent heat release is crucial for ensuring efficiency and performance when designing and optimizing phase change processes in supersonic flows.

This section highlights the distinctions between the ideal gas and real gas models concerning condensation characteristics, mass transfer, and heat transfer in supersonic flows. Particularly under high-pressure inlet conditions, the ideal gas model exhibits limited simulation capabilities, which pose significant constraints. The ideal gas model often fails to accurately capture the complex behaviors and interactions that occur in real gas scenarios, especially under high-pressure conditions. This limitation becomes evident in the simulation of condensation processes, where the assumptions of the ideal gas law do not hold true. Consequently, the ideal gas model may produce inaccurate results, leading to erroneous predictions of condensation characteristics, mass transfer rates, and heat transfer efficiencies. In contrast, the real gas model accounts for the deviations from ideal behavior by incorporating factors such as intermolecular forces and the actual volume occupied by gas molecules. This makes the real gas model more suitable for simulating high-pressure conditions, providing a more accurate representation of the physical processes involved in condensation, mass transfer, and heat transfer. Therefore, relying on the ideal gas model for high-pressure supersonic flow simulations can lead to significant inaccuracies, affecting the design and optimization of systems that depend on precise predictions of these phenomena. The real gas model, with its enhanced simulation capabilities, offers a more reliable alternative, ensuring better performance and efficiency in practical applications.

In conclusion, understanding the distinctions between ideal and real gas models is crucial for accurately simulating condensation, mass transfer, and heat transfer in supersonic flows, particularly under high-pressure conditions. The ideal gas model's limitations highlight the need for adopting the real gas model to achieve more accurate and reliable simulation outcomes.

9.2.2 *Effect of Inlet Conditions on Condensation Characteristics of Wilson Point*

After demonstrating the superiority of the real gas model in analyzing CO_2 supersonic flow from various perspectives, this section will continue utilizing this model to explore the condensation mechanism of CO_2 within a supersonic nozzle and the condensation characteristics at the Wilson point. Special attention will be given to the variation of condensation parameters at the Wilson point under different inlet conditions. Furthermore, a predictive model will be developed to describe the relationship between the degree of supercooling, pressure, and expansion rate at the Wilson point during the supersonic flow of CO_2.

Different inlet temperatures and pressures have a significant influence on the condensation characteristics at the Wilson point during CO_2 supersonic flow. In this section, we examine three groups of inlet pressures: 30, 60, and 100 bar. The inlet superheat (ΔT_H) is defined as $\Delta T_H = T_{CO_2} - T_s$, where T_s is the saturation temperature. For inlet pressures p_{in} of 30 and 60 bar, the values of ΔT_H are set at 10, 15, and 20 K, while for an inlet pressure p_{in} of 100 bar, the values of ΔT_H are 31, 36, and 41 K. Figure 9.10 illustrates that at the same inlet pressure, an increase in inlet superheat leads to higher degrees of supercooling and latent heat at the Wilson point.

Tables 9.2, 9.3 and 9.4 provide a quantitative analysis of condensation parameters under different inlet conditions, comparing the degree of supercooling, supersaturation, pressure, and expansion rate k_p at the Wilson point within and between groups. The data indicates that with increasing superheat, the pressure and expansion rate at the Wilson point decrease, while the degree of supersaturation increases, similar to the degree of supercooling. Supersaturation S is defined as $S = p_{CO_2}/p_s$ and represents the extent to which steam continues to expand in a metastable supersaturated region. Greater supersaturation results in deeper continued expansion.

Figure 9.11 shows that at the same superheat, higher inlet pressures cause the condensation position to move forward. For instance, with an inlet superheat of 25 K, the condensation onset position (Wilson point) shifts from $x = 38.57$ mm at 30 bar, to $x = 13.96$ mm at 60 bar, and to $x = 0.827$ mm at 100 bar. This shift occurs because higher inlet pressures increase the partial pressure of CO_2 and the supersaturation,

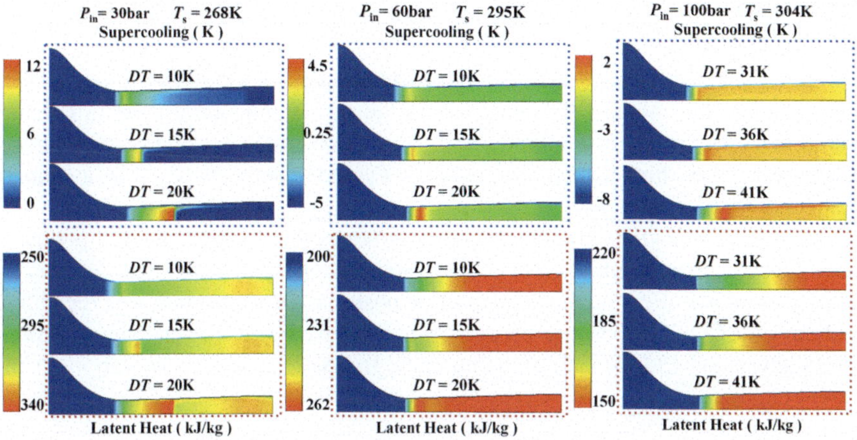

Fig. 9.10 Comparison of supercooling and latent heat parameters of CO_2 based on different inlet superheat under different inlet temperatures and inlet pressures

Table 9.2 Quantitative analysis of condensation parameters with $p_{in} = 30$ bar

$p_{in} = 30$ bar, $T_s = 268$ K				
	Supercooling (K)	Supersaturation (-)	Pressure (bar)	k_p (1/s)
$\Delta T_H = 10$ K	7.70	1.29	16.79	7486
$\Delta T_H = 15$ K	9.88	1.42	13.43	4664
$\Delta T_H = 20$ K	11.94	1.58	10.82	3182

Table 9.3 Quantitative analysis of condensation parameters with $p_{in} = 60$ bar

$p_{in} = 60$ bar, $T_s = 295$ K				
	Supercooling (K)	Supersaturation (-)	Pressure (bar)	k_p (1/s)
$\Delta T_H = 10$ K	1.92	1.05	40.72	10,911
$\Delta T_H = 15$ K	2.90	1.08	34.32	8918
$\Delta T_H = 20$ K	4.04	1.12	28.66	3497

Table 9.4 Quantitative analysis of condensation parameters with $p_{in} = 100$ bar

$p_{in} = 100$ bar, $T_s = 304$ K				
	Supercooling (K)	Supersaturation (-)	Pressure (bar)	k_p (1/s)
$\Delta T_H = 31$ K	0.67	1.02	53.85	8430
$\Delta T_H = 36$ K	1.22	1.03	47.04	4087
$\Delta T_H = 41$ K	1.92	1.05	40.79	1753

leading to a larger expansion degree and making spontaneous condensation easier under the same superheat.

Figure 9.12, which uses ΔT_w to represent the degree of supercooling at the Wilson point, corroborates this finding. As the inlet temperature rises, a greater degree of supercooling is required for condensation, making condensation more challenging. Thus, higher inlet temperatures T_{in} and superheats result in a further back condensation position.

The qualitative and quantitative analyses demonstrate that both inlet temperature and pressure exert substantial influence on condensation characteristics at the Wilson point, particularly affecting the condensation position. These insights are crucial for predicting the CO_2 condensation process within supersonic nozzles and play a pivotal role in advancing CO_2 separation technology. By understanding how variations in

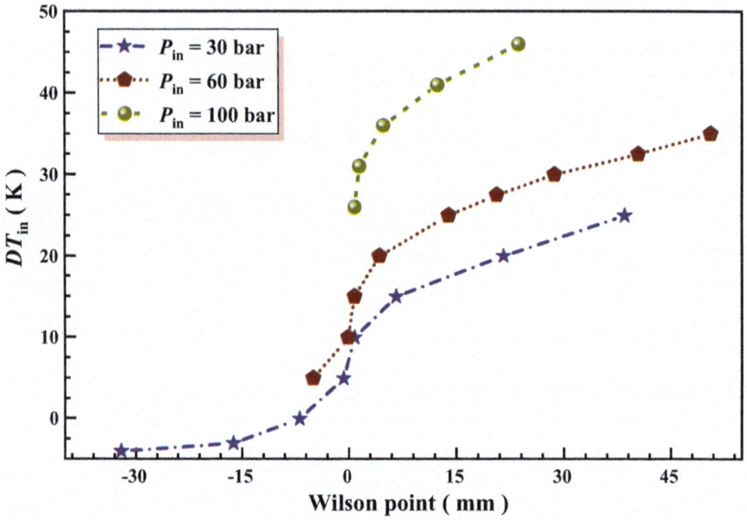

Fig. 9.11 Relationship between Wilson point location and inlet superheat of CO_2

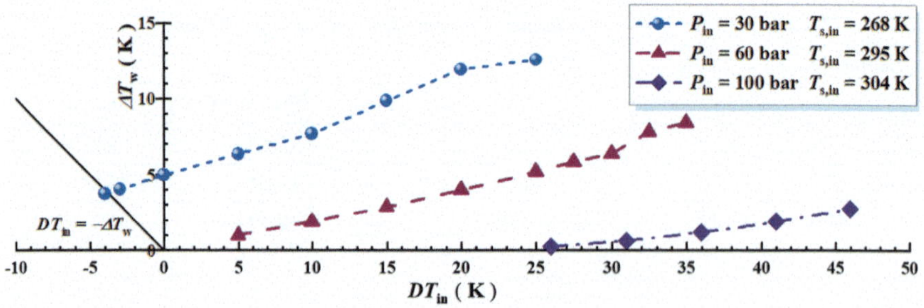

Fig. 9.12 Relationship between the degree of supercooling at Wilson point and the inlet superheat of CO_2

inlet temperature and pressure impact condensation, researchers can better predict where and how efficiently CO_2 condenses within supersonic separators. This knowledge enables the optimization of separation processes, potentially leading to more effective and energy-efficient CO_2 capture methods. Furthermore, these findings contribute to ongoing efforts aimed at enhancing CO_2 separation technologies, which are critical for mitigating greenhouse gas emissions and combating climate change. By improving the understanding and predictive capabilities regarding condensation behavior in supersonic flows, we can further refine and innovate CO_2 capture technologies, making them more reliable and economically viable for widespread adoption.

In summary, the detailed analysis of inlet temperature and pressure effects on condensation characteristics provides valuable insights into CO_2 separation processes. These insights not only advance the fundamental understanding but also pave the way for practical improvements in CO_2 capture technology, supporting global efforts to achieve environmental sustainability.

9.2.3 Prediction Model for Wilson Point

In this section, a prediction model was established for parameters at the Wilson point. This model is designed to predict the relationship curves among the degree of supercooling, pressure, and expansion rate under the corresponding dry gas conditions at the Wilson point.

In this section, a prediction model for parameters at the Wilson point was proposed to evaluate the relationship curves among the degree of supercooling, pressure, and expansion rate under the corresponding dry gas conditions at the Wilson point: denotes the expansion rate k_p at the Wilson point.

$$\begin{aligned}\Delta T_w = b_1 + b_2 k_p + b_3 p_w + b_4 p_w^{b_5} k_p^{b_6} \\ + b_7 e^{b_8 p_w k_p} + b_0 (p_w k_p)^2\end{aligned} \tag{9.3}$$

To ensure the applicability and feasibility of the proposed prediction model, we employ composite evaluation indexes based on the mean relative error (MRE) and relative root mean square error (rRMSE).

$$\text{rRMSE} = \sqrt{\frac{1}{n}\sum_{i=1}^{n}\left(\frac{\hat{y}_i - y_i}{y_i}\right)^2} \tag{9.4}$$

These metrics were chosen due to their sensitivity to the data characteristics and their ability to accurately reflect the overall data trends in prediction and regression models. The calculation methods for MRE and rRMSE are given below. The MRE and rRMSE are used to assess the agreement between the predicted values from

Eq. (9.3) and the simulated values obtained through CFD. In Eq. (9.4), $\hat{y}_i$ represents the simulation value by CFD, and y_i is the predicted value by Eq. (9.3). When MRE and rRMSE are approximately 0, it indicates good agreement between the predicted and simulated results. In this study, the MRE was calculated to be 0.176%, and the rRMSE was 2.275%.

Comparative analysis between prediction results and CFD simulation data shows a consistent relationship, specifically an inverse proportionality between pressure and the degree of supercooling. This inverse relationship is due to the continuous reduction in temperature and pressure as CO_2 expands in supersonic flow. The higher the inlet superheat, the deeper the continued expansion, thus achieving a greater degree of supercooling necessary for spontaneous condensation.

Figures 9.13 and 9.14 illustrate the relationship curves between the predicted and simulated values, validating the accuracy and effectiveness of Eq. (9.3). When the degree of supercooling at the Wilson point reaches a certain value, the expansion rate decreases, as shown in Fig. 9.14. Figure 9.15 further demonstrates that the error between simulation and prediction results is within ± 5%, with an MRE of 0.176% and an rRMSE of 2.275%, indicating a high degree of accuracy in the prediction model.

This prediction model offers a straightforward method for predicting simulation data trends and identifying patterns within limited datasets. Its flexibility allows the independent variables to be adjusted to other condensation parameters at the Wilson point, accommodating a variety of scenarios. The calculation methods for mean relative error (MRE) and relative root mean square error (rRMSE) provide quantitative measures of prediction accuracy. These metrics are essential for comparing the prediction model's accuracy against experimental data and other simulation results, ensuring its reliability and robustness across different applications. By employing MRE and rRMSE, researchers can quantitatively evaluate the model's performance,

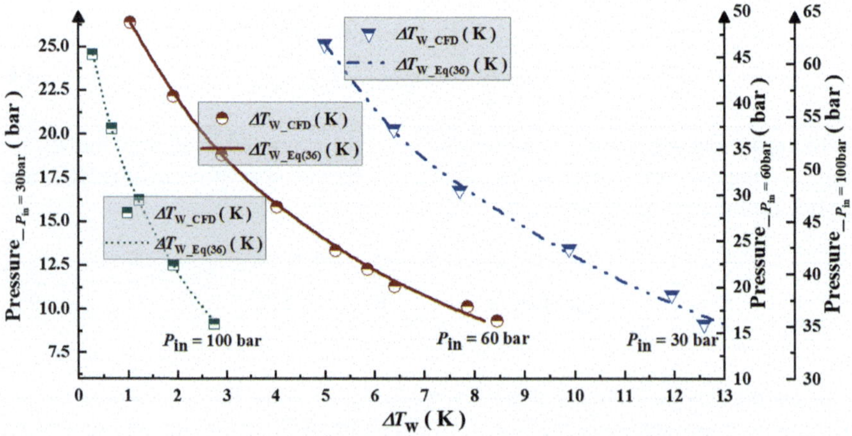

Fig. 9.13 Analysis prediction curves of pressure and supercooling at Wilson point

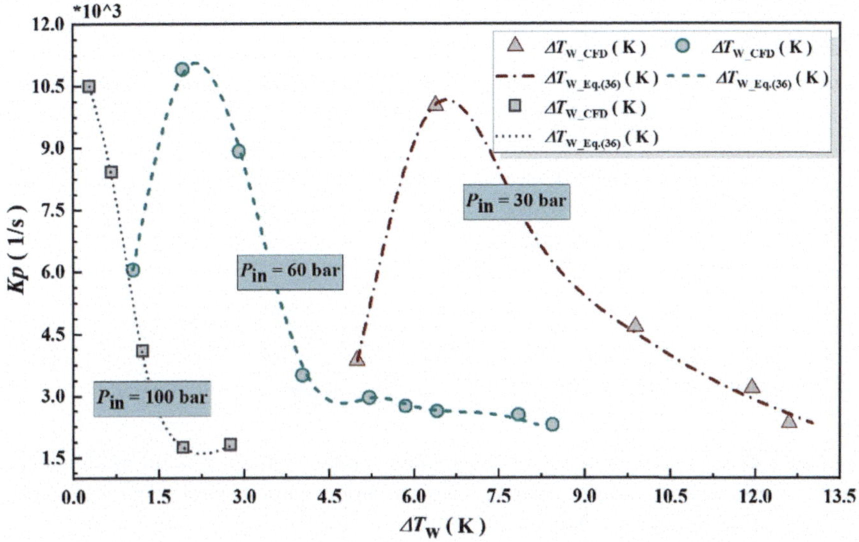

Fig. 9.14 Analysis prediction curve of expansion rate and supercooling at Wilson point

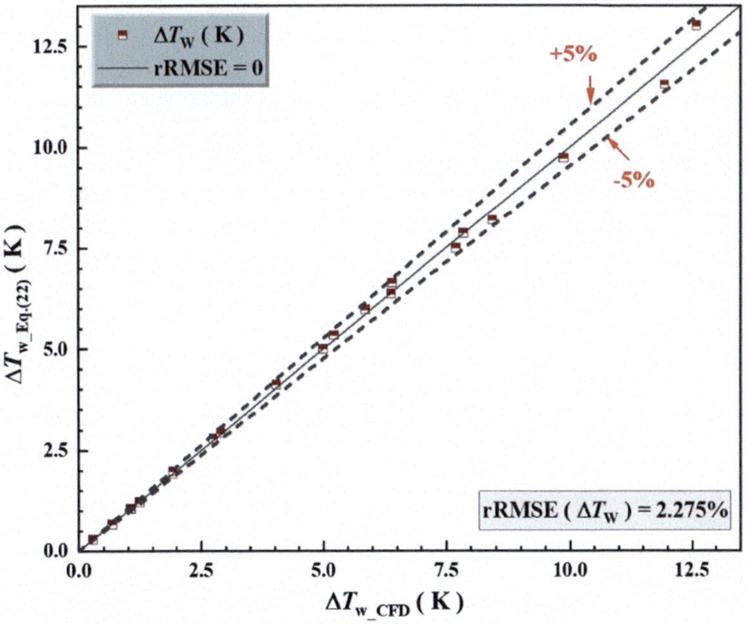

Fig. 9.15 Error analysis of CFD value and predicted value of supercooling at Wilson point

identifying discrepancies between predicted and actual values. This is useful for validating the model and ensuring its effectiveness in diverse research settings.

In summary, this prediction model simplifies the process of predicting simulation data trends and identifying patterns in limited datasets. Its flexibility in adjusting independent variables and the quantitative measures of prediction accuracy, such as MRE and rRMSE, make it an invaluable tool for comparing experimental and simulation data in various studies. This enhances its applicability and usefulness in advancing research and development in fields reliant on accurate modeling and simulation of condensation processes.

9.3 Condensation Performance in Supersonic Separator

The supersonic separator needs to be validated before the condensation and separation characteristics can be studied. The validation aims to assess the prediction accuracy of the model for the separation performance in a two-component system using a supersonic separator, as depicted in Fig. 9.16. Given the lack of experimental data on CO_2 condensation and separation in this context, we conducted experiments using an air–steam system with detailed device and methods previously described. Experimental conditions included an inlet pressure p_{in} of 3 atm, outlet pressure p_{out} of 1 atm, inlet temperature T_{in} of 300 K, and 100% inlet humidity Φ_{in}. The diameter and mass concentration of heterogeneous droplets at the inlet were 2.2 μm and 0.005 kg/m^3, respectively.

In the simulation, the same conditions were applied, and a three-dimensional structured mesh with 557,560 grid cells, determined through GCI tests, was used. Figure 9.17 compares experimental results with simulation results. Dew-point depression (ΔT_d) and vapor separation efficiency (η_v) are used to evaluate separation performance, with calculations as follows:

$$\Delta T_d = T_{d,in} - T_{d,out} \tag{9.5}$$

$$\eta_v = \frac{y_{v,in} - y_{v,out}}{y_{v,in}} \times 100\% \tag{9.6}$$

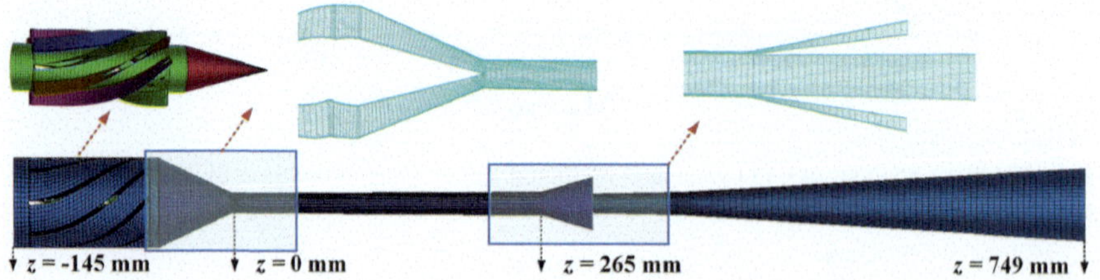

Fig. 9.16 Structure, size, and mesh of the supersonic separator

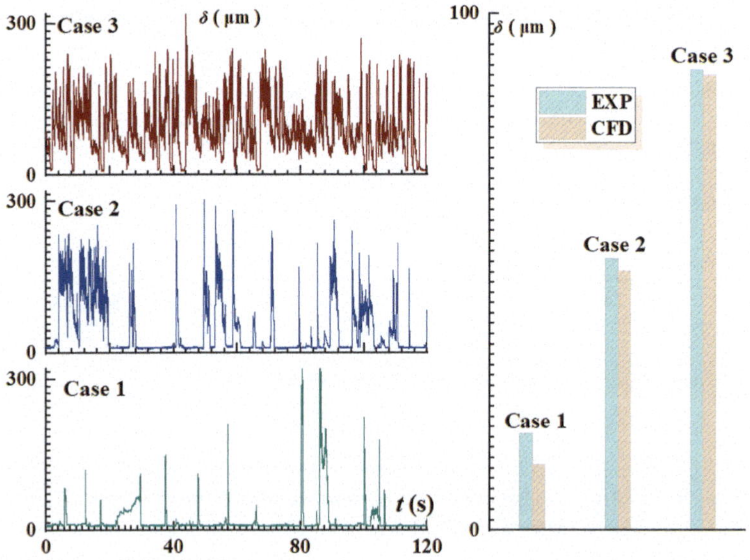

Fig. 9.17 Model validation: prediction of condensation and separation performance

where $T_{d,in}$ and $T_{d,out}$ are the dew point at the inlet and gas outlet, and $y_{v,in}$ and $y_{v,out}$ are the vapor mole fraction at the inlet and gas outlet.

The experimental and CFD results for ΔT_d were 20.04 K and 22.95 K and for η_v were 73.04% and 77.27%, respectively, which can be seen in Fig. 9.17. The relative errors were 12.67 and 5.47%, which are within acceptable ranges, indicating a strong agreement and validating the model's accuracy in predicting separation performance. Additionally, three cases comparing liquid film thicknesses at varying inlet mass concentrations ρ_{in} (0.0001, 0.0010, and 0.0020 kg/m^3) (in Fig. 9.17) also showed that the model's predictions were generally in reasonable agreement with experimental results, despite minor deviations due to flow fluctuations at the liquid outlet. In summary, the established model is reliable and can serve as a foundation for further research.

Then this section uses the developed Euler–Euler–Euler (E-E-E) model to study the condensation separation performance inside the supersonic nozzle and inside the supersonic separator, which distinguishes this chapter from other literature related to supersonic separators. The section compares different thermodynamic models to evaluate their accuracy in predicting the behavior of pure CO_2 in a supersonic nozzle. In Fig. 9.18, the compressibility factor, a measure of deviation from ideal gas behavior, varies from 0.65 at the inlet to 0.80 at the outlet, indicating that an ideal gas equation of state (EOS) is inadequate for accurate heat and fluid flow predictions in such scenarios.

Figure 9.19 illustrates discrepancies between real gas EOS and ideal gas EOS predictions under varying conditions. In Case 1 (subcritical state), the ideal gas EOS overestimates the Mach number at the nozzle outlet by 11%. This overestimation

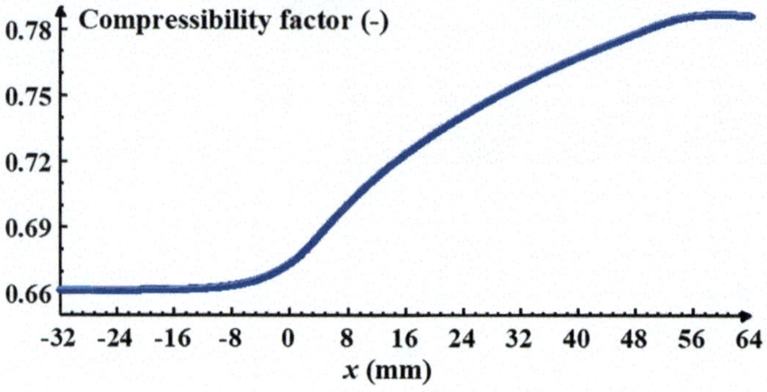

Fig. 9.18 Compressive factor distribution of ideal gas EOS

increases to 17.4% and 19.4% in Cases 2 and 3 (near-critical and supercritical states), respectively, indicating that higher pressures amplify this error. Additionally, the ideal gas EOS significantly overestimates the heat transfer characteristics at the Wilson point, with errors increasing from 9.6 to 25.1% as pressure rises. Moreover, the liquid fraction at the nozzle outlet is grossly underestimated by the ideal gas EOS; for instance, in Case 3, the ideal gas EOS predicts a liquid fraction of 0.00723 compared to the real gas EOS's 0.03557, resulting in a 79.7% error. Real gas EOS, therefore, provides more accurate predictions of expansion characteristics and mass-heat transfer in supersonic condensation flow, with discrepancies between the two models increasing with pressure.

Figure 9.20 quantifies CO_2 condensation under different conditions. Higher inlet temperatures decrease both the condensed mole fraction (x_{con}) and condensed CO_2

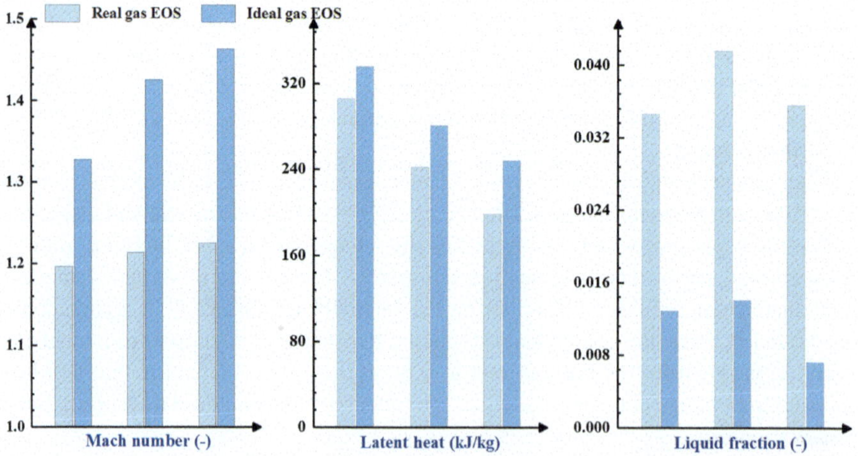

Fig. 9.19 Prediction properties of thermodynamic models in supersonic flow under different conditions

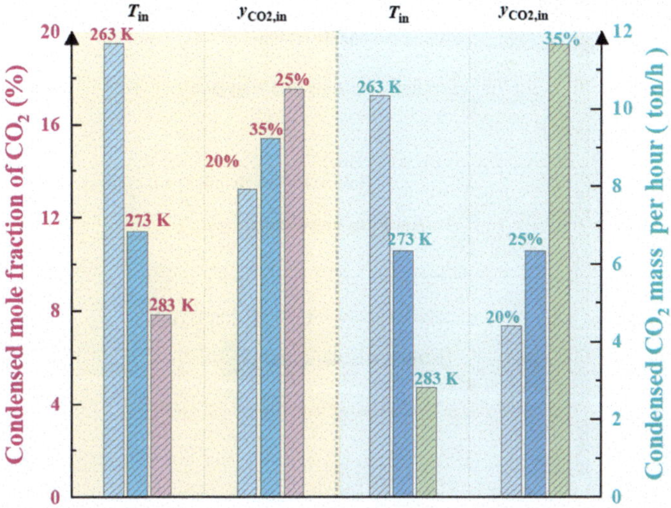

Fig. 9.20 Condensation amount of CO_2 under different conditions

mass per hour, whereas increasing the inlet CO_2 vapor mole fraction enhances these parameters. For instance, raising the inlet temperature from 263 to 283 K decreases x_{con} from 19.47 to 7.84% and condensed mass from 10.33 to 2.84 ton/h. Conversely, increasing $x_{CO_2,in}$ from 20 to 35% raises x_{con} from 13.22 to 17.53% and condensed mass $\dddot{m}_{con}$ from 4.41 to 11.67 ton/h. Lowering the inlet temperature or increasing the CO_2 vapor mole fraction thus improves CO_2 condensation, aiding the advancement of supersonic decarbonization technology. Here, the condensation amount is evaluated by two indexes: the condensed mole fraction of CO_2 (x_{con}) and the condensed CO_2 mass per hour $\dddot{m}_{con}$ (ton h^{-1}), namely

$$x_{con} = \frac{x_{CO_2,in} - x_{CO_2,out}}{x_{CO_2,in}} \quad (9.7)$$

$$\dddot{m}_{con} = \frac{x_{CO_2,in}q_{in} - x_{CO_2,out}q_{out}}{3.6} \quad (9.8)$$

where q_{in} and q_{out} are the mass flow rate (kg s^{-1}) at the nozzle inlet and outlet.

9.4 Separation Performance in Supersonic Separator

Then this section investigates the impact of foreign droplets on mass transfer and separation efficiency within a supersonic separator, focusing on both homogeneous and heterogeneous condensation processes. The cloud chart in Fig. 9.21 provides detailed insights into the behavior characteristics of vapor, droplets, and liquid film

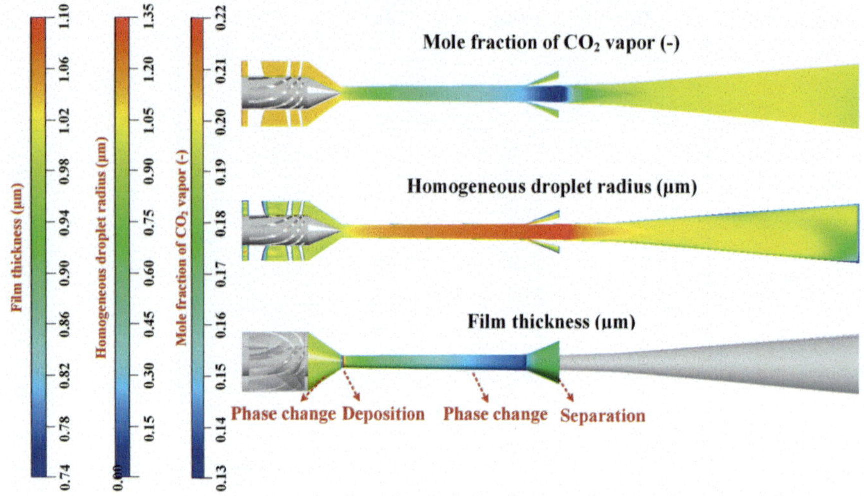

Fig. 9.21 Behavior of gas, liquid droplets, and liquid film in supersonic separator dominated by homogeneous condensation

within a supersonic separator, specifically under conditions where only homogeneous condensation is present. The simulated inlet conditions for this analysis are the inlet pressure p_{in} of 150 atm, a temperature T_{in} of 258 K, and a CO_2 mole fraction $x_{CO_2,\text{in}}$ of 23.2%. The gas outlet condition is set at a pressure $p_{\text{g,out}}$ of 80 atm, while the liquid outlet condition is at a pressure $p_{\text{l,out}}$ of 60 atm.

As illustrated in Fig. 9.21, the CO_2 vapor initially undergoes condensation followed by a phase of evaporation within the supersonic separator. During this process, the mole fraction of CO_2 vapor can decrease to as low as 0.13. This indicates significant initial condensation. In Fig. 9.21, the formation of homogeneous CO_2 droplets during the condensation process is shown, with these droplets being very small in size. The largest droplet formed has a radius of 1.35 μm. Due to the swirling flow within the separator, these CO_2 droplets are driven toward the separator wall, where they contribute to the formation of a liquid film.

However, because of the small size of these homogeneous droplets, they are more prone to being carried away by the gas phase rather than coalescing into a substantial wall film. Figure 9.21 details the characteristics of the liquid film, showing that it forms a very thin layer with a maximum thickness of just 1.10 μm. This thin film results from the condensation of CO_2 vapor near the wall, along with the deposition of a small number of CO_2 droplets.

Next, the separation process dominated by heterogeneous condensation is discussed. Keeping the above simulation conditions unchanged, heterogeneous droplets with a radius of 10 μm and a mass concentration of $\rho_{\text{het,in}} = 7.5$ kg/m^3 are continuously injected into the inlet. The changes in gas, droplet, and liquid film are shown in Fig. 9.22a, and compared to Fig. 9.21, the lowest value of the mole fraction of CO_2 vapor can be reduced to 0.10. Additionally, the particle size of homogeneous droplets increases to 1.62 μm. At the throat of the separator, as shown in Fig. 9.22b, homogeneous droplets collide with heterogeneous droplets and aggregate into larger heterogeneous droplets. When the heterogeneous droplets reach their maximum size, they can grow to about 0.1 mm in diameter. This larger particle size means that there are more opportunities for vapor condensation. The separation process is completed when the liquid film on the wall flows out of the liquid outlet. Therefore, more liquid film content is beneficial for the separation. As illustrated in Fig. 9.22c, the presence of heterogeneous droplets significantly increases the number of large droplets in the flow field, which in turn increases the number of droplets that can be directed to the wall surface. Figure 9.22d shows that the liquid film thickness in this case ranges from 13 to 121 μm, indicating better separation capacity.

In summary, the introduction of heterogeneous droplets into the supersonic separator markedly improves the condensation and separation process by reducing the CO_2 vapor mole fraction more effectively and increasing the size and quantity of droplets that form the liquid film. This results in a more substantial and thicker liquid film, enhancing the overall separation efficiency.

Figure 9.23 illustrates the proportion of different forms of carbon dioxide within the supersonic separator for the given case, with all data sourced from the axis of the separator. The figure reveals that the mole fraction of CO_2 vapor at both the inlet and outlet of the supersonic separator remains nearly equal. This observation

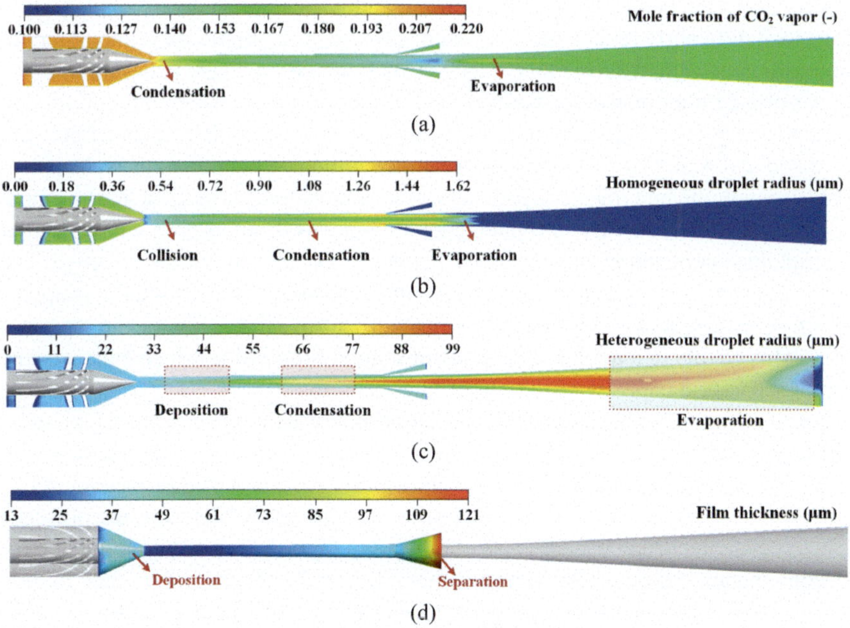

Fig. 9.22 Behavior of gas, liquid droplets, and liquid film in supersonic separator dominated by heterogeneous condensation: **a** CO_2 mole fraction distribution profile; **b** homogeneous droplet radius of CO_2; **c** heterogeneous droplet radius of CO_2; **d** CO_2 film thickness distribution

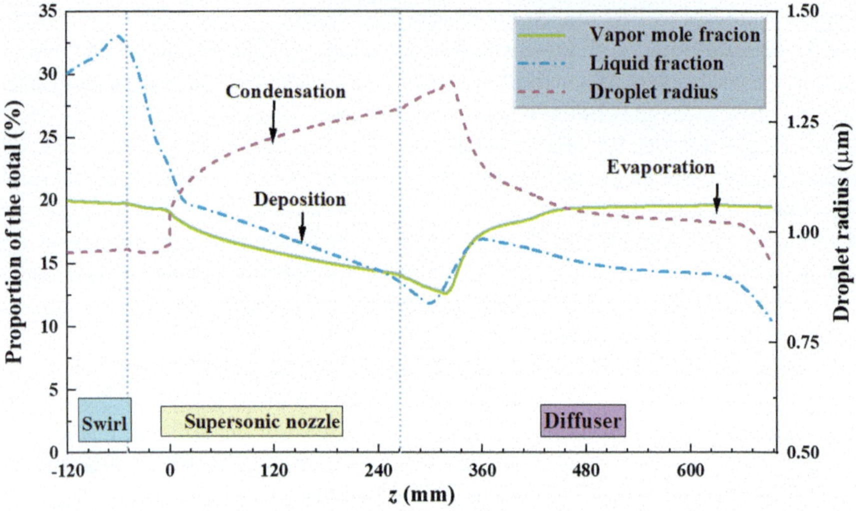

Fig. 9.23 Proportion of different forms of carbon dioxide

suggests that the separation efficiency of the process dominated by homogeneous condensation is negligible, effectively approaching zero.

This negligible separation is due to the inability of homogeneous condensation alone to significantly reduce the CO_2 vapor content. The small size of the homogeneous droplets formed during this process makes them more likely to be entrained by the gas phase rather than contributing to the formation of a substantial liquid film on the separator walls. Consequently, the overall amount of CO_2 that can be effectively separated and removed from the gas stream is minimal. In contrast, the subsequent analysis involving heterogeneous condensation demonstrates a marked improvement in separation efficiency. By introducing heterogeneous droplets with a larger initial size and mass concentration, the condensation process becomes more effective. The larger droplets have a greater capacity to collide and coalesce with smaller droplets, forming even larger particles that can more readily form a liquid film on the separator walls. This results in a thicker and more substantial liquid film, which enhances the overall separation capacity of the supersonic separator.

Further, Fig. 9.24 illustrates the CO_2 vapor mole fraction profile along the axial direction for varying heterogeneous droplet concentrations $\rho_{\text{het,in}}$. In scenarios with only homogeneous droplets, the CO_2 vapor mole fraction at the dry gas outlet of the supersonic separator is nearly equal to that at the inlet. However, the presence of heterogeneous droplets causes part of the CO_2 vapor to condense on the heterogeneous nuclei. These nuclei typically have larger diameters and greater masses than homogeneous nuclei, increasing the likelihood of centrifugal separation. Consequently, heterogeneous droplets are more prone to contacting the wall, depositing, and separating, thereby effectively removing vapors from the supersonic separator in droplet form. For example, when $\rho_{\text{het,in}} = 0.1\ \text{kg/m}^3$, the mole fraction of CO_2

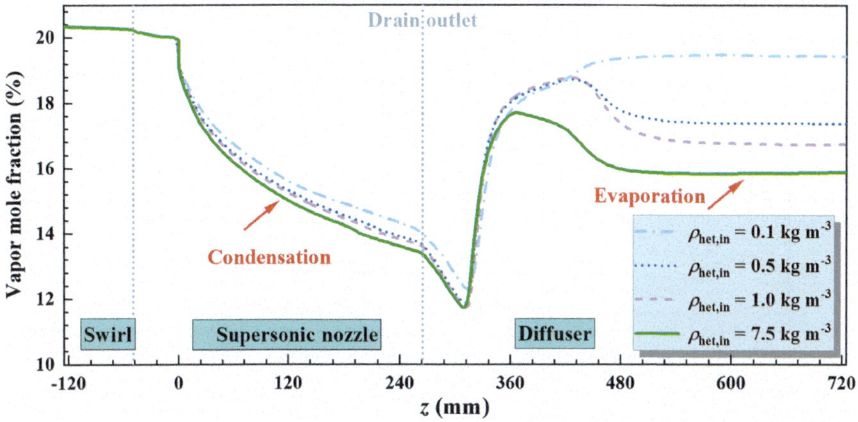

Fig. 9.24 Effects of different heterogeneous droplet mass concentrations on CO_2 vapor molar fraction

vapor at the outlet is 19.4%, and when $\rho_{\text{het,in}} = 7.5$ kg/m^3, this value decreases to 15.9%. In the diffuser, the mole fraction of carbon dioxide vapor rises.

Thus, in the presence of heterogeneous droplets, the CO_2 vapor mole fraction at the dry gas outlet is generally lower than in the case of homogeneous condensation alone. Furthermore, as the concentration of heterogeneous droplets at the inlet increases, the CO_2 vapor mole fraction at the outlet decreases, indicating enhanced separation of CO_2.

The evolution of liquid film thickness across varying heterogeneous droplet concentrations is depicted in Fig. 9.25. When swirling droplets enter the convergent section of the nozzle, centrifugal force drives them toward the wall, forming a liquid film. As gas velocity increases to supersonic levels at the nozzle throat, the liquid film velocity also increases, albeit at a slower rate, leading to film stripping due to high shear forces. In the divergent section, continuous droplet deposition and vapor condensation cause the liquid film thickness to increase gradually. At the liquid outlet, around 270 mm into the system, the film thickness increases significantly, reaching up to 168 μm for an inlet heterogeneous droplet concentration of 10 kg/m^3. This suggests a higher potential for effective separation, as a thicker liquid film enhances the likelihood of successful separation. Expanding on this, the centrifugal forces in the convergent section initiate liquid film formation, influenced by initial droplet concentration. At the nozzle throat, shear forces strip the liquid film, reducing its thickness momentarily. In the divergent section, decreasing gas velocity and continuous droplet deposition lead to a gradual increase in film thickness. At the outlet, the significant increase in film thickness, especially at higher droplet concentrations, indicates better separation efficiency. This emphasizes the importance of controlling initial droplet concentration to optimize separation efficiency and enhance liquid film behavior under various conditions.

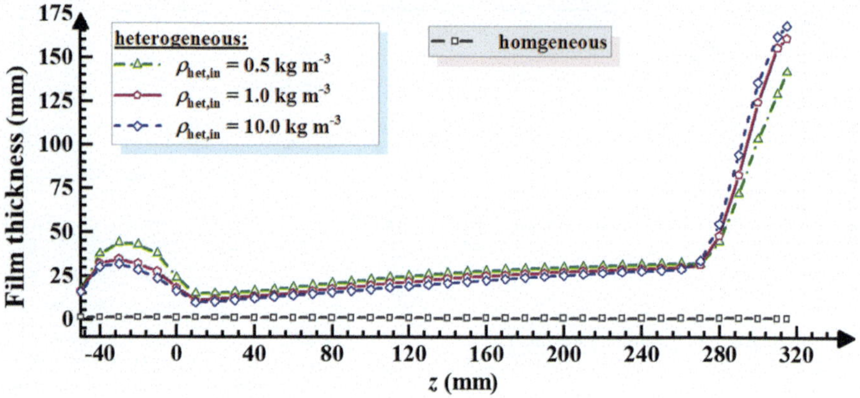

Fig. 9.25 Film thickness distribution under different $\rho_{het,in}$

In summary, the introduction of heterogeneous droplets significantly enhances the separation efficiency in a supersonic separator. This improvement is achieved by reducing the CO_2 vapor mole fraction at the outlet and increasing the liquid film thickness. The thicker liquid film facilitates better mass transfer, leading to more effective separation processes. This approach not only improves overall separation efficiency but also optimizes the performance of the supersonic separator, making it a more viable solution for industrial applications requiring efficient CO_2 removal.

9.5 Energy and Economic Analysis

In carbon capture, utilization, and storage (CCUS) technology, the efficiency of the supersonic separator in the separation process is of paramount importance. Equally crucial to the separation quality is the evaluation of energy consumption and economic viability. The separation process in a supersonic separator involves complex thermodynamic and fluid dynamic interactions that can significantly impact energy usage. Accurate assessment of these energy requirements ensures that the process is not only technically feasible but also sustainable over the long term. Moreover, the economic evaluation of the separation process is vital. It involves analyzing the costs associated with energy consumption, maintenance, and operational efficiency. A comprehensive economic assessment helps in identifying the most cost-effective strategies and optimizing resource allocation. In the competitive landscape of carbon capture technologies, balancing the separation efficiency with energy and economic considerations can determine the overall success and scalability of the CCUS implementation.

Therefore, integrating thorough energy and economic evaluations into the development and operation of supersonic separators is as important as achieving high separation quality. This holistic approach ensures that the technology remains viable and effective in real-world applications, promoting sustainable and economically sound carbon capture solutions.

9.5.1 Entropy and Exergy Assessment

The entropy generation of the supersonic separator is explored in this section. The total entropy generation consists of four components ($s_{gen,D}$, $s_{gen,C}$, $s_{gen,L}$, $s_{gen,A}$), representing the entropy generation due to viscosity, heat conduction, phase change, and aerodynamic losses. For the case of $\rho_{het,in} = 7.5$ kg/m^3, the entropy generation data of different cross sections on the divergent section of the supersonic nozzle were analyzed and compiled into Fig. 9.26. It can be seen from the data that with the increase of axial position, the values of total entropy production and four kinds of entropy production all increase, among which viscous entropy generation $s_{gen,D}$, phase change entropy generation $s_{gen,L}$, and aerodynamic entropy generation $s_{gen,A}$ occupy the main components. The entropy generation data four types regarding the z = 260 mm cross-section was selected for further illustration. The total entropy generation of this cross-section is 0.529 kJ/(kg k), of which $s_{gen,A}$ accounts for 33.28%, $s_{gen,D}$ accounts for 48.60%, $s_{gen,L}$ accounts for 18.07%, and $s_{gen,C}$ is only 0.05%. This indicates that fluid viscosity, phase change, and aerodynamic losses are the main factors leading to entropy generation in the nozzle expansion section, causing irreversible energy loss.

The entropy generation due to phase change is mainly caused by homogeneous condensation and heterogeneous condensation, and the values of entropy generation along the supersonic nozzle due to homogeneous and heterogeneous phase change are shown in Fig. 9.27. From the graph, it can be analyzed that both $s_{gen,hom}$ and $s_{gen,het}$ are showing a growing trend. But $s_{gen,hom}$ fluctuates between 0.001 and 0.002 kJ/

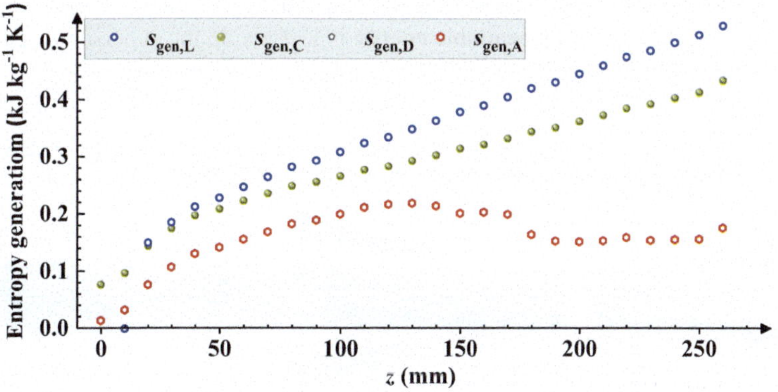

Fig. 9.26 Four types of entropy generation along the nozzle

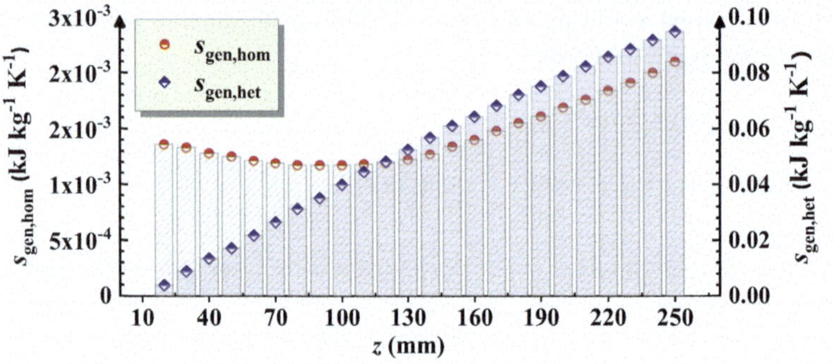

Fig. 9.27 Entropy generation due to homogeneous and heterogeneous phase change along the nozzle

(kg K), while $s_{gen,het}$ grows from zero all the way up to 0.095 kJ/(kg K). Numerically speaking, $s_{gen,het}$ is two orders of magnitude larger than $s_{gen,hom}$, indicating that heterogeneous condensation is the dominant factor in phase change entropy production, while homogeneous condensation is almost negligible.

Figure 9.28 illustrates the distribution of the mean value of viscous dissipation $Y_{gen,\overline{D}}$ across the flow field. Four significant regions with prominent viscous dissipation include the sharp corner of the cone, the near-wall region, the drain outlet region, and the diffuser section. This distribution highlights potential inefficiencies in the current design of the supersonic separator, suggesting that optimizing the design of components such as the drain outlet and cone could improve the separator's performance.

Additionally, two measures were selected to assess energy efficiency, which are the exergy destruction and the total entropy generation at the outlet. In Fig. 9.29, the impact of increasing $\rho_{het,in}$ ranging from 0.1 to 10 kg/m^3 becomes evident. There is a clear tendency for both indicators to increase and then decrease, with peaks reaching 179.58 kJ/kg and 0.66 kJ/(kg K), respectively, when $\rho_{het,in}$ was 5 kg/m^3. This indicates that the more available energy is lost inside the supersonic separator due to irreversible processes, an indication of reduced energy efficiency.

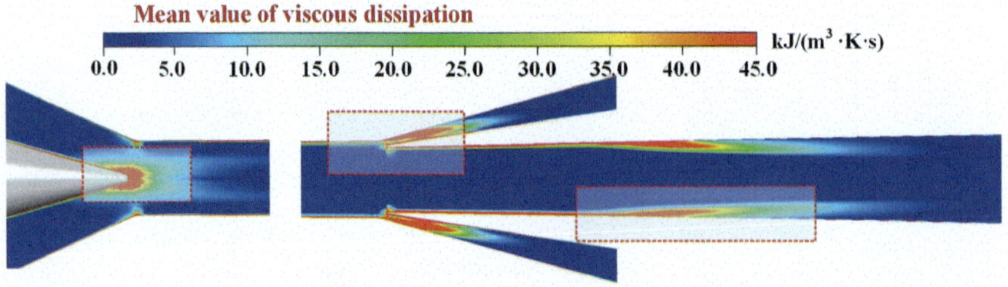

Fig. 9.28 Distribution of the mean value of viscous dissipation

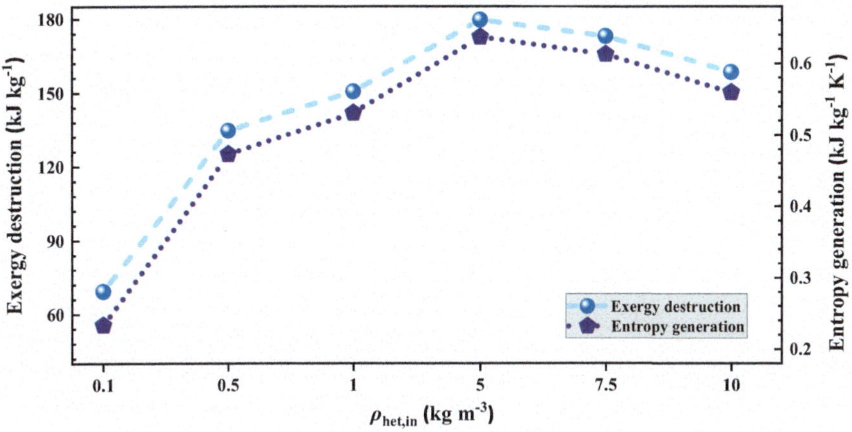

Fig. 9.29 Energy utilization under different $\rho_{het,in}$

After calculation, when $\rho_{het,in}$ is 5 and 7.5 kg/m^3, the maximum values of 477.29 and 451.75 kW can be reached, which also proves that the energy loss is the largest at these two concentrations as shown in Fig. 9.30. It can be seen that mass separation performance and energy utilization performance are two contradictory evaluation parameters. Specifically, improving mass separation performance usually requires more energy consumption, while pursuing lower energy utilization may sacrifice mass separation performance. The combination of these two factors is crucial for the design and optimization of supersonic separators.

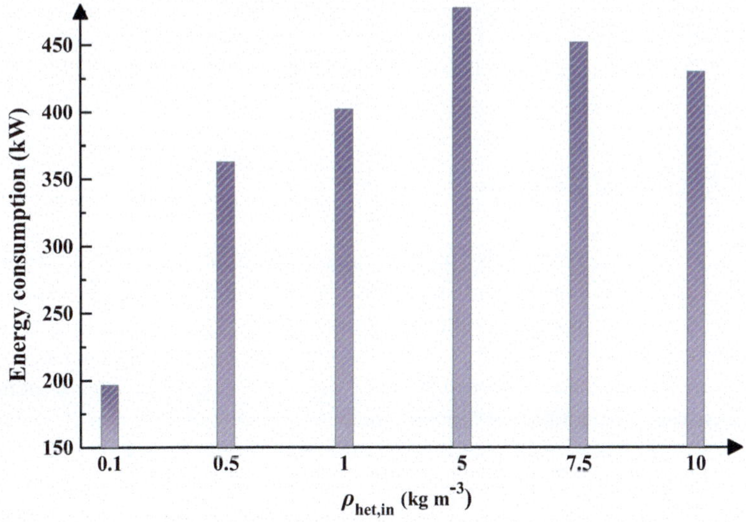

Fig. 9.30 Energy consumption under different $\rho_{het,in}$

Understanding the contributions of different entropy generation mechanisms is crucial for making targeted improvements in supersonic separators. For instance, reducing fluid viscosity dissipation can lower overall entropy generation, while optimizing phase change processes can minimize the entropy generated during phase transitions. By addressing these specific areas, the total entropy generation and associated energy losses can be significantly reduced. This not only enhances the efficiency of the separator but also minimizes energy losses, thereby increasing economic benefits. A detailed analysis of entropy generation and available energy losses in supersonic separators can identify the primary sources of energy loss. This foundational knowledge is essential for future design optimizations. By focusing on targeted improvements, it is possible to maximize energy efficiency and economic benefits while maintaining high separation quality. This is particularly important for the advancement and application of carbon capture, utilization, and storage (CCUS) technologies. Optimizing entropy generation mechanisms involves several strategies. For example, improving the aerodynamics of the separator can reduce frictional losses and fluid viscosity dissipation. Enhancing the efficiency of the phase change processes can be achieved by better controlling the conditions under which condensation and evaporation occur, thus reducing phase change entropy generation. These targeted improvements not only make the supersonic separator more energy-efficient but also enhance its economic viability. By reducing operational costs through lower energy consumption, the overall cost-effectiveness of the separation process is improved. This is critical for the broader implementation of CCUS technologies, which are essential for reducing greenhouse gas emissions and combating climate change.

In conclusion, a thorough understanding and optimization of entropy generation mechanisms in supersonic separators can lead to significant improvements in efficiency and economic benefits. These advancements are vital for the successful deployment and operation of CCUS technologies, ensuring they can meet the demands of reducing carbon emissions while being economically sustainable.

9.5.2 Economic Analysis

In addition, economic analysis is crucial for the application of supersonic separators in carbon capture, utilization, and storage (CCUS) technology. As a key separation device, the design and operating efficiency of the supersonic separator directly affect the economic feasibility of the entire carbon capture process. Through economic analysis, the cost-effectiveness of the supersonic separator can be evaluated, including comprehensive consideration of factors such as equipment investment, operating costs, energy efficiency, and carbon capture efficiency. In addition, economic analysis can also help determine the optimal operating conditions and technical parameters to maximize the carbon capture rate and optimize energy utilization efficiency, thereby reducing the cost of carbon capture. This information is crucial to attracting investment, promoting technology commercialization, and achieving economic feasibility in large-scale applications. In summary, economic analysis is not only an important part of the CCUS technology development process, but also a key factor in promoting its market promotion and sustainable development.

According to a report released by the Global CCS Institute in September 2021, the number of commercial CCUS facilities in the world has now increased to 135, with a total average CO_2 capture capacity of 149.3 Mtpa of which 27 are in operation, 4 are under construction, 58 are in advanced development, 44 are in early development, and 2 have been suspended in Fig. 9.31a. Most of the plants implementing CCUS technology are located in North America, Europe, and East Asia and the Pacific, accounting for 63, 22, and 9% of the world's maximum capture capacity as shown in Fig. 9.31b. These values are completely different compared to 2020 and 2019, when the number of large-scale CCUS facilities was 65 and 51, respectively, with a maximum CO_2 capture capacity of approximately 114.3 and 96 $MtCO_2$/year. Nevertheless, even if the current performance of CCUS systems allows for a gradual reduction in CO_2 emissions to the atmosphere, more than 2000 facilities would be needed to achieve the 2050 Paris Agreement climate goals outlined in the International Energy Agency's Sustainable Development Scenario. This would mean that 70–100 new facilities would need to be built each year [25]. Therefore, as more facilities are put into use, the economic costs have to be taken into account.

For cost analysis, this section synthesizes the CO_2 purity and recovery rates of currently commonly used CO_2 capture and separation technologies [15, 17]. As a new green capture technology, supersonic low-temperature phase change separation technology stands out for its significant potential and application value. This technology operates without the need for chemical reagents, thus avoiding additional pollution. Its innovative approach relies on the principles of phase change at supersonic speeds to separate CO_2 from natural gas efficiently. One of the notable advantages of this technology is the use of a cascade mode, which can significantly enhance the recovery rate of CO_2. Current calculations on mass and energy utilization in supersonic separation technology indicate that a single-stage separator can achieve a CO_2 removal rate of approximately 25% while incurring only a 46.7% pressure loss. This demonstrates a balanced trade-off between efficiency and operational cost in terms of pressure drop. However, the implementation of a multi-stage cascade separator, while increasing the overall pressure loss, offers a considerable boost in the CO_2 recovery rate. This multi-stage approach can address the limitations of a single-stage system by sequentially removing more CO_2, thus improving the overall effectiveness of the process. The enhanced CO_2 recovery rate achieved through the multi-stage cascade configuration makes this technology a promising solution for the natural gas industry, where efficient and sustainable CO_2 removal is crucial.

In summary, supersonic low-temperature phase change separation technology represents a cutting-edge solution with significant environmental and operational benefits. Its ability to operate without chemical additives and its potential for high CO_2 recovery rates position it as a valuable tool in the pursuit of greener and more efficient natural gas processing. Table 9.5 summarizes the economic parameters under different CO_2 separation modes.

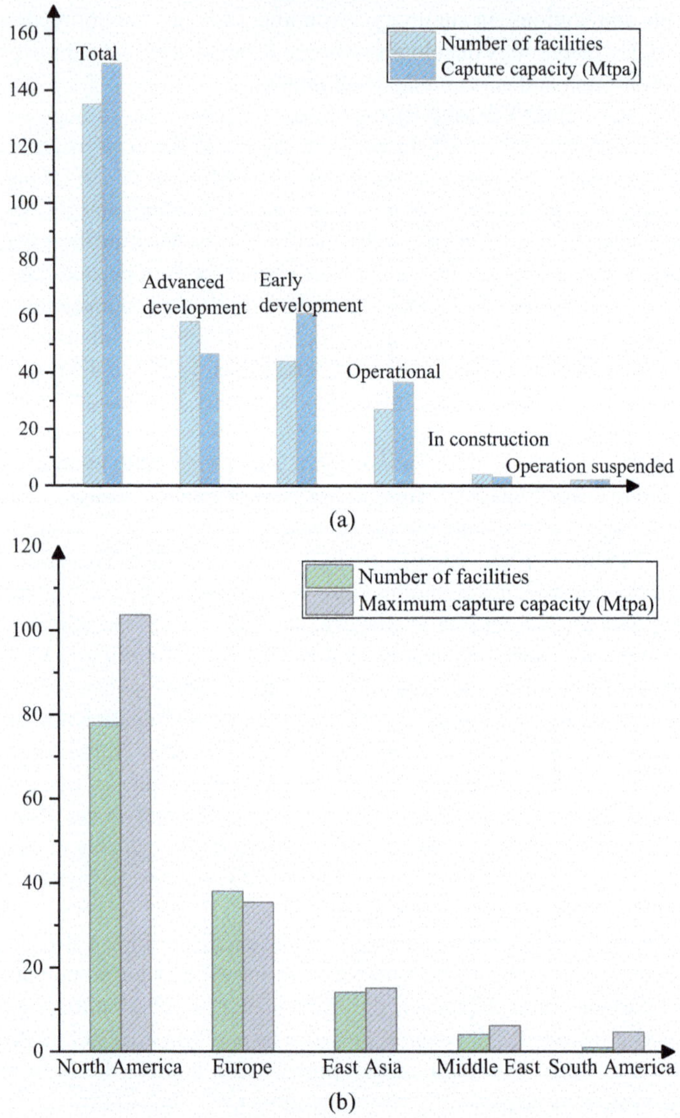

Fig. 9.31 Summary of the number of commercial CCUS facilities worldwide and their CO_2 capture capacity: **a** numbers of global commercial CCUS facilities and their mean CO_2 capture capacity by different stages of development in 2021; **b** numbers of global commercial CCUS facilities and their maximum CO_2 capture capacity by world region in 2021

Table 9.5 Evaluation of important technical and economic parameters under different technologies

CO_2 separation technology	CO_2 recovery (% vol)	CO_2 purity (%)	Energy consumption (GJ/t CO_2)	Capture cost ($/t CO_2)
Chemical absorption	85 ~ 90	> 95	1.83–4.0	62.80
Physical absorption	88 ~ 95	> 99	2.4–9.0	89.66
Membranes	70 ~ 90	> 95	1.1–1.9	80.46
Biological	10 ~ 40	> 99	–	793
Cryogenic	90 ~ 99	> 99	0.85–3	52
Supersonic separator (SS)	60 ~ 80	> 99	0.85–2	38
SS + absorption	90 ~ 95	> 99	1.34–3	50.4

Note $ = USD
USD 1 = EUR 0.93

References

1. Castier M (2014) Modeling and simulation of supersonic gas separations. J Nat Gas Sci Eng 18:304–311
2. Castro-Munoz R, Ahmad MZ, Malankowska M, Coronas J (2022) A new relevant membrane application: CO_2 direct air capture (DAC). Chem Eng J 446:137047
3. Chen J, Jiang W, Han C, Liu Y (2021) Numerical study on the influence of supersonic nozzle structure on the swirling condensation characteristics of CO_2. J Nat Gas Sci Eng 88:103753
4. de Oliveira Arinelli L, Teixeira AM, de Medeiros JL, Ofélia de Queiroz FA (2019) Supersonic separator for cleaner offshore processing of natural gas with high carbon dioxide content: environmental and economic assessments. J Clean Prod 233:510–521
5. Delale CF, Schnerr GH, Zierep J (1993) Asymptotic solution of transonic nozzle flows with homogeneous condensation. I. Subcritical flows. Phys Fluids A Fluid Dyn 5(11):2969–2981
6. Deng Y, Zhang L, Hou H, Yu B, Sun D (2019) Modeling and simulation of the gas-liquid separation process in an axial flow cyclone based on the Eulerian–Lagrangian approach and surface film model. Powder Technol 353:473–488
7. Ding H, Wang C, Zhao Y (2014) An analytical method for Wilson point in nozzle flow with homogeneous nucleating. Int J Heat Mass Transf 73:586–594
8. Ding H, Dong Y, Zhang Y, Yang Y, Wen C (2022a) A potential strategy of carbon dioxide separation using supersonic flows. Sep Purif Technol 303:122153
9. Ding H, Sun C, Wen C, Liang Z (2022b) The droplets and film behaviors in supersonic separator by using three-field two-fluid model with heterogeneous condensation. Int J Heat Mass Transf 184:122315
10. Ding H, Zhang Y, Sun C, Yang Y, Wen C (2022c) Numerical simulation of supersonic condensation flows using Eulerian–Lagrangian and Eulerian wall film models. Energy 258:124833
11. Ding H, Zhang Y, Dong Y, Wen C, Yang Y (2023a) High-pressure supersonic carbon dioxide (CO_2) separation benefiting carbon capture, utilisation and storage (CCUS) technology. Appl Energy 339:120975
12. Ding H, Zhang Y, Yang Y, Wen C (2023b) A modified Euler–Lagrange–Euler approach for modelling homogeneous and heterogeneous condensing droplets and films in supersonic flows. Int J Heat Mass Transf 200:123537

13. Dobbins RA (1983) A theory of the Wilson line for steam at low pressures. ASME. J. Fluids Eng 105(4): 414–422.
14. dos Santos Cruz WR, dos Santos FP, de Andrade Medronho R (2021) Uncertainty quantification of real gas models in CO_2 supersonic flow. J Comput Sci 56:101479
15. Dziejarski B, Krzyżyńska R, Andersson K (2023) Current status of carbon capture, utilization, and storage technologies in the global economy: a survey of technical assessment. Fuel 342:127776
16. Han J, Feng J, Hou T, Chen W, Peng X (2021) Numerical and experimental study on gas-water separators for a PEMFC system. Int J Green Energy 18(5):490–502
17. Hong WY (2022) A techno-economic review on carbon capture, utilisation and storage systems for achieving a net-zero CO_2 emissions future. Carbon Capture Sci Technol 3:100044
18. Huang L, Young JB (1996) An analytical solution for the Wilson point in homogeneously nucleating flows. Proc R Soc Lond Ser A Math Phys Eng Sci 452(1949):1459–1473
19. Jiang W, Bian J, Wu A, Gao S, Yin P, Hou D (2018) Investigation of supersonic separation mechanism of CO_2 in natural gas applying the discrete particle method. Chem Eng Process Process Intensif 123:272–279
20. Karasu S, Altan A (2022) Crude oil time series prediction model based on LSTM network with chaotic Henry gas solubility optimization. Energy 242:122964
21. Lettieri CLME, Yang DDME, Spakovszky ZZME (2015) An investigation of condensation effects in supercritical carbon dioxide compressors. J Eng Gas Turb Power 137(8):082602
22. Lettieri C, Paxson D, Spakovszky Z, Bryanston-Cross P (2018) Characterization of nonequilibrium condensation of supercritical carbon dioxide in a de laval nozzle. J Eng Gas Turb Power 140(4):041701
23. Liu Y, Cao X, Yang J, Li Y, Bian J (2021) Energy separation and condensation effects in pressure energy recovery process of natural gas supersonic dehydration. Energy Convers Manage 245:114557
24. Niknam PH, Mortaheb HR, Mokhtarani B (2018) Effects of fluid type and pressure order on performance of convergent–divergent nozzles: an efficiency model for supersonic separation. Asia-Pac J Chem Eng 13(2):e2181
25. Ratner M, Glover CF (2014) US energy: overview and key statistics, vol 40187. Congressional Research Service, Washington, DC, p 27
26. Shen M, Tong L, Yin S, Liu C, Wang L, Feng W, Ding Y (2022) Cryogenic technology progress for CO_2 capture under carbon neutrality goals: a review. Sep Purif Technol 299:121734
27. Shoghl SN, Naderifar A, Farhadi F, Pazuki G (2021) A novel strategy for comprehensive optimization of structural and operational parameters in a supersonic separator using computational fluid dynamics modeling. Sci Rep 11(1):21850
28. Shooshtari SR, Shahsavand A (2018) Optimal operation of refrigeration oriented supersonic separators for natural gas dehydration via heterogeneous condensation. Appl Therm Eng 139:76–86
29. Song C, Liu Q, Deng S, Li H, Kitamura Y (2019) Cryogenic-based CO_2 capture technologies: state-of-the-art developments and current challenges. Renew Sustain Energy Rev 101:265–278
30. Wen C, Li B, Ding H, Akrami M, Zhang H, Yang Y (2022) Thermodynamics analysis of CO_2 condensation in supersonic flows for the potential of clean offshore natural gas processing. Appl Energy 310:118523
31. Yamasaki H, Wakimoto H, Kamimura T, Hattori K, Nekså P, Yamaguchi H (2022) Visualization and measurement of swirling flow of dry ice particles in cyclone separator-sublimator. Energies 15(11):4128
32. Yang Y, Wen C, Wang S, Feng Y (2014) Numerical simulation of real gas flows in natural gas supersonic separation processing. J Nat Gas Sci Eng 21:829–836
33. Yue T, Chen J, Song J, Chen X, Wang Y, Jia Z, Xu R (2019) Experimental and numerical study of upper swirling liquid film (USLF) among gas-liquid cylindrical cyclones (GLCC). Chem Eng J 358:806–820
34. Zhang Z, Cai J, Chen F, Li H, Zhang W, Qi W (2018) Progress in enhancement of CO_2 absorption by nanofluids: a mini review of mechanisms and current status. Renew Energy 118:527–535

35. Zhang Z, Wang T, Blunt MJ, Anthony EJ, Park AHA, Hughes RW et al (2020) Advances in carbon capture, utilization and storage. Appl Energy 278:115627
36. Zhao N, You F (2020) Can renewable generation, energy storage and energy efficient technologies enable carbon neutral energy transition? Appl Energy 279:115889
37. Zhao R, Zhao L, Deng S, Song C, He J, Shao Y, Li S (2017) A comparative study on CO_2 capture performance of vacuum-pressure swing adsorption and pressure-temperature swing adsorption based on carbon pump cycle. Energy 137:495–509
38. Zhou L, Hang J, Bai L, Krzemianowski Z, El-Emam MA, Yasser E, Agarwal R (2022) Application of entropy production theory for energy losses and other investigation in pumps and turbines: a review. Appl Energy 318:119211

The manufacturer's authorised representative in the EU is Springer Nature Customer Service Centre GmbH, Europaplatz 3, 69115 Heidelberg, Germany. If you have any concerns regarding our products, please contact ProductSafety@springernature.com

Printed and bound by CPI Group (UK) Ltd, Croydon, CR0 4YY
01/10/2026
02225958-0009